John Bachman

The Publications of the Southern Texts Society

SERIES EDITOR

David S. Shields, University of South Carolina

EDITORIAL BOARD

Richard J. M. Blackett, Vanderbilt University

Susan V. Donaldson, College of William and Mary

Fred Hobson, University of North Carolina

Anne Goodwyn Jones, Allegheny College

David Moltke-Hansen, Historical Society of Pennsylvania

Michael O'Brien, University of Cambridge

Steven M. Stowe, Indiana University

John Bachman

Selected Writings on Science, Race, and Religion

EDITED BY GENE WADDELL

THE UNIVERSITY OF GEORGIA PRESS *Athens and London*

Publication of this book was made possible in part by
a grant from the Watson-Brown Foundation.

Paperback edition, 2016
© 2011 by the University of Georgia Press
Athens, Georgia 30602
www.ugapress.org
All rights reserved
Set in Minion by Graphic Composition, Inc., Bogart, Ga.

Most University of Georgia Press titles are
available from popular e-book vendors.

Printed digitally

The Library of Congress has cataloged the hardcover
edition of this book as follows:
Bachman, John, 1790–1874.
[Selections. 2011]
John Bachman : selected writings on science, race, and religion /
edited by Gene Waddell.
viii, 380 p., [16] p. of plates : ill. (some col.) ; 24 cm. —
(The publications of the Southern Texts Society)
Includes bibliographical references and index.
ISBN-13: 978-0-8203-3818-7 (hardcover : alk. paper)
ISBN-10: 0-8203-3818-4 (hardcover : alk. paper)
1. Mammals—North America. 2. Birds—North America.
3. Agriculture. 4. Monogenism and polygenism.
5. Religion. I. Waddell, Gene. II. Title.
QL715.B12 2011
599.097—dc22   2010052714

Paperback ISBN 978-0-8203-4983-1

British Library Cataloging-in-Publication Data available

Contents

Preface   vii

Introduction: John Bachman's Works and Life   1

Funeral Discourse of the Rev. John G. Schwartz   22

Address Delivered before the Horticultural Society of Charleston   41

Experiments Made on the Habits of the Vultures   70

Migration of North American Birds   82

Species of Squirrel Inhabiting North America   105

Changes of Colour in Birds and Quadrupeds   134

Benefits of an Agricultural Survey   177

American Beaver: A Chapter from Audubon and Bachman's *Quadrupeds*   217

Generation of the Opossum   238

Unity of the Human Race   251

Defense of Luther and the Reformation   263

Address on Education   279

Vindication   299

Selected Letters, 1831–1871   317

Bibliography   359

Index   375

Preface

John Bachman was a Lutheran clergyman who by 1854 was also the leading authority on North American mammals. He was the author or coauthor of three major books and more than sixty shorter publications on science, religion, and other subjects. His writings are incisive, erudite, thorough, honest, and lucid, and they continue to be worth reading.

For inclusion in this volume, I have selected representative portions of his books, ten of his articles, and about one-tenth of his surviving letters. Of the thirteen selections, six are about zoology (three on mammals, two on birds, and one on mammals and birds). Two articles are about religion, and two others are about agriculture. One article is on anthropology, another on education, and yet another on history. The letters included are primarily about the preparation of Audubon and Bachman's monograph on viviparous quadrupeds (mammals).

An introduction indicates the scope and importance of Bachman's writings and provides biographical details. Headnotes indicate each selection's contents and significance and include other needed information.

No attempt has been made to update Bachman's scientific findings. For those in need of current information about mammals, the staff of the American Museum of Natural History has updated Bachman's scientific nomenclature (Cahalane 1967). The Smithsonian Institution and the American Society of Mammalogists have produced a more comprehensive study of North American mammals than was possible a century and a half ago (Wilson and Ruff 1999).

Bachman's writing needed only minimal editing to make his meaning clear on a first reading. I have retained his spelling, but silently corrected obvious misprints. I have updated his punctuation (chiefly by omitting commas when they interrupted the flow of his sentences and by changing dashes at the ends of sentences to periods). Scholars who need more exactness can consult facsimiles of Bachman's shorter publications online at the Special Collections Web site of the College of Charleston Library. In some cases I have identified individuals more specifically by inserting names and relationships in brackets within Bachman's text, but ordinarily the index and bibliography made this unnecessary. All footnotes are Bachman's.

I am grateful to David Shields for asking me to edit this volume, to John Poindexter and Katina Strauch for suggestions for improving it, and to Marie Ferrara for supporting this project and the online edition of Bachman's writings. Accurate transcriptions of Bachman's writings were prepared by Todd Hagstette.

Lester Stephens generously shared his knowledge of Bachman in a review of the manuscript for this volume, and he characterized Bachman as an "exemplary scientist." The soundness of Bachman's approach to the solution of problems is the reason his work still deserves the most careful consideration.

Gene Waddell
Archivist Emeritus
College of Charleston

John Bachman

Introduction

John Bachman's Works and Life

John Bachman prepared the text for John James Audubon and his *Viviparous Quadrupeds of North America* (1845–54). In his *Doctrine of the Unity of the Human Race Examined on the Principles of Science* (1850), he established that all races were a single species equivalent to the varieties of any species of domesticated animal. In *A Defence of Luther and the Reformation: Against the Charges of John Bellinger, M.D., and Others; To Which are Appended Various Communications of Other Protestant and Roman Catholic Writers Who Engaged in the Controversy* (1853), he used primary sources to prepare the first American monograph on the doctrines of Luther and on the history of the Reformation.

In professional journals, Bachman published comprehensive studies of several genera of North American mammals. He specialized in the smaller and neglected types of mammals such as rabbits, shrews, and squirrels. He produced the first scientific descriptions for twenty species, and he and Audubon together described eleven other species. These thirty-one species represented about one-fifth of all the species that were later included in Audubon and Bachman's *Quadrupeds*.

The *Quadrupeds* was published in installments. The first set of lithographs was issued in 1842; the three folio volumes of illustrations eventually consisted of 150 folio plates made from paintings by Audubon and his son John Woodhouse Audubon. The three text volumes were prepared by Bachman.

According to the biography compiled by members of Bachman's family, the renowned nineteenth-century scientist Louis Agassiz stated that the *Quadrupeds* "has not its equal in Europe" (Bachman 1888: 252). The twentieth-century zoologist Victor H. Cahalane wrote that the *Quadrupeds* was recognized as "the unquestioned authority in its field. In coverage, scientific accuracy and popular interest, it had no equal at the time of its publication and for a half-century thereafter. Even today the anecdotal flavor of the text and the feel of a lost century make fascinating reading" (Cahalane 1967: xiv).

Audubon and Bachman's monograph went far beyond previous studies of North American mammals. For example, "in the last *general* work on American quadrupeds by an American author, published by Dr. Godman in 1826, only two hares were admitted," and one was European and the other was misnamed. While John Godman had included no valid species of hare, Audubon and Bachman described twelve species. Similarly, Richard Harlan's *Fauna Americana* was a one-volume compilation with largely European names that did not apply to American species (Stephens 1999: 832). The most accurate study of North American mammals had been John Richardson's *Fauna Boreali-Americana* (1829), but it was limited to Canada.

Bachman was also deeply interested in problems relating to the whole of natural history such as seasonal changes in the color of birds and mammals, the effects of environments, and the migration of birds. He was one of the few American naturalists of the nineteenth century interested in animals and plants, in how they interacted, and in conducting controlled experiments. Among his earliest scientific publications was a list of plants (Bachman 1834B); he later applied his knowledge of botany and entomology to agriculture.

In the field of natural history, Bachman made acute observations and kept careful notes, and his conclusions accounted for all relevant evidence. He wrote, "we are perhaps too prone to build our theories first, and afterwards seek for the facts which are to support them. Hence, naturalists, having the same field of inquiry before them, and reading from the same book of nature, which is open to all, are very apt to be swayed by their preconceived notions, and thus retard the progress of science by unprofitable disputes" (Bachman 1838B: 198). He considered animals and plants infallible sources of information in their own right, and he had a low opinion of naturalists who argued on the basis of authorities: "evils result from a blind reliance on authorities, whilst we have much better authority before us in the book of nature" (Bachman 1850A: 142). Since all organisms were believed to have been individually created by God, the "book of nature" had been considered by Francis Bacon to have equal authority with the Bible. "Like other contemporary naturalists, Bachman subscribed to the Baconian philosophy of inductive inquiry that avoided hypothesizing" (Stephens 2000: 2, 19).

As a Lutheran clergyman, Bachman believed that every individual should follow his or her conscience in seeking truth. He sought truth above all else and had no fear of the consequences.

In 1833, two years after meeting Audubon, Bachman began regularly to publish pamphlets and articles on natural history. In 1834, he published a defense of Audubon and a catalog of approximately one thousand plants growing in the vicinity of Charleston. He wrote several articles on plants, birds, and insects before specializing in mammals.

Prior to working on the *Quadrupeds*, Bachman had written scientific descriptions of forty-seven species of mammals, and this represented about one-third of the total number known at the time. These descriptions included the thirty-three new species that he and Audubon first identified, and the remaining fourteen species had not previously been distinguished adequately from one another. By considering all species of a genera at the same time, Bachman was able to eliminate much duplication and many mistakes that had been made in previous writings on separate species. By 1842 Bachman had produced scientific descriptions of more species of North American mammals than any other author.

Among Bachman's most significant accomplishments was his demonstration that the varieties of races had been produced through the same processes as the varieties of other kinds of domesticated animals. Amassing a comprehensive body of evidence, he greatly improved upon the distinctions between species, varieties, and hybrids of all kinds in his *Doctrine of the Unity of the Human Race*.

Bachman's *Unity* went against the prevailing views published by many American scientists, including Agassiz. In order to persuade them, Bachman attempted to use arguments based wholly on scientific evidence. He established that in nature, as with domesticated animals, various types of "permanent varieties" could remain unchanged in isolation but would be reabsorbed into the original type if they interbred, and he showed that the same was true for every race. By identifying ways to distinguish varieties from species, he was able to establish that many supposed species were in fact varieties of other well-known species.

Bachman considered in detail the origins and developments of the principal species of domesticated animals and plants, and he concluded that the variety of races had been produced by equivalent causes. Aristotle had listed human beings among the domesticated animals and had noted that all domesticated animals were represented in the wild, but he had not realized that all domesticated animals had wild ancestry. Bachman identified the wild ancestor for every species of domesticated animal.

Bachman defined a species as a type of organism capable of producing offspring that could also produce prolific offspring. He demonstrated that every type of hybrid was not a separate species. He discussed every variety of race, showed in detail what they had in common, and indicated how their differences had been exaggerated. His comprehensive examples and persuasive evaluations constitute a model of inductive reasoning.

Bachman's *Unity* best reflects his insight, the breadth and depth of his knowledge, his powers of reasoning, and his ability to write clearly and effectively. This book was a largely successful attempt to persuade his contemporaries that they were wrong to think that some races were a different species.

As a historian, Bachman prepared the first comprehensive and accurate record of Martin Luther's accomplishments to be published in the United States. Using German primary sources written by Luther and others, he showed how Luther destroyed intolerance of dissent and promoted education, and he revealed how Luther had been intentionally and continually misrepresented. Bachman's *Defense of Luther and the Reformation* (1853) provided a carefully reasoned defense of freedom of speech and freedom of religion. He wrote that Luther had "always been an advocate for unrestrained freedom of thought as well as liberty of speech" (Bachman 1888: 358). Much of his *Defense of Luther* dealt in detail with incorrect assertions that had been made in local publications. By dealing with specific inaccuracies, he was able to generalize effectively to demonstrate the need to maintain the separation of church and state as required by the Constitution of the United States.

Johannes Bachman, the son of Jacob and Eva Bachman, was born on February 4 (or 7) in 1790 in Rhinebeck, New York (Bachman 1888: 11; Bost 1963: 3, n. 6). He was of German-Swiss descent:

> My paternal ancestor was a native of the Canton of Berne, Switzerland. After visiting England, he came to America as private Secretary to William Penn. Finally he settled near Easton, Penn. As a reward for faithful services rendered to the infant Colony, the Government granted him two Townships of land called Upper and Lower Sackeny, which are now settled by his numerous descendants. He was the seventh generation from the above [Bachman himself]. My ancestors on my mother's side were from the kingdom of Würtemberg, Germany. (Bachman 1888: 9)

Bachman began to study natural history intensively around 1800, when he was about ten years old; he later referred to the studies of natural history that he made in New York State as "fifteen years of close observation" (Bachman 1839b: 91). Some of the memoranda he cites in the *Quadrupeds* were from notes made in his youth:

> From my earliest boyhood ... I had an irrepressible desire for the study of Natural History. At the age of fourteen, I made an extensive collection of plants, birds and quadrupeds of my native State. I was intimate with Alexander Wilson, the pioneer of American Ornithology, and furnished him with the rare birds existing in the Northern parts of New York. In Carolina, I was enabled to compare the native productions of a Southern climate with those of the State of my nativity. (Bachman 1888: 94)

Having grown up on a farm, Bachman had the opportunity to study plants and animals continually. His father's farm was located seven miles north of Lansingburg (present-day New City), New York (Bost 1963: 3). He made "occasional botanical excursions among the Catskill mountains, and those of Vermont and New Hampshire," and in 1835 he made a "botanical excursion" in the Allegheny mountains (Bachman 1836A1: 82). Even as a boy, he wanted to know how and why some animals changed colors with the seasons, and he kept a weasel and an ermine in captivity to observe them. He credited a slave named George with teaching him about animals and encouraging his interests in natural history.

Bachman had innumerable pets, and he kept notes on their characteristics and behavior. For example, while a boy, he raised Northern Gray Squirrels to maturity to see if tufts would appear on their ears, and he found that squirrels with and without tufts were the same species (Bachman 1888: 17, 120). Throughout most of his life, he continued recording observations on small mammals and birds that he raised and that he hunted in the wild. As he later wrote Audubon, "all that you write on the spot, I can depend on, but I never trust to the memory of others, any more than to my own" (Herrick 1938: vol. 2, 271).

As a youth, Bachman was raised on a farm and spent much of his time outdoors. He trapped animals to acquire pets and to raise money to buy natural history books. He traveled widely in the woods of New York State and was "Knickerbocker's" secretary on a three-month expedition to the Oneida Indians. Diedrich Knickerbocker was the pseudonym Washington Irving used for his *History of New-York from the Beginning of the World to the End of the Dutch Dynasty* (1809), in which he described the customs of the Oneida in detail. The Oneida named Bachman "Big Moose" for killing a large moose with a single shot. Like Audubon, Bachman enjoyed hunting. He wasted no game and took no "unfair advantage" (Bachman 1888: 56, 15, 17–19, 253).

As a young man, Bachman moved to Philadelphia to study religion, and while there he taught in several small schools to earn a living. He initially learned about the scientific aspects of natural history through his friendships with Alexander Wilson and William Bartram. Wilson's *American Ornithology* was published from 1804 to 1814, and Wilson was the ornithologist Audubon relied upon and set out to surpass. Bartram's famous *Travels* had been published in 1791. His vivid natural descriptions influenced Samuel Taylor Coleridge.

Bachman was in Philadelphia when Alexander von Humboldt visited the United States in 1804. Although Bachman was only 16,

> it being known that I was occasionally in the habit of accompanying Wilson in his researches in Ornithology, and of spending my vacations

and Saturdays in Bartram's garden, the usual resort of botanists, I was honored with an invitation to meet those who were about to welcome this eminent philosopher and naturalist to our country. . . . I saw him every day during the few days he remained in Philadelphia. . . . for the past sixty years we corresponded at long intervals. His publications, as they successively appeared, mostly in the French language . . . were regularly sent to me. (Bachman 1888: 173, 391)

In addition to learning from Wilson and Bartram, Bachman learned much from Audubon, and Audubon learned much from him. Bachman was living in Charleston in 1831 when Audubon came there, and he immediately invited him and his two assistants to be guests in his house. Audubon recorded their meeting: "When I first saw this excellent man, he was on horseback, but upon my being named to him, he leaped from his saddle, suffered his horse to stand at liberty, and gave me his hand with a pressure of cordiality that electrified me. I saw in his eyes that all he said was good and true" (Herrick 1938: vol. 2, 9). Audubon wrote his wife, Lucy, that these were "some of the happiest weeks of my life" (Shuler 1995: 78). After Audubon and his assistants (the landscape artist George Lehman and the taxidermist Henry Ward) had left for Florida, Bachman wrote Lucy and echoed Audubon's sentiments: "the last has been one of the happiest months of my life . . . besides receiving many lessons from him in Ornithology, he has taught me how much can be accomplished by a single individual who will unite enthusiasm with industry. . . . we were inseparable." He wrote Audubon, "your visit to me gave me new life, induced me to go carefully over my favorite study, and made me and my family happy" (Bachman 1888: 94–95, 101). Their friendship flourished, and they began to exchange letters, information, and visits. Bachman also wrote to Audubon, "I liked Wilson because he studied nature; I like you because you give theory to the dogs; because you give to the opinions of others just as much as they are worth; because you will examine and judge for yourself, and because you study, where every naturalist ought, in the wide field of Nature." Bachman was always an adversary "when nature is distorted for the purpose of advancing a theory" (Bachman 1888: 130, 314).

The first volume of Audubon's *Birds of America* had been hailed throughout Europe when it was published in Britain in 1827. In 1831 Audubon was preparing the second volume of illustrations. Over a period of years, Bachman furnished Audubon with information for the five volumes of the *Ornithological Biography* that accompanied the four volumes of plates, and in his text, Audubon mentions Bachman's assistance 134 times (Happoldt 1960: 63). Audubon named two birds for him: Bachman's Warbler (*Vermivora bachmanii*) and the American Black Oystercatcher (*Haematopus bachmani*) (Sanders and Ripley 1985: 125). During the 1830s, Audubon lived primarily in Edinburgh and London while his *Birds* was being published. As a consequence, "Bachman's house became, in effect, Audubon's American home where he was always welcome, where he stored his treasures, where a studio awaited him, and where he frequently returned to work and plan the moves necessary to complete *Birds of America*" (Shuler 1995: 79).

Despite the great acclaim he was receiving for his *Birds of America*, Audubon was attacked by several men intent on destroying his reputation in favor of Alexander Wilson's. Bachman urged Audubon to ignore "smoke from a Dung Hill" and to let his friends defend him (Herrick 1938: vol. 2, 142). In 1834, after an American writer sent unfounded criticism to Loudon's *Magazine of Natural History*, Bachman quickly and effectively defended Audubon in the same journal. An item of contention was whether vultures located their prey by sight or smell, and Bachman devised additional experiments to confirm Audubon's findings that the two American species in question relied on sight. His conclusions have been questioned, but his approach to solving this problem is worthy of consideration.

While writing the text of the *Birds*, Audubon acknowledged that he gained many insights as well as much information from Bachman. Bachman persuaded Audubon, for example, that every species of birds has a distinctive song and that consequently a different song indicated a different species. Audubon first noticed, for example, that the Grey Kingbird was a different species by its song (Shuler 1995: 92, 95). During his 1838 trip to Europe, Bachman spent two weeks in Edinburgh helping Audubon revise a volume of his *Ornithological Biography*. After Bachman left, Audubon wrote, "the days which we enjoyed together were few, but delightful; and when at the end of a fortnight my friend left us, I felt as if almost alone, and in the wilderness" (Shuler 1995: 159).

In 1838 Bachman traveled throughout Europe from July 2 through November 11. He carefully examined the mammal collections in many European museums and learned about American species from the far west, extreme north, and deep south, areas that Audubon and he had not visited. Although he went to Europe primarily for his health, he took full advantage of the opportunity to do research and to visit leading naturalists. He met Charles Darwin, who attended Bachman's lecture on squirrels for the Zoological Society of London and who asked him numerous questions afterward and recorded the answers in detail in his notebooks of research. Later, Darwin frequently cited Bachman's text for the *Quadrupeds*, and he cited four of Bachman's separately written publications. Bachman probably also met Charles Lyell in London, and Lyell later visited Bachman twice in Charleston and recorded what he learned from him, especially about the limits imposed by geography on the distribution of species. Bachman visited Humboldt repeatedly in Berlin and Paris. In Freyburg, he summarized the progress of natural history in the United States in an address to 650 delegates attending the international meeting of naturalists. He also attended a meeting of the Academie des Sciences Naturelles in Paris (Happoldt 1960). He returned from this European trip even better prepared to be the coauthor of the *Quadrupeds*.

In 1839 Audubon planned to prepare the text as well as the illustrations for a study of mammals as he had done for his *Birds*. Bachman urged him to examine museum collections before leaving Britain: "It would be wise for you to study the skulls & teeth a little—you take these things by intuition." He further wrote, "I promised Harris & others that I would give a full synopsis of American Quadrupeds. I have done no more than make pretty full notes. These and all the information I have to give are at your service" (Shuler 1995: 168–69). Audubon's knowledge of mammals was limited primarily to the larger species, which were a small part of the total number. He soon realized that only Bachman could prepare the needed scientific descriptions of all types of mammals.

In his *Ornithological Biography* (the text that accompanied his *Birds*), Audubon had included much information about mammals in his "delineations" of American scenery, which were inserted to add variety to the scientific text. The information related mostly to incidents that occurred while hunting cougar, moose, pole-cat, deer, bear, raccoon, and opossum. He had been interested in the life histories of birds far more than mammals, but had made numerous notes about mammals in his journals and memorandum books. As important, literary, and interesting as these descriptions were, Audubon provided little scientific information that was required for distinguishing and defining species and rarely mentioned the smaller, often nocturnal mammals that form the overwhelming majority of mammalian species. Bachman knew the smaller mammals uniquely well, and most of the information about the majority of species represented in the *Quadrupeds* was adapted from his observations. The most vivid incidents about larger animals were edited from Audubon's notes.

During the early years of their friendship, Bachman's and Audubon's families grew continually closer. Audubon and his younger son, John Woodhouse Audubon, visited Bachman for much of the winter of 1833–34. Within a year, John Woodhouse was engaged to Bachman's elder daughter, Maria Rebecca, and in 1837 Bachman married them. This union of the two families produced descendants of Audubon and Bachman. During the winter of 1836–37, John and Lucy Audubon stayed with the Bachmans. In 1839 Audubon's other son, Victor Gifford Audubon, married Bachman's daughter Mary Eliza. In response to a letter of congratulations from Audubon on "this double union of our families," Bachman wrote that he would miss his daughters, whose "happiness requires a removal," but he was glad to have two additional sons, who would doubtless help greatly with the planned study of quadrupeds. When Audubon later became mentally incapacitated, John Woodhouse prepared approximately half the illustrations for the *Quadrupeds*, and Victor supplied Bachman with requested information from published sources unavailable in Charleston. After Maria Rebecca and Mary Eliza died from tuberculosis in 1840 and 1841, Bachman continued to treat Audubon's sons as his sons-in-law and referred to them as such even after they remarried, and he continued to help them and his and their descendants in every way possible (Bachman 1888). He accepted no remuneration for writing the three volumes of text for the *Quadrupeds* so that all proceeds would go to members of the Audubon family.

For decades Bachman was at the center of a group of Charleston naturalists, and he was one of two who contributed comprehensive and internationally significant publications. The importance of Bachman's role in the group was fully revealed by Lester D. Stephens in *Science, Race, and Religion in the American South: John Bachman and the Charleston Circle of Naturalists, 1815-1895* (2000). The other most significant naturalist in the group was the herpetologist John Edwards Holbrook. Holbrook's *North American Herpetology* was originally published in four volumes from 1836 to 1840, and a second edition in five volumes was published in 1842. Agassiz also praised this study as finer than anything Europe had produced on the same subject. In addition, Holbrook wrote a less comprehensive study of fishes, the *Ichthyology of South Carolina* (1855, with a second edition in 1860).

The naturalists in Charleston helped one another in many ways, particularly in keeping their enthusiasm high. They included the polymath Lewis Reeve Gibbes, the paleontologist Francis Simmons Holmes, the zoologist John McCrady (who became Agassiz's successor at Harvard University), the botanist Francis Peyre Porcher, and the conchologist Edmund Ravenel (Sanders and Anderson 1999; Stephens 2000). Five of the seven principal naturalists in Charleston had an interest at the professional level in zoology and two in botany. Gibbes's research extended to chemistry, physics, and astronomy. During the antebellum period, only Philadelphia, Boston, and New York had more naturalists of the same caliber, and although a smaller city, "Charleston produced a group of naturalists equal in ability and accomplishment to any elsewhere in the nation" (Stephens 2000: 9).

Charleston's naturalists and others met regularly to discuss presentations at meetings of the Literary and Philosophical Society of South Carolina. Bachman presented numerous lectures to its members, and several were published by the organization. He became its president in 1836 (Stephens 2003: 156, 164-70). Stephen Elliott had been its president from 1814 until his death in 1830, and his two-volume *Sketch of the Botany of South-Carolina and Georgia* (1816-24) had provided a basis for Bachman's list of plants found in the vicinity of Charleston (Bachman 1834). Bachman was later an active member of the separate Elliott Society (Stephens 1989).

Louis Agassiz is best known for discovering that much of Europe had been covered by glaciers, and he later became the leading authority on invertebrates worldwide. From 1847 to 1853 he spent a portion of almost every year in Charleston, and he had much influence on its naturalists. Bachman had thought highly of Agassiz from the time he first heard him speak in 1838 at the meeting of European naturalists in Freyburg. In 1850 Agassiz helped persuade the American Association for the Advancement of Science to hold its third annual meeting in Charleston (Stephens 2000: ix). During this meeting, Agassiz proposed the creation of a museum of natural history in Charleston. He urged that the natural history collections that had been begun in 1773 by the Charleston Library Society in emulation of the British Museum and that had been greatly enlarged by the Literary and Philosophical Society be further augmented with contributions from naturalists to create a professionally staffed museum with a full-time curator. Agassiz, Bachman, Holmes, and Michael Tuomey (the state geologist) contributed large parts of their collections to help create the College of Charleston Museum. In a few years, under the curatorship of Holmes, this organization became by far the largest museum in the South. Its collection included many of Bachman's type specimens and skins collected by Audubon for the *Quadrupeds*. During the second half of the nineteenth century, the museum filled one-third of the college's main building, and the need for more classroom space resulted in the relocation of the museum off campus. A separate board of trustees was created to oversee what became the independent Charleston Museum (Mazyck 1908).

As interested as Bachman was from early in his life in natural history, he seems never to have considered becoming a full-time naturalist like Audubon or Agassiz. He planned to be a lawyer, but upon reading Luther's *On Galatians*, he felt called to the ministry.

About his religious education, he wrote, "I was especially indebted to Dr. Quitman, of Rheinbeck; to the Rev. A. Brown [Braun], of New York; and to Dr. Mayer of Philadelphia, for their instructions while I was a student in Theology" (Bachman 1888: 21, 25). The Rev. Frederick Henry Quitman was president of the New York Synod and was widely known as a rationalist. He had attended the University of Halles when Johann Salomo Semler was creating modern biblical criticism there. "Few men in the history of the Lutheran Church in America have been so prominently caught up in the turbulence of an intellectual revolution as was Quitman" (Bost 1963: 58). Quitman and the Rev. Anthony Theodore Braun were both professors at Hartman Seminary, which also had an academy and later became Hartman College (Bost 1963: 5, 24, 49). Bachman was evidently a student first at its academy and later at its seminary. For a time, he lived with Braun, who had been a Catholic missionary among the Indians of Canada before converting to Lutheranism. The third teacher who influenced Bachman most was the Philadelphia clergyman Philip Frederic Mayer. Bachman had been in Philadelphia as early as 1804, and he moved there by 1810. His initial Philadelphia connection seems to have been Mayer, who was Quitman's stepson. Mayer was a graduate of Columbia College (later University) in New York City and had studied under the Rev. John Christopher Kunze, a former professor of classics at the University of Halle. Later, Mayer became renowned for biblical criticism. Thus, Bachman studied under three theologians with exceptional educations and unusual qualifications.

Beginning around 1810, Bachman taught for about a year each in three different schools while completing his own education. He wrote, "first I went to Frankfort, Penn., where I remained nearly a year" (Bachman 1888: 22). In Frankfort, a suburb of Philadelphia, Bachman taught at a school created by Saint John's Lutheran Church, of which Mayer was pastor (Bost 1963: 64–66). Next he was persuaded by Alexander Wilson and Wilson's nephew William Duncan to teach at a school where both of them had formerly taught, Elwood School in Milestown (Bachman 1888: 22). Afterward, he returned to Philadelphia to study more closely with Mayer.

When Braun died in 1813, Bachman was asked to succeed him as pastor of the three churches of Gilead Parish in upstate New York, including the church he had attended as a youth. Consequently, the Lutheran Synod of New York licensed him to preach in 1813. Bachman served

Gilead Parish for about a year and a half until his health failed. While in college, Bachman had had a "hemorrage of the lungs" and was forced to withdraw before graduation, and when the hemorrhaging recurred in 1814, he "took a sea voyage to the West Indies, and recuperated greatly." Although Bachman intended to remain in New York State, his lungs continued to hemorrhage, and his physician advised him "to seek relief in a more southern climate" (Bachman 1888: 26). St. John's Evangelical Lutheran Church in Charleston had been without a Lutheran minister since 1811, and when its vestry asked Dr. Quitman and Dr. Mayer to recommend someone to be its minister, both recommended Bachman. Quitman ordained Bachman in December 1814 so he could accept the position.

Bachman arrived in Charleston on January 10, 1815, and lived there during the remainder of his life. He delivered most of his sermons in English, which the majority of his congregation spoke, but from 1815 to 1841 he also delivered one sermon a month in German, which he had spoken fluently from childhood (Bachman, 1888: 201).

From 1811 to 1814, Episcopalian ministers had preached at St. Johns, and the number of communicants had fallen to four, but others were waiting to be confirmed (Bost 1963: 49). About eight months after Bachman's arrival, the church had so many communicants that a larger building was needed. A cornerstone was laid on August 8, 1815; the large brick building was constructed entirely by members of the congregation; and the extant church was dedicated in January 1818. During the church's construction, Bachman asked that provisions be made for Blacks to attend services, and a balcony was added for them. By 1825 the congregation had grown to 290 Whites and 91 Blacks, and in 1862, there were 310 Whites and 200 Blacks. The membership continued to grow despite the creation of two additional Lutheran churches in Charleston (Bost 1963: 124–26). During his ministry at St. John's from 1815 to 1871, Bachman baptized over 2600 Whites and 2000 Blacks. He instructed, baptized, married, and buried four generations of its members.

Bachman was much beloved for his fairness and earnestness. His sermons emphasized the New Testament rather than the Old, and his principal subjects were "truth, integrity, justice, and mercy" (Bachman 1888: 57). His sermon on the life of the Rev. J. G. Schwartz reveals as much about Bachman as it tells about Schwartz. Some other sermons were published, and a few survive in manuscript, but most were destroyed in 1865.

> He wrote a manly and pure English.... his style and manner were indescribably earnest.... He was a man always 'fully persuaded in his own mind,' and therefore did not fail to persuade others.... Everything he did was recounted. They [the members of his congregation] were proud of his fame.... Even pain for a moment ceased at the sound of his voice. His tenderness comforted. His prayer seemed more certain to be answered.... Dr. Bachman's power lay not in what he did or said, nor in his manner, but in himself. (Horn, 1884: 272–73)

In an introductory note to Bachman's last sermon, which was published in 1870 in the *Charleston Daily Courier*, the editor wrote, "he stands forth at once the type and embodiment of all that can attract and concentrate human affection. Probably, no man in this community is more tenderly beloved" (Bachman 1870). A friend and colleague wrote that Bachman was "genial, observant, of vast and varied knowledge and experience, among those whom he had known from babyhood and on whose whole life, even their schooling and their business, he had exercised much influence, in ardent sympathy with their institutions and manner of thought, and keenly appreciative of their daily affection, it is no wonder he was beloved as well as admired" (Horn 1884: 272–73).

During his years in the ministry, Bachman's religious influence was so great that ten members of his congregation became ministers (Bachman 1888: 318, 402–10). He devoted the greater part of his time to ministering individually to the members of his congregation, but he also helped others regardless of denomination. Every morning he visited the sick and needy with vegetables and flowers from his own garden. He treated all Christians as "brothers" and found numerous positions of employment for Catholics as well as Protestants (Bachman 1888: 286).

Bachman usually managed to remain on good terms with most of the persons he disagreed with. In one case, for example, he wrote, "we regret that we are obliged to differ from an author who is generally accurate, and who is always courteous in his language towards other naturalists, but in this case we must do so" (Audubon 1989: 82). In 1837, he published a pamphlet that included a statement directed only against a group of Lutherans who believed in transubstantiation, which he dismissed as irrational, that inadvertently offended Charleston's Catholics. Bishop John England wrote to him at the time,

> I have always regarded you, and I trust I ever shall regard you, as a lover of literature and a man of mind who has successfully cultivated a taste naturally correct, a man who deserves to be considered a scholar and a friend of science; in society, a gentleman, and with me, this word means much. Differing widely from you in religion, I have, however, esteemed you, not only as a man possessing much candor and honesty of purpose, and having as little of the bigot in your composition as most men I know. (Bost 1963: 299–300)

Later, however, Bachman was personally and repeatedly attacked in a local Catholic journal, and he responded by being explicit about why Luther broke with the Catholic Church.

Belief in an omniscient God gave Bachman an unshakable faith that any apparent contradictions in the Bible could be reconciled. He had no doubt that God created human beings separately from all other animals. He seems nowhere to have recorded his reaction to the *Origin of Species*, but he had a low opinion of "the absurd theory of La Mark [Jean-Baptiste Lamarck]" (Bachman 1850a: 37).

In 1833 Bachman wrote Audubon, "I have read the speculations of men, I have listened to the tales of the ignorant traveler, and it seemed as if there were defects in all the works of God. Then I have turned to the fields and woods; to the air, the earth, and the sea; and I perceived that all was order, harmony, and beauty, and I have acknowledged that all the defects were in the short-sightedness of man" (Bachman 1888: 131). Although his religious convictions rarely affected his conclusions, they greatly influenced his interest in natural history. He believed that nature had been created in ways that would not require further change to remain in balance and that everything had been created as it needed to be.

There is some uncertainty about the extent of Bachman's education, but none about his erudition. As noted, he definitely attended Hartman Seminary, and its successor, Hartman College, lists him as an alumnus.

Some early sources mention that he attended Williams College but withdrew before graduation because of his health, and one source mentions Union College, but no record of his enrollment has been found at either institution (Bost 1963: 10, n. 20; cf. Bachman 1888: 20; Peattie 1928: 466). He received informal training at the college level from at least two professors in religion and from Wilson, Bartram, and Audubon in natural history, and he further trained himself through reading and research to produce writing at a professional level in his two principal fields of study.

From 1834 to 1848, Bachman served as a trustee of the College of Charleston and from 1848 to 1853 as the college's first professor of natural history. In 1835, Pennsylvania College at Gettysburg awarded him the Doctor of Divinity degree, and the letter from Dr. S. S. Schmucker informing Bachman of this degree has survived (Bachman 1888: 92). After his name, he sometimes added "D.D., LL.D." (as on the title page for his *Defence of Luther and the Reformation*). In 1838 South-Carolina College (later the University of South Carolina) awarded Bachman a doctor of letters (Horn 1884: 271; cf. Stephens 2000: 59), and in 1841 its trustees asked him to become its president. He declined, preferring to continue to minister to his congregation, to remain in Charleston, and to continue his researches. As he wrote to Audubon, "such a step would moreover have put an end to my amusements in Nat. History" (Bost 1963: 359). Writing during Bachman's lifetime and undoubtedly with his assistance, James Wood Davidson asserted that Bachman had been awarded a Ph.D. in Berlin, and he mentions having frequently visited the University of Berlin (now Humboldt University). Although no record of this degree has been found, Wilhelm von Humboldt (Alexander's brother) is likely to have conferred it (Davidson 1869; cf. Bachman 1888: 200; Stephens 2000: 277, n. 31). In 1839 Bachman was made a member of the Board of Visitors of West Point to examine the education of its cadets, and he wrote a detailed report for the secretary of war, his friend Joel Roberts Poinsett. In 1857, he was elected as chairman of the board of the newly created institution that became Newberry College (Bost 1963: 357–58, 367).

For his work in natural history, Bachman was the recipient of numerous honors. He was president of the Literary and Philosophical Society of South Carolina from 1836 to 1839 (later renamed the Elliott Society of Natural History and called by various other names) (Stephens 2003: 165, 174). On the title page of his *Doctrine of the Unity of the Human Race*, he noted that in addition to being "Prof. Nat. Hist. College of Charleston," he was a "Corresponding Member of the Zool. Soc.; Hon'y Member of the Entomol. Soc., London; Cor. Memb. Royal Botanical Soc., Saxony; Royal Soc., St. Petersburgh; R. S. A., Copenhagen; Acad. Nat. Sciences, Philad.; N. Y. Lyceum; N. H. Soc., Boston, N. Haven, and Toronto; National Institute; American Assoc., etc., etc."

Ordinarily, Bachman went out of his way to avoid controversy, but being a man of high principle and strict integrity, he was not infrequently at the center of controversies. He rarely defended himself, but he often defended his principles and his friends. He angered slave owners (even though he was one) by making an impregnable scientific argument against the assertions of Agassiz and others that the Negro race was a separate species. He successfully defended Audubon against a campaign by unscrupulous naturalists who attempted to discredit him in a total of nineteen publications (Shuler 1995: 122). He exposed frauds and entered numerous sectional and religious controversies when he felt truth was at stake. A number of scientists attempted to discredit him and failed, and he exposed their shoddy scholarship.

The historian Frederick Porcher, a fellow professor at the College of Charleston, wrote that Bachman's "well known reputation as a naturalist gave him a position of eminence and his genial manners and benevolent spirits made him loved by his associates.... He had a very tolerant spirit for every thing except spirits, both material and moral, and against anyone who doubted the unity of the human race" (Porcher 1946: 218). Bachman and Audubon had serious disagreements about Audubon's excessive drinking and about Audubon's attentions to Maria Martin, whom Audubon had taught to paint and whom Bachman later married, but their friendship survived this and other disagreements and misunderstandings (Shuler 1995: 149).

Much of Bachman's life was devoted to his large immediate family. He married his first wife, Harriet Martin, in 1816, and they had fourteen children, eight of whom died as children or young adults (Shuler 1995:

87; Bost 1963: 141). In 1846 Harriet died after a long illness; the following year, Bachman married her sister, Maria Martin, the talented artist who painted a number of backgrounds for Audubon's *Birds of America* and reptiles for Holbrook's *Herpetology*. Maria died in 1863. Bachman was survived by five of his children (two sons and three daughters).

During his lifetime, Bachman had seventeen grandchildren and two great-grandchildren, and they and his remaining children were a great solace for all the family members he had lost. Many grandchildren lived nearby and often joined him for breakfast. He was especially fond of his namesake John Bachman Haskell: "he calls himself '*Dr. Bachman*'; he is willful and full of fun; he amuses me and vexes me by turns; yet, I confess that I do not feel quite satisfied if he is not sitting at table on his high chair at my side." He was later gratified when this first grandson became a minister. Before his untimely death, the Rev. J. B. Haskell compiled the material that his aunt Catherine L. Bachman used to complete her father's biography. It is almost an autobiography, for Haskell left these instructions: "Let him speak for himself, whenever it is possible" (Bachman 1888: 275, 420, 5).

Although Bachman was extraordinarily vigorous, he suffered serious illnesses at intervals throughout his life (Bachman 1888: 56, 189). He contracted what was evidently tuberculosis in his youth, which at one point confined him to bed for six months. Twice he nearly died from internal bleeding, and on two other occasions he nearly lost his sight (once from a high fever and later from an accidental gunpowder explosion caused by the ignorance of a slave). He contracted cholera while ministering to the sick. His 1838 trip to Europe was made primarily to recover from an incapacitated arm and leg. In 1865 a Union soldier permanently injured one of his arms. In old age he suffered severely from rheumatism and in 1871 had a paralyzing stroke, but he continued to correspond with his friends. His final stroke was early in 1874, and he died soon afterward (Shuler 1995).

The illnesses Bachman endured and the losses in his family were severe tests that at least once made him question his faith. After a lifetime of attempts to understand why so many losses had been necessary, he concluded that "the ways of God are dark and incomprehensible to us poor short-sighted mortals. It is our duty not to murmur, but to pray for submission." In 1864 he wrote, "God may have something more for me to do, if it be only to suffer" (Bachman 1888: 197, 221, 240, 377). Despite his often poor health and many crushing disappointments, he lived a long and productive life that enabled him to complete nearly every project he began. His scientific work ended after his library was burned in 1865 in the fire that destroyed Columbia, South Carolina, where his books happened to be on their way to Newberry College.

Bachman died on February 24, 1874, soon after his eighty-fourth birthday. "Thousands came to view his body, especially Negroes who had known his kindness, and hundreds of children were lifted to kiss the face of a leader especially beloved" (Peattie 1928: 467). His body was placed in a metal casket and buried in a brick vault in front of the chancel of the church his congregation had built soon after his arrival and in which he had served for six decades (Bachman 1888: 434).

Upon Bachman's death, Benjamin Nicholas Martin, a professor at New York University, wrote a "review of his life and labors" that included the following passage: "He was the first to give any general and trustworthy account of the origin of animal varieties; and his work [*Unity of the Human Race*], though subsequently thrown into the shade by the more elaborate treatises of Darwin, is yet a most valuable repository of accurately observed facts, and illustrates some important laws nowhere else so distinctly stated" (Martin 1874: 118).

Everything Bachman wrote is still well worth reading, and there is much to be learned from it. Most worth reading is his *Unity*, in which he presented conclusive evidence that all races were a single species and demonstrated that the selective evidence presented by polygenesists was incomplete, inaccurate, and misleading. Not many people have gone against the thinking of nearly all their contemporaries and been right, but Luther, Darwin, and Bachman did. This is not to equate Bachman's accomplishments with theirs, but to recognize the value of his similar methods of thought.

The text of the *Quadrupeds* is, as Cahalane wrote, a great pleasure to read as well as a great storehouse of fascinating information on all types of mammals. It has been surpassed, but not superseded.

Bachman's history of the Reformation was written to defend Luther against recent attacks by local Catholics, but since his defense was based on primary sources, it has permanent value. It is strongly biased in favor of the Lutheran point of view but well documented. It is also of interest as the record of a religious controversy in which free speech was a major issue.

Primarily for how he approached the solution of major problems, Bachman's writings continue to be worthy of careful consideration. His articles are characterized by sound methodology and great common sense. His essays and letters make good reading. His scientific writings are not primarily important for their literary merit, but his prose is clear and to the point, and more importantly, his insights were often profound. His insight, erudition, and integrity and the breadth of his knowledge and understanding enabled him to create a large body of work with permanent value.

Funeral Discourse of the Rev. John G. Schwartz

John G. Schwartz was born in Charleston in 1807, and when his father died in 1819, Bachman assumed responsibility for his education. Schwartz graduated from South-Carolina College in 1826 and briefly taught classics at the College of Charleston, but resigned to become a minister.

In 1827 when Bachman was seriously ill for several months, Schwartz preached in his stead and wrote to him, "to every member of your congregation your illness has been an affliction, and your recovery a blessing. I think that I could die easy and happy if I had such a congregation weeping for me and praying for my welfare" (Bachman 1888: 64). Schwartz died when he was only twenty-four, but even in so short a life, he had become a highly regarded scholar and had created a college that still exists.

The Synod of South Carolina unanimously selected Schwartz to create a Lutheran theological seminary. At the time, a seminary was widely opposed as unnecessary, but Bachman successfully argued that ministers were "bound to keep pace with the intelligence of the people."

"Schwartz was a man of great promise and of extraordinary devotion, and the story of his election to the professorship and self-denying labors, as detailed by Dr. Bachman, is very affecting" (Horn 1884: 269–70). In his eulogy, Bachman praised Schwartz for his "mildness, forbearance and charity." He showed that all Schwartz had learned and accomplished in so short a life was an "example [that] shall remain."

The / Funeral Discourse / of the / Rev. John G. Schwartz. / Delivered / September 11, 1831. / by the Rev. John Bachman / Charleston: / Printed by James S. Burges / 1831

At a Meeting of the Congregation of the Lutheran Church of German Protestants of this City, held on the 11th September the following Preamble and Resolution were adopted:

Whereas, It has pleased the Almighty Disposer of Human Events, to remove from society, in the morning of his life, and from a scene of active and useful exertion in the Church of Christ, the Rev. John G. Schwartz, who was brought up in the Congregation under the kind instructions of their Pastor, who was for a length of time an active and successful teacher in our Sunday School, who frequently officiated in this Church, who was engaged as a Missionary by a Society formed in this Congregation, and who was endeared to us by his talents, his virtues, and piety.

Sympathising as we do with his widowed Mother and Relatives for the loss of one who was so dutiful a son and so affectionate a brother and friend, duly sensible of the affliction of his Congregations, in being deprived of an intelligent and pious Pastor, and above all deploring the irreparable loss which the Theological Seminary, over which he presided as Professor, has sustained in the death of one who was so well qualified to advance the interests of our beloved Church:

Resolved, therefore, That the Congregation of the Lutheran Church of German Protestants in Charleston, deeply deplore the loss which the Lutheran Church of South-Carolina, and the Seminary of the Lutheran Synod of South-Carolina, and the adjacent States, have sustained in the death of their lamented friend, the Rev. John G. Schwartz.

Resolved, That they deeply sympathise with the Congregations of Bethlehem Church, St. Luke's Church, and St. Matthew's Church of Newberry District; also with the Congregation of St. Mark's Church of Edgefield District, in the irreparable loss which the Lutheran Church in South-Carolina, and they in a more particular manner have sustained in the lamentable death of their Pastor.

Resolved, That they fully commiserate the heavy affliction of his bereaved Mother and family and also of our beloved Pastor in the lamented death of his pupil and friend.

Resolved, That the Church be clad in the space of mourning for three months, as a testimony of the deep regret of this Congregation for the great loss which the Lutheran Church in South-Carolina has experienced.

Resolved, That a copy of these Resolutions be transmitted to the Congregations of Bethlehem, St. Luke's, and St. Matthew's Churches of Newberry District; to St. Mark's Church of Edgefield; also to the Students of the Seminary of the Lutheran Synod of South-Carolina and the adjacent States; to his bereaved mother and family and also to our distressed Pastor.

Resolved, That the Rev. Mr. Bachman be requested to give a copy of the Sermon delivered by him this Morning on the Lamented death of the Rev. J. G. Schwartz for publication, and that 600 copies be printed for distribution.

Jacob Sass, *President*,
Jacob F. Schirmer, *Secretary*.

Funeral Discourse.
Revelations ii. 10
"Be thou faithful unto death, and I will give thee a crown of life."

The life of man is so full of difficulties and trials—his happiness is so liable to interruption, either by his own losses and pains or by the afflictions and death of those in whose fate he is most interested—his own period of existence is so uncertain, and may so unexpectedly be brought to a close that he is reminded of the necessity of looking beyond the confines of this uncertain world for consolation and happiness.

On those solemn and mournful occasions when it has pleased a wise and holy God to remove from us the objects of our affection and love, we have insensibly yielded to a law of our own nature, and shed the tear of sorrow; we dwelt in fond affection on their virtues; we threw the mantle of charity over their faults, and we endeavoured to fortify ourselves by the religion of hope and of mercy to be enabled to bow with submission to the will of heaven. But above all, we have felt, when the poignancy of our grief was assuaged, that a lesson of instruction could be derived on reviewing the characters of the lamented dead and that our eulogies though they fell unheeded on the graves of the departed might yet prove incentives to the living. It is on such an occasion, my brethren, that we are assembled in this sacred place to cherish the remembrance of a dear departed friend, whose early life promised so fair and who as a man and a minister was every day growing in our affections.

The task of portraying his amiable and pious qualities of heart and life has fallen upon one whose fond hopes never allowed him to indulge the thought that such a duty would ever devolve upon him, and whose feelings for his own private loss render him but imperfectly qualified for the task.

John G. Schwartz was born in this city on the 6th of July, 1807. His father, a native of Germany, was a truly exemplary and pious man, a strenuous supporter of our church, and a communicant with us till the day of his death, which occurred in 1819, when the subject of this address was not quite twelve years of age. His mother is still living. How faithfully she discharged her duty to her household, when the husband and father was removed, is known to you all. The reflection that some of the most talented and pious men, the ornaments and benefactors of society and the world, were indebted for those early incentives to virtue and goodness which rendered them so conspicuous in after life, to a mother's watchful care and pious example, cannot fail to assuage the acuteness of sorrow and offer supports to many a mourning widow whilst anxiously revolving in her mind the fate of her now fatherless children.

But although the formation of the mind and the direction of the life of this young man in the most critical period of his existence (from his twelfth year) devolved solely on his mother, yet the principles of a noble emulation, and the seeds of virtue and religion were instilled into his mind at an earlier period. Like Zechariah and Elizabeth of old, the piety of parents was rewarded by the blessings of heaven, and like them, God encouraged them with the hope, in answer to prayer, that their son should also one day preach repentance to the people in the sanctuary of the Lord.

It is now more than twelve years (although it seems but as yesterday, so indelibly are the circumstances impressed on my mind) since the father of Mr. Schwartz, reduced almost to infant weakness by a lingering disease, called me to his bedside one evening and said, "This boy has given me such proofs of possessing talents. He seems to be so virtuously and religiously disposed that I thought if I should live I would try to give him an education so that by God's blessing he might become a minister in our church; but as I feel that I am but a short time for this world, I will take it kind in you if you will encourage him should he continue to feel so inclined, and I trust God will bless you for this act of kindness to one who will soon be fatherless." Among the few acts of an imperfect life that afford consolation in the review, the belief that this promise has been faithfully fulfilled, now comes in to cheer me under this unexpected and heavy bereavement.

At the death of his father, when he found the family dissolved in tears, he in the most feeling and affectionate manner comforted his mother with an assurance that he would do all in his power to support her in life and cheer her in her journey to the grave. That a mother's love has been requited with no common filial attachment is well known to most of those who hear me.

His grief for his recent affliction had not yet subsided when he came to express a desire that I would advise him in his studies, as he was anxious, by the blessing of God, if his life should be spared, to devote himself to the ministry. I gave him such advice as I thought suited to a person of his tender years, and fearing that his resolutions might have been hastily made under peculiar excitements, and perhaps by the persuasions of his relatives, I requested him to reflect on the subject for one year, and if his resolutions still continued firm, I would then give him farther directions. I met him frequently; but although he was as studious and correct as before, yet he said nothing farther on the subject, and our conversation had almost escaped my recollection, when he one day presented himself before me. It appeared an unusual visit at an unusual hour, for it was early in the morning of a rainy day. He told me that he had come to remind me of his promise to give me the result of another year's reflection; that it was that day a year since I encouraged him to call, and he had come punctually to say that his feelings and wishes were still the same, and that his resolutions to devote himself to the service of the church remained unchanged.

From that day till the time of his entering college he recited to me his lessons on Saturdays, and became like a member of my family—the pride and joy of my house; and from thence commenced an attachment which he never weakened by saying or doing any thing that ever gave me offence or a moment's pain.

He received his academical education first at the school of Dr. Jones of this city, but principally at the school of the German Friendly Society, where he uniformily held the first rank; and it must be remarked that so well was his reputation established that all of his classmates cheerfully accorded to him the praise of deserving the highest honours in the school.

In 1824, steady to his purpose of devoting himself to the service of our sanctuary, he applied to become a member of this church, and was publicly confirmed. And here let it be observed to the shame of many professors, that he carried into practice and exhibited in after life to the world, the vows and resolutions which he made at this altar. He never neglected his communion. He never lost his attachment to the church or her ordinances. It grew with his growth and strengthened with his strength. It was but a short time before his death he expressed himself on this subject in these words: "The book you have sent me on the purity and orthodoxy of our church places the subject in a clearer light than I ever beheld it before. I wish we could have it translated for the benefit of our people. The more I study the principles of our church, the more am I convinced that they contain the true doctrines of the Bible; and that the more they are studied, the more they will be admired."

He entered the Junior Class of the South-Carolina College in the autumn of 1824, and graduated in 1826, having throughout his collegiate course conducted himself with such propriety that he was greatly beloved by the members of his class. He received a high honor when he graduated, and a letter from one of the professors stated, "He is not only among the best scholars, but one of the very best young men that graduated here for many years past."

He had already turned his attention to the study of Theology whilst in the Senior Class at College. He was unremitting in his studies during that winter and the following spring, and in the summer of 1827, when he had not yet attained to the 20th year of his age, he preached his first sermon in this church. This congregation will not soon forget the interest which was then excited. We congratulated each other that another minister, and one who promised so fair, was to be added to the church. Duty called me away from the congregation as we then supposed for a few weeks, but a severe illness detained me at the North for a whole summer, and when I returned all had been gratified with the services of their young minister. He had preached twice on the Sabbath and frequently lectured once a week, and attended to many parochial duties, besides teaching every day in a grammar school. All this he had been able to effect by a careful distribution of his time, and by devoting a part of each night to his studies.

In the autumn of that year he was licensed as a minister by the Synod of our church and was immediately engaged as a missionary. In this capacity he visited nearly all the middle and upper districts of the State, preaching frequently every day in the week. He brought to us an account of the state and wants of the country far more correct than any we had before received, and he thus enabled us to see the necessity of renewed exertions to procure ministers and encourage our brethren in the interior to build churches and organize themselves into congregations.

His anxiety to remain in the city where he might have recourse to books and more thoroughly ground himself in the study of theology induced him to accept of the situation of classical teacher in the Charleston College, which duty he discharged with satisfaction to the Trustees and to the young gentlemen who were entrusted to his charge. But he soon discovered that the fatigues incident to the faithful discharge of his duties at College left him little leisure for study; this added to a strong conviction that he was not on the post of duty, as long as he was not preaching the tidings of salvation to others, caused him to be uneasy and unhappy in this secular employment, and his health and spirits sunk together. He relinquished this situation, to the great regret of his employers and spent a short time in the summer of 1829 in travelling through the Northern States. In this journey his health was once more restored, and he hastened to present himself as a missionary to "the Society for the promotion of Religion," formed principally of members of this congregation. His services were gladly accepted, and with renewed energy he recommenced his missionary labors. Several congregations in Newberry and Lexington were committed to his care. The number of his hearers gradually increased. They soon became acquainted with his talents and his worth, and when the Synod met in Newberry, in November, 1830, they petitioned that his services might be continued to them and stated that such had been the increase of their congregations that they hoped to be able to support him without farther aid from our Society.

At a meeting of the Synod of our church in November last, he, together with three other young ministers, was solemnly ordained and set apart according to the forms of our church to the work of the ministry. It was one of the most impressive sights that I have ever witnessed; many tears were shed by individuals in the concourse of people who were present on that occasion. They were tears of joy: God seemed to visit our Zion with his smiles, and we anticipated a long train of blessings to our beloved church. But alas! what a sad reverse have we been doomed to witness. One of those young men of early promise has removed to New-York, and we shall see him only at very long intervals, and the grave has closed over two others.

Mr. Schwartz took a lively interest in the transactions of the Synod. His modesty, his good sense and his correct view of ecclesiastical matters caused him to be looked up to with deference, and his sentiments were generally considered as correct and were sanctioned by the Synod. He was of a very conciliatory spirit, and contributed very materially to preserve that harmony which has reigned among the members of our ecclesiastical body.

It is well known to most of my hearers that the great obstacle to the increase of our church in this State was the want of ministers. Our doctrines were generally approved, new congregations were forming and new churches building, but still the number of ministers did not keep pace with the wants of the church. It was found impossible to procure clergymen from the North. Although the Theological Seminary at Gettysburgh now contains fifty students, yet we are informed that they can give us no assistance. Our church in the Middle and Northern States has increased so fast within a few years as to require all the assistance of those who are now entering into the ministry. There appeared no other way opened before us than that of establishing a Theological School of our own and affording the means of instruction to pious young men of our own State who were habituated to our climate and institutions, and accustomed to the people to whom they were expected to officiate. There appeared to be many difficulties in the way; we had no funds; few who either had the ability to become, or could be spared as professors, and we were doubtful whether we could obtain young men of high talents who would feel themselves urged by a sense of love to the Redeemer's cause to enter on the arduous duties of the ministry. Yet the object was so important, no less a one for the preservation of our church south of North-Carolina, that we determined prayerfully and zealously to enter on the work. Many difficulties had to be surmounted in the very commencement. A prejudice existed in the country against all institutions of this kind. Many of our people had imbibed the ideas which were entertained by other denominations by whom they were surrounded, that an intelligent ministry was not necessary, that no such institutions existed in the early ages of the Church, and that as our Saviour chose his successors from among the fishermen and tent makers, God would carry on his work under the preaching of unlearned men at the present day. They did not recollect that the Apostles of our Lord were for three years under the tuition of no less a personage than the Son of God himself, before they entered into the ministry, that the world has since undergone a material change, that the days of inspiration and miracles have gone by, and that the ministers of the Gospel are bound to keep pace with the intelligence of the people and must be ready to parry the weapons so successfully wielded by the advocates of infidelity. By a judicious explanation of these subjects, the prejudices against the institution were removed, and an unusual zeal was manifested in its support. Liberal subscriptions were set on foot, and such promises of pecuniary aid given, as authorized us to enter on work immediately.

It was necessary that a professor to the institution should be elected, and that he should enter at once upon the discharge of his duties. Every eye among the clergy and laity was immediately directed to Mr. Schwartz. They knew his education, his talents, and piety. Although but twenty-three years of age, he had made the best use of his short life. They were few better Greek and Latin scholars in our country. He had attended considerably to the Hebrew language. He was proficient in the French, and he was studiously directing his attention to the German, and read and translated that language with considerable ease. He had made an equal proficiency in the other sciences. In theology he was probably as well read as any young man of his age. He had attentively read all the most important writings on the subject; and although he preferred the doctrines of our church to all others, yet his soul was the seat of Christian liberality, it should be spoken to his praise, that although surrounded by Christians of other denominations, yet he never gave them offense, and they generally attended with satisfaction and improvement of his ministrations. The objections to his youth were every day removing. He received a unanimous vote as Professor of Theology. After the election there was a pause of many minutes, when he arose to address us. For a time his feelings almost prevented the power of utterance. He at length proceeded to thank us for our favourable opinion; stated his sense of his incapacity to discharge the duties of the station to which he had been appointed; pointed out its difficulties, but signified his willingness to undertake it by the help of God, and entreated our prayers and intercessions, and those of all Christians in his behalf. The use of the individual, the occasion, the importance of the subject, and the feeling and eloquent address moved the audience into tears, and I am sure that few who were then present will ever forget that impressive scene.

There was yet one difficulty to overcome. His congregations implored us not to deprive them of the services of their minister for this year. He had organized and built them up, and they thought without him they could not prosper. He immediately stated that he could not conscientiously leave the people of his charge, and although it was making a great sacrifice, yet he must remain with them another year, and would attend to such students as might be offered to him, provided they would consent to remove with him to the place where he should be located. His friends consented to this arrangement under an impression that during the sickly months he would comply with their request and remove to a more healthy situation.

He soon after entered upon his arduous and important duties. And it seemed as if Providence at once gave him the fairest prospects of success. A library was fast accumulating under the patronage of the Synod and the liberality of individuals. Five students immediately entered the institution; three more were promised on the first of January next, and in one of his letters he states that his fear now began to be that more would offer than we could accommodate.

Of the character of the individuals who were to be instructed for the ministry, he had a very correct conception. In a letter dated May the 9th, he thus writes: "You perceive, my dear friend, that we have at length taken a decided step, it may be said fairly to have made a start. God grant his blessing to us in the commencement and throughout the continuance of our institution. How fervently should we all pray to the great Head of the Church, that he may give into our hands pious and zealous young men, and enable us to send them forth into season such workmen as need not be ashamed. I feel, my dear friend, that I have taken on myself a burden of responsibility almost greater than I can bear, yet God's grace is sufficient for me, and I trust that under his blessing I shall at least perform my duty faithfully and conscientiously. I stand in need, however, of the prayers of my friends, and I call upon those at whose request I consented to accept a situation. I call upon my brethren in the ministry, to aid me by their prayers and by their counsel."

He continued: "All the young men now with me are promising, and if their hearts be right in the sight of God, I have no doubt that they will prove a blessing to our church. The heart is known, however, only to God. We can judge only by the outward appearance; but did I think that any of these students were deficient in proper use of religion and of the ministerial office, I should feel it my bounden duty to advise them at once not to enter this institution. I dread the idea of being instrumental in educating any one for the holy office of the ministry, who through a want of personal religion may bring disgrace upon our sacred calling. Whilst I can testify to the consolations and encouragements which the Christian minister will receive at the hands of God, in the midst of the peculiar discouragements and difficulties which belong to his profession, I believe those difficulties and discouragements to be of such a character as to drive any one from his office who does not feel the supporting comforts of God's presence. I could not, therefore, advise any person to enter upon this office without being convinced that he realizes the balm of religion to his own soul, and the importance of that duty which commands him to preach this religion to others."

Shortly afterwards he announced his resolution of not removing from Newberry during the summer, as his friends had advised. He writes in these words: "I know it will distress you to hear that I have determined to remain in this district during the present summer, inasmuch as I cannot otherwise attend as I ought to my congregations. Some of my people are increasing in their attention and seriousness, and I trust I shall yet be permitted to see the fruit of my labours. My students also preferred being in the neighborhood of their homes for the present year. I may be incurring some risk, but am I not in the hands of God? Has he not hitherto helped me? If it please him to remove me by any means from the church, will he who is the Head of the church permitt it to suffer thereby? I would not be presumptuous and my confidence, but am I not authorized to commit myself and all my concerns into the hands of him who hath said, 'Lo I am with you always?' Happy shall I be if I be made to the humblest instrument of glorifying his Almighty name."

To the further remonstrances of his friends and family he returned no answer. We knew that his resolutions were taken, and we committed him to the care of Providence.

He every day grew in the affections of his people. One of them wrote to me: "We are just beginning to worship your young friend." He wrote me on the 10th of August, detailed his labours, gave an account of the Sunday school he was establishing, and asked advice about the best books he could introduce. Of his studies he says: "Should Providence grants me the continuance of the health I present enjoy, I hope to note this summer as the most profitable season of my life. I have a good course marked out for my endeavors and noble subjects for the exertions of my powers. With the return of my health too, I feel my mind invigorated, and I begin to realize that the country is my home. My health is remarkably good. I feel myself in the hands of Providence, and I find that the more I can realize my dependence upon God, the more cheerful, contented and happy I am."

Shortly afterwards, however, these bright prospects were clouded with anxiety and sorrow. A letter from his physician came to state that he had been ill with fever, but it was hoped that his symptoms were favourable to his recovery. To this letter were added a few fond lines dictated by himself to which he wrote his name in a weak and unsteady hand. He stated that he trusted in and committed himself to his God and begged me to comfort his family. This was probably the last time he put his hand to paper. The next mail informed us that he was worse—the three physicians were in attendance—and that there was no hope of recovery. The next brought the sorrowful tidings that their friend and pastor had died on 26 August—that our fond hopes were all prostrate in the dust—and that Zion mourned the loss of one of her most faithful and devoted sons.

You will naturally expect me to give a short detail of the last hours of our departed young friend. Happily I am prepared to do this for he was blessed with the full powers of his mind till the last hour; but if even his intellects had been clouded, and his life had terminated in delirium, he would have still left behind him a higher consolation than a dying testimony, even the memory of a blameless, pious, well-spent life. As it is, we have the additional satisfaction of knowing that death had no terrors for him and of believing that he has gone to his rest above. One of his physicians writes: "His mind was calm, cool and deliberate throughout his painful illness, particularly so a few moments before he died, when he said, 'I shall shortly enjoy the glorious light of heaven, happiness and immortality.' 'I am not afraid to die, for I know that my Redeemer liveth.' He expired without a struggle and without a groan, having the faculties of mind and speech to the last." Another friend writes that when he visited him on his sick bed, he told him not to be distressed on his account for "whether he lived or died, all would be well." Another writes: "he declared himself ready and willing to go, and if he had any wish to live, it was on account of his dear mother, who might perhaps stand in need of him. But after that he never mentioned any thing more about his relations and friends, but was altogether engaged with himself and his own soul and the souls of others. One night he called them all into his room, and after one of his students had read a chapter out of the Bible, he prayed in his bed with a loud voice for nearly half an hour. To one of his physicians he said, 'See, Doctor, how much better it is to make our peace with God in the time of health than to wait till we are laid on a bed of sickness; for repentance on a sickbed is seldom of any avail.'"

Another friend writes of the calmness of his last moments, from which we learn that the nearer he approached to his end, the more composed and resigned he became that his face for the expression of calmness and submission and that it seemed as if the bright anticipations of the soul were settled on the lineaments of his countenance.

The shock given to the people among whom he lived by this event was unusual, and the calamity was heightened by its bereaving them of their fondest hopes. A gentleman who attended the funeral writes: "no tongue can express, no pen can describe the feelings of the people on this melancholy occasion. The remains of our dearly beloved friend were interred this morning in Bethlehem church yard; the largest concourse of people that were ever assembled in this country attended the funeral. The sad looks, the loud sobs, and tears shed on this mournful occasion amply testified the high esteem in which he was held by all, rich and poor, old and young, white and black — pardon me for introducing the word black but I must say that even the poor Africans sympathized and sorrowed, saying, 'Dear Mr. Schwartz.'" Three of the ministers of our church officiated at his funeral, and all bore testimony that never had an individual departed in that community who was more beloved, or whose loss was more sincerely lamented than was that of our departed friend.

Should it be asked what was the peculiar trait in the character of Mr. Schwartz, I would say that it was a solemn determination conscientiously to discharge his duty to his God. For this he left his peaceful home and the friends of his youth and retired into a sickly part of our country; and from thence he wrote: "Here in the woods of Carolina I suspect my lot is cast. Here I shall live and here I shall die. To be instrumental in doing good in enlarging the Redeemer's kingdom is all I ask."

That a man, who was so devoted to the duties of the Christian, should possess the amiable graces of benevolence, we cannot wonder. He felt it his duty to exert all his powers to do good to the bodies and souls of men. The great maxim, no man will liveth to himself, was engraven on his mind. Without profession or show, he engaged in and ardently devoted himself to every work of benevolence. He became a member of the German Friendly Society, and whilst in the city was an active member of the School Committee. He became a teacher in our Sunday

School, and if he was but one Sabbath in our city, he took his seat among our little children, and inquired of their progress and improvement. He formed similar schools in his own neighbourhood and interested his young friends to become teachers in them. He was active in the distribution of the Scriptures and was at the time of his death the President of the Bible Society of Newberry District. He saw the evils of intemperance in the country. He aided in the formation of Temperance Societies, and resolved to abstain even from wine, lest his example might be an encouragement to others. He contemplated, if his life should be spared, aiding in several works of a practical nature for the use of our people, and he had already commenced collecting the materials for an improved catechism for our church; and a benevolent individual from the country, who is better known by his deeds of benevolence than by his professions, had offered to defray the expenses of this publication.

Few persons discharged their duty to society more faithfully and conscientiously than the subject of this discourse. There could not have been a more affectionate son and brother, and he strove in his frequent letters and by every fond and endearing act to direct his kindred to the true sources of happiness and to convince them of his affection of love. He who addresses you can testify how firm and unaltered a friend he was and grateful for the slightest favors. From his boyhood he was entrusted with every secret and every feeling of his heart, and never in a single action, word, or thought did he evidence any other conduct than that of respect, gratitude and affection. To all around him he uniformly evidenced a spirit of mildness, forbearance and charity. No feelings of unkindness could keep possession of his soul beyond a single hour. The language of acrimony was entirely foreign to him, and if he could not speak well of another, he remained silent.

One great peculiarity of Mr. Schwartz, which was the admiration of his friends, consisted in a well regulated mind. His feelings, though warm, were under the control of his judgment. He loved study, but was also an interesting companion in society. His mind was harmonious, and from his looks and demeanor one could discover that it was at all times under a religious influence. There was nothing like religious cant in his conversation, yet his thoughts were continually recurring to the mercy and goodness of God, and when he was with his intimate friends it formed the frequent subject of remark even whilst in cheerful conversation. A beautiful flower or bright day, a cool breeze, the beauties of nature which he witnessed in his rides to his churches in the country were often the subjects of his remark in his conversations and letters, and he always spoke and wrote as if he be held in every new beauty and charm of the world a subject of love and gratitude to his Maker. In one of his last letters, whilst speaking of the happiness of his mind and contemplating these gifts of God's goodness, he says: "Excuse me for expressing so frequently the feelings of my heart on the subjects, but I cannot help it. It is almost like being again in your study. I seem to be talking with my friend who is near me."

As a minister, those who hear me had no fair opportunity of judging of the extent of his powers. He was under many restraints whilst preaching in a city and among his early associates. Unaccustomed in the country to preach written sermons, he felt embarrassed when he had to resort to it in cities. Those, however, who saw him and heard him among his own people were struck with the clearness of his mind, with the arrangement of his thoughts, the fervor of his eloquence, and the deep feeling of piety which pervaded his discourses; and it was remarked to me by one of our most eminent Judges that he was one of the most interesting and impressive speakers that he had ever heard in that part of the country.

On the whole, I feel confident that I will be borne out in my assertions by all who were intimately acquainted with the character of that lamented young man, when I state that in the faithful discharge of his duties as a son, a brother and friend—in the improvements of his mind—in every kind of knowledge that might render him an ornament to his profession—in all the various occupations of a herald of the Gospel of peace, he was an example and a model to all who duly appreciate the importance of the ministerial character.

To his young friends who now hear me, the high incentives to the improvement of the mind, the practice of virtue and piety, and the solemn admonitions given by his death cannot fail to be deeply impressive. Behold how much he effected in the short life. Yes! it was short to us, but he died at twenty-four. But still in the best sense of the word his life was long. He lived to discharge with faithfulness the various relative duties of life. He lived to awaken the sentiments of affection and esteem in the breasts of thousands who cherish his worth as long as feeling and memory shall last. He lived to place before the young an example of youthful character which could serve as a model of their own lives. He lived to direct thousands of his fellow creatures in the way of salvation. He died sincerely and deeply lamented by all who knew him, and the tears of this audience testify how much his character was venerated and how fondly he was loved. This then is an evidence that although existence may be brief, yet the great object of man's creation may be effected even in early life. "Honorable age is not that which standeth in a length of time, nor that which is measured by number of years, but wisdom is the gray hair unto men, and an unspotted life is old age."

Permit me to close this part of my address by reading an extract of one of the last letters of Mr. Schwartz, addressed to a very young man, and one to whom he was fondly attached; and it is to be hoped that in this last advice he may be considered, although dead, as still speaking.

"You are now, my dear ———, upon the wide world, as it were, and at liberty to do as you please. This with all of us is a dangerous situation, more particularly with the young and inexperienced. I think of you often and of the dangers to which your soul may be subject, however good your intention, and I now sit down to address you a few thoughts that may prove useful to you.

"1st. Never forget that you have a soul that must live after the body is dead, that is capable of eternal happiness at God's right hand, or may be banished forever from the presence of God, and is consigned to darkness and everlasting despair. The thought of this will help you to deny yourself sinful gratification and sensual indulgence.

"2dly. Endeavor to keep the fear of God constantly before your eyes. Remember the Searcher of hearts is always looking down upon you, that you are in his hands, and that he is able to raise you to heaven or to sink you down to hell. Remember that his eyes are always upon you, and you will 'learn to do well in fear to do evil.'

"3dly. Make it a rule, wherever you are, to let nothing keep you from the house of God on the Sabbath except it be actual sickness. When we neglect the church our souls begin to be in awful danger.

"4thly. Make it a rule never to lie down at night, nor to commence the labors of the day, without thanking God for his mercies and praying to him for his protection and favor. This will be of immense advantage to you in assisting you to do good and in helping you to avoid sins.

"5thly. Keep out of the way of temptation. It is the part of a wise man to keep at a distance from danger. We are so weak that if we give the least opportunity to our besetting sins, they soon get the better of us. Always recollect them to avoid that kind of company and those places where you know there is danger.

"6th. Seek good company, and avoid the society of such as show themselves to be the enemies of God, by cursing, profaning God's holy Sabbath, and by other immoral practices.

"Lastly. Think always that you as well as all men are fallen creatures, a rebel against God, and that you can only be saved through the merits of that Savior who loved us and gave himself for us. Oh! never forget that salvation is by the cross of Christ. Pray to God to help you to believe in Jesus, and give your heart to him to be renewed and to be sanctified."

Brethren of this congregation! He who was brought up in the midst of us, who ministered often at our altar, who would become the instructor of a number of pious young men who we hoped would in a very few years go forth into every corner of our land and carry there the tidings of salvation, he who we believed was raised up by Providence to be glory of our church is by the wise but mysterious directions of Heaven removed from us. Our fond hopes have been disappointed. The fair prospects which were before us are gone, and we are left in perplexity and sorrow.

Shall we then despair? God forbid. The dying words of our friend who said "he hoped the church would gain more by his death than by his life," forbid it. That pure and gentle spirit which may now be hovering around us would chide us for such despondency. Thanks be to God, the church is in his hands. He bringeth light out of darkness, and He can cause these conflicting dispensations of his Providence to be the commencement of a long train of mercies and blessings to our church. Let us now, even now, whilst our tears are not yet dried for this heavy bereavement, rise up and call upon God to help us. Let the two Societies which belong to this church, in which I conscientiously believe have mainly contributed to the excitement and exertions in the country which led to the formation of so many new congregations and to the establishment of a school of instruction which promised so much good to our beloved church, be up and doing. My brethren, try to increase the number of subscribers to our Society and add the names of your sons. My female friends I trust will be equally vigilant. God is still on our side. He has promised to be a wall of fire round about his church, and a glory in the midst of her, and if God be for us, who can be against us.

Young men, we go and imitate his bright example for you too may be cut off in early life like a flower in its bloom. Ye middle aged and old, let the wisdom and piety of that youth chide you for negligence in the cause of duty; may all of us here learn the importance of an early devotion to the improvement of the mind in the exercise of every Christian duty in laying the foundation for the esteem of men, for a peaceful death and a blessed immortality.

And now, the love of pupil and friend, thou faithful minister of the Gospel, we bid thee farewell. Thy early death was as full of regrets as thy youth was a promise. Thou art removed from us, but the influences of thy life and example shall remain. From thy deportment we will learn the excellency of piety and virtue. From thy death we will learn the power of religion in preparing the soul for the regions of the blest, and thy memory long remain as an example of goodness on earth whilst thou art enjoying thine everlasting reward. *Amen.*

Address Delivered before the Horticultural Society of Charleston

The Latin word *hortus* means garden, and horticulture as distinct from agriculture is generally done by hand and involves science or art. The aspects of the subject that Bachman emphasized most were the development and introduction of more useful varieties of plants. He discussed "in what way ornithology, chemistry, entomology, and physiological botany are closely allied to and inseparably connected with the science of horticulture."

He also considered the development and cultivation of vegetables and fruits for food, plants for medicines, and ornamental plants to provide "shade, fragrance, or beauty." He gave examples of scientific knowledge that had practical applications for agriculture, and he pointed out how useful knowledge made the work of planting more profitable, interesting, pleasurable, and more challenging for the laborer. He urged that fields of study with direct relevance to agriculture be made an integral part of curriculums. He recommended that streets, gardens, and cemeteries be enhanced with plants that were readily available in local forests.

Bachman noted that the cultivation of vegetables, fruits, herbs, medicinal plants, and flowers had been praised for many reasons since antiquity, but that systematic experiments to improve existing crops and to produce new crops had begun only about sixty years earlier. For example, he noted that Lavoisier had doubled the amount of produce through chemical experiments. He reminded his audience of discoveries of plants worldwide that a century or two earlier had been unknown in Europe and the United States, but had since become widely distributed. He noted the recent establishment of nurseries, the creation of organizations and prizes to promote improvements, the increasing willingness to "diffuse botanical and scientific knowledge," and particularly the potential for further improvements through continued experimentation and acclimatization. He emphasized that much work remained to be done and could be done by individuals.

He pointed out the widespread need for at least "sufficient knowledge" of science to take advantage of its potential for improving agriculture by individual farmers. He noted the need to know enough about ornithology to be able to recognize and distinguish birds that devour insect pests from birds destructive of crops, to know enough about chemistry to determine which types of soils are best suited for which crops and what kinds of additions to soil increase their fertility, to know enough about entomology to be able to defeat insects by planting before or after their usual time of hatching, and especially to know enough about botany to recognize plants and to understand their potential.

This speech indicates the breadth of Bachman's interest in and knowledge about natural history, related sciences, and agriculture and his desire to apply his knowledge for the public good. His address contained so much useful information and was so persuasive that it was published as a pamphlet by the Society and then reprinted in a journal to give it still wider circulation.

An / Address / Delivered Before the / Horticultural Society of Charleston, / at the / Anniversary Meeting, / July 10th, 1833 / by the Rev. J. Bachman / Published by Request of the Society. / Charleston: / Printed by A. E. Miller, / No. 4 Broad-st. / 1833.

Address.

Mr. President, and Gentlemen of the Horticultural Society.

At the last anniversary of this Society, the high and honourable duty was assigned me of addressing you on this occasion. I feel and I acknowledge most gratefully this proof of your kindness and favourable opinion; but I have to lament that however zealously and ardently I am attached to the cause which is espoused by this Society, I am unable to bring before you the results of much practical knowledge on the subject of horticulture, and that, therefore, some of the theories which I am about to advance may not, in the end, bear the test of experiment. I am encouraged, however, to trust in your indulgence whilst I attempt to discuss a subject on which I may not be practically as familiar as some other members of this Society under a belief that many important discoveries in horticulture still remain to be made—that, whilst many theoretical speculations may be demonstrated to be futile, and many experiments may fail of producing the desired effect; yet, that this failure may serve as a beacon to future travellers, and that every successful experiment will, when recorded, confer a benefit on mankind for ages to come.

Horticulture has two objects in view.

First: The introduction and cultivation of such vegetables and fruits as may serve for the food or medicine of man.

Secondly: The cultivation of trees, shrubs, and flowers, which by their shade, fragrance, or beauty, may serve to refine and purify his mind, add to his pleasures, and awaken in his bosom sentiments of admiration to that being who in mercy to man has promised that "while the earth remaineth, seed-time and harvest, and summer, and winter, shall not cease."

Time will not permit me to enter into a detail of the history of this art. Suffice it to say that several of the ancients and particularly Cicero enumerated this as among the most pleasing occupations of the mind, as particularly adapted to the aged, and calculated to give health to the body, and afford agreeable exercise to the mental faculties. From this exhaustless store of human happiness, the poet has derived some of his greatest beauties, the philosopher some of his most interesting disquisitions and the philanthropist some of the noblest plans for the amelioration of our race. How greatly is our pleasure in reading the works of Homer and Virgil, of Milton, Thompson and Cowper, enhanced by the continual references to this delightful theme. But not even the charms which Milton and Homer and Lucullus have sung or the descriptions which Lord Walpole and Sir William Temple in more modern times have left us on this delightful subject shall prevent me from entering into those more humble practical details with which the present age furnishes us. I proceed to remark that comparatively little was done in the science of gardening till within the last sixty years. Since that period, the justly celebrated national establishment of France under the auspices of Desfontaine, Jussieu and Thouin has arisen, which contains every thing directly and remotely connected with this department of knowledge. It was not until 1804, that Sir Joseph Banks aided by Sir

James Edward Smith, Mr. Thomas Andrew Knight, and a few others instituted the Horticultural Society of London. Five years afterwards the Caledonian Horticultural Society was formed in Edinburgh, and from thence a fondness for the studies and labours of this art was by the aid of similar institutions defused over all Europe. In some parts of Germany, the culture of a garden and fruit trees forms a part of the education of the ordinary seminaries, and no schoolmaster is permitted to exercise that function without a certificate of his capacity to teach the management of the garden and the orchard. Travellers inform us that in Seville and Cadiz, the windows and balconies are every where filled with pots containing a great variety of the beautiful Amaryllis, with the favourite Polyanthus and Narcissus, the gaudy Tulip and other bulbs, and with ornamental jars of the Geranium and the Jessamine. The Pink is there, as in every part of the world, a favourite flower, and even the lowest cottages have a few pots of the Sweet Basil, the Daisy or the Violet.

In France, among that gay and luxurious nation, the science of horticulture is cherished with the greatest enthusiasm. In Paris, alone, three courses of rural botany are delivered gratis every year. Several classes are composed of upwards of two hundred individuals, some of whom are soldiers and cottagers, and men who are moving in the humbler walks of life. The garden of the Tuilleries are invaluable from their situation in the very centre of Paris and from their being open at all times to all the world. Their walks are shaded with beautiful and airy groves, bordered with a constant succession of showy, flowering plants. In England, such have been the improvements since the establishment of these Societies that no one can gaze upon their beautiful, well trimmed lawns, their gay parks, and the flowers that bloom around many a cottage, and climb over the lattice of the poor man's dwelling, without being convinced that a love for the beauties of nature rises spontaneous in the human heart; and that the more we cultivate it, the more we will be led to admire the works of God. An intelligent German traveller, speaking of the habits of the English people, makes these remarks: "We have visited the celebrated flower market of London, of which no German who has not seen it could have formed a proper idea. What chiefly struck us is that the greatest rarities and the most trifling articles are here exposed for sale together and are both eagerly bought. The wealthy and respectable

Englishman who is a connoisseur will purchase nothing that is common for if pretty he has it already in his garden; and the poor Londoner who cannot afford to buy what is beautiful will still obtain something green to decorate the window of his dark little attic and give his last farthing for a bit of verdure." He speaks of the wonderful improvement in horticulture and ascribes it to the influence of the Horticultural Societies, declaring that although he had been forty years conversant with the raising of fruit, he had never beheld finer Peaches, Nectarines, Plums, Melons, Grapes and Pine Apples than he saw there. Agents from some of the Societies have been sent into distant lands to enrich their native country with the beauties of Pomona and Flora, and such has been their success under these exertions that some of the most delicious fruits and beautiful shrubs and flowers that are known in the world have been introduced, naturalized, and cultivated and are now ministering to the wants or adding to the gratification of man.

In 1818, a small number of enterprising and intelligent practical gardeners and nursery-men in the city of New York formed themselves into an association for the purpose of introducing such improvements in the cultivation of vegetables as they were competent to effect. Premiums were offered by the Society, and in a very few years there was such an improvement in the products of their gardens that the vegetable markets of New-York, which before had been very indifferent, may now vie with those of any other city in our country. In the month of May last, I had the pleasure of attending their exhibition of flowers, which, for abundance, variety, and beauty exceeded any thing that I have witnessed. Similar institutions have arisen in Pennsylvania, Maryland and New-England, and it is probable that Massachusetts now possesses the most flourishing and useful institution of the kind in our country.

The establishment of nurseries is but of a comparatively recent date, but still there are few valuable trees, shrubs, bulbs or flowers, that may not be found in the gardens of Prince at Flushing; of Floy, Wilson, Hogg or Thorburn in New-York; of Buel and Wilson in Albany; of Landreth or Carr in Philadelphia; or of Noisette, who long has devoted himself to the cause in the neighbourhood of our own city.

The reason why horticulture, until very recently, is with few exceptions limited to the culture of common culinary vegetables and fruit is very evident. The wants and necessities of a young nation are generally so imperious that they have little time to attend to the ornamental and scientific departments of gardening. The introduction of luxuries requires time, leisure, and wealth.

Our own institution is but of very recent origin. We have met to celebrate our second anniversary. We have had to encounter difficulties on all sides. Some prejudices at first existed among some of our gardeners, who seemed to fear that we were associating for the purpose of obtaining and publishing to the world the secrets of gardening, which they had acquired after many years of experience, and which by increasing competition might eventually prove an injury to their business. These individuals could scarcely have recollected that this Society was principally instituted for their benefit, that scarcely an individual member was engaged in cultivating vegetables for the market, that the knowledge which we acquired, that the improvements which were made on the subject of horticulture were all at their service, and that by improving the articles we increased the demand for their consumption. We believe that these prejudices have, in a great measure, been removed.

Another difficulty with which we have to contend it is that those members of our Society who are planters by profession and could aid us by their experience are absent from us a portion of the year and attend but little to their gardens, on their plantations, during the winter and spring since they are aware that in summer and autumn, when the fruits and most of our vegetables are in perfection, they would be absent from their plantations and therefore could not enjoy them.

And we may further add that the fierce political contest [Nullification] in which our people have been engaged had, in some measure (for a time, at least), diverted many active and good men from these useful and pleasing employments to the study of the principles of government and have sometimes led to those who breathed the same air and who once admired each other's gardens—interchanged for delicious fruits of the season, and presented each other with the Rose, the Pink and the Violet—to regard each other as the enemies of their common country. These dark and unpleasant scenes, we trust, have now all passed away and will soon be buried in oblivion; and nature, that is so full of harmony and love, that has covered the earth with fragrance and with beauty invites us to repose and friendship together on the green lap of earth beneath the shade of her majestic trees; and the Father of the universe seems to say "Let there be no strife between you for ye are brethren."

Come then, let us unitedly engage in studies and employments which will not be confined to the sweets of Flora or the apples of Pomona; our views will embrace a wider field, a more extended sphere of public utility. Whilst we are introducing new objects of horticultural industry, we may be able to diffuse botanical and scientific knowledge, contribute something to ameliorate the condition of the poor, add to the morals and the virtue of our people, and lead the contemplations of man from "nature up to nature's God."

The science of horticulture has not heretofore been held in that estimation to which it was certainly entitled. It was formerly pursued principally by persons in the humble walks of life, persons possessed of but little scientific knowledge, who obeyed the first impulse of nature and procured the bread of life by labour and toil. No wonder then, that nothing very interesting or attractive could be found either in the life or the employment of such an obscure uncultivated being. To my view, there are few states of existence less enviable than that of an ignorant man or woman working hard on the farm or garden, without having knowledge or science enough to be interested in their occupation and in the scenes around them.

But it cannot fail to awaken pleasure in every virtuous and reflecting mind to observe how generally a taste for rational enjoyments, as exemplified in the growing partiality for the study of natural history and in the encouragement given to the various branches of horticulture, is superseding the sports of the field and the revels of the banquet. The eager search after truth in the present age has, in some measure, redeemed the supineness of former times. The tree of knowledge, whose fruit was heretofore so inaccessible to man in the humbler walks of life, has been freely plucked by all who choose to gather it. The obstructions which were thrown in the way by the ancient languages and by the pretended hidden secrets of the art have all been stript of their mysterious covering. A more general knowledge of what the soil is capable of producing is diffused among the cultivators, a taste for reading various valuable productions upon horticultural subjects has increased. A majority of the articles contained in the horticultural publications of England and Germany are written by professed gardeners, who labour in the garden and green-house, and we trust that the time is not far distant when our own excellent publication on Southern agriculture will be enriched with the productions of the scientific and practical gardener. Although we are yet sadly deficient in our knowledge on these subjects, yet there are improvements of a very gratifying character in many portions of our land, and we hope that before many years the sciences of chemistry, botany, entomology, ornithology and conchology will be as regularly taught in our schools and private families as our music and the French language at the present day; and this is certainly calculated to open a great source of pleasure and advantage to the rising age.

The advantages of science in horticultural pursuits do not appear to be sufficiently estimated, and in order to elucidate the subject, I beg leave to invite your attention to the observations and facts I am desirous of bringing to your notice. I would endeavor to show you in what way ornithology, chemistry, entomology, and physiological botany are closely allied to, and inseparably connected with the science of horticulture.

The study of ornithology which is least allied to the subject still presents strong claims. Man is known to look with a jealous eye upon all who oppose his interests. In obedience to this natural dictate of the passions, he not only grapples with him of his own species whom he views as his enemy; but he wages war on the beasts of the field, on the fowls of the air, and the insect world, and all that he believes is about to endanger the prospect of his success. In this way the innocent often suffer for the guilty, and the harmless bird that comes to add to our pleasures by warbling its sweet noises in our gardens and on our housetops, or who is a positive blessing to us, by lessening the number of depredating insects falls indiscriminately with the crow and the grackle at the sound of the murderous gun. Now all this does not usually proceed from a natural disposition to cruelty, but from ignorance. Without a suitable knowledge of the science of ornithology, we are unable to know which birds are injurious, and which are a positive benefit to the farmer, which ought to be banished from our fields, orchards and gardens, and which ought to be encouraged there by all the allurements in our power.

Kalm tells us that when a bounty was set on the head of the little crow in Virginia (meaning probably some of the genus *Quiscalus* and *Icterus*, which go under the common name of black-bird), which were destroyed at an enormous expense to the state, the insects so increased that they would have bought them back again at any price. The purple grackle in New-England was destroyed in consequence of the governor's offering three-pence a head, and the result was that insects multiplied so rapidly, that the herbage was destroyed, and the inhabitants were obliged to import hay from Pennsylvania and England. The poor wood-pecker is shot by every idle boy because he is said to extract the juices of apple trees, when in most cases he is allured thither by the worm which is perforating the tree; and thus the bird on which we pronounce sentence of death as on an enemy has come to save the tree by feeding on its destroyer. The tyrant flycatcher (*muscicapa tyranus*) is called the bee-bird and is slaughtered when for one bee that he destroys, he relieves the farmer of a thousand insects that were depredating on his fields. Of the large family of flycatchers (*muscicapa*), warblers (*Sylvia*), and thrushes (*turdus*) that constitute three-fourths of our land birds, scarcely one is in any respect a depredator on the property of man; but on the contrary, all greatly aid him in preserving his fields and fruits from devouring insects. Let then a sufficiency of ornithology be known by the cultivators of the soil to distinguish in the feathered race an enemy from a friend; and if the hawk, the crow, and the starling are deserving of death for their depredations, let us spare the beautiful warblers, the thrushes, and the wrens that come to our gardens to claim the worm that is injuring us and who are ready to reward us with a song.

The science of chemistry advances no inconsiderable claims to the attention of the horticulturalist. In order to the successful rearing of plants, we must place them in soils adapted to their natures. It is well known that the soil calculated for the growth of one plant is often destructive to the life of another. The experience of the members of this Society can testify, that the plants which flourish in the garden of one will not succeed in that of another. The okra, the tomato, and the watermelon succeed well in some soils, whilst in others they struggle through a sickly existence and die before they bring their fruits to maturity. The nettle haunts, as it were, the footsteps of man and claims, as poetry might urge, in the very sociality around his dwelling. This plant will not flourish but in a soil containing nitrate of potash (*salt petre*), a salt always abounding in the neighbourhood of places where there is calcareous matter. Chemists inform us that every soil is composed of silica, alumina, oxide of iron, salts, and animal and vegetable remains. The most important consideration is, in what proportions these must be mixed, in order to constitute a fertile soil. Alumina or clay imparts tenacity to a soil when applied. Silica or sand diminishes that power whilst chalk and lime have an intermediate effect: they render heavy soils more porous and light, soils more retentive. These simple facts are all important. Two neighbouring fields by an interchange of soils being often rendered fertile, one of which had been forbidden too tenacious, and the other too porous. The experiments of Sir Humphrey Davy on the subject of soils are full of instruction. He found that a rich black mould containing one-fourth of vegetable matter had its temperature increased in an hour from 65 to 88 degrees by exposure to the sunshine whilst chalk soil was heated to only 69 degrees under similar circumstances. But the first, when removed into the shade, cooled in half an hour 15 degrees, whereas the latter only 4. This explains why the crops on light coloured soils are in general so much more backward in the spring, but are retained longer in verdure during autumn than those in black coloured soils; the latter obtain a general warmth more readily, but part with it with equal speed. Coal ashes sewn on beds cause beans and peas and many other vegetables to come up two or three days earlier than where no such application is made; it being a well-known fact that dark coloured bodies absorb caloric more readily and in larger proportions than those of a brighter hue. As an evidence of what can be effected by a combination of chemical and practical knowledge in the cultivation of the earth, it is only necessary to mention the experiments of the great chemist Lavoisier in order to impress on the minds of his neighbours, the people of Levandee in France, the advantage of combining chemical and practical knowledge. He cultivated two hundred and forty acres on scientific principles. In nine years his produce was doubled and his crops afforded one-third more than those of ordinary cultivators. I trust that these few hints will suffice to show how much may be gained in horticulture by a knowledge of chemistry.

Entomology too, a science but little known until very recently, lays weighty claims to the attention of the horticulturalist. Wherever we go, we find the earth, the trees, the shrubs, and the air filled with thousands of living beings, assuming the most wonderful changes and gifted with the most surprising instincts. Some of these like the silk worm, the cochineal, and the cantharides add to the wealth or luxury of man or minister relief to his diseases. Others are destructive of his prospects and the enemies of his repose. Some attack the roots of his trees and plants which soon wither and die, whilst others fasten upon the blossoms, or upon the fruit, and all his bright prospects are blighted. The fair one who has reared with care and perseverance some favourite plant, finds it drooping and decaying in spite of all her vigilance and is not aware that a worm may be at its root or that some insect may visit it at night and deprive it of its buds and leaves; but she knows not the characters of either. She knows not where its eggs are deposited, at what season of the year she may apprehend its attacks, and is utterly unable to guard against it.

When the insect called the Hessian fly made its appearance on Long-Island in 1776, it was wrongly conjectured that the Hessian soldiers under the pay of the British government had conveyed this evil along with them in their straw from Germany. The British government feared that it might be introduced into England, and took measures to prevent it. Information was sought by government from practical men in America, some of whom had lost their entire crop by the insect; and yet they were ignorant of whether it was a moth, a fly, or what they term a bug. Expresses were sent to ambassadors in France, Austria, Prussia, and America. The information obtained was so voluminous as to have filled two hundred octavo pages, yet still so little science was possessed by the persons who gave information about the insect and by those who met to ward off its ravages that it was impossible to form any idea of its genus or character till Sir Joseph Banks, an eminent naturalist, lent his aid in the investigation, and gave the nation the only information that could be relied on. An insect with a somewhat similar character actually made its appearance in England some time afterwards. It threw the country into great consternation as they feared that it might prove destructive to the staff of life; when Mr. Marsham, by tracing out the species proved the alarm to be unfounded. Pursuing the history of this insect again in America, entomologists discovered its character and habits, and by sowing their wheat at a particular time in autumn, when it was too late for the insect to multiply before the cold weather set in and when the plants would be too much forward to sustain much injury in the spring, the cultivators have, in a great measure, arrested its distractive progress, and thus science has lent her aid to agriculture and averting evils which at one time threatened to banish from our land the culture of the finest grain, with the exception of rice, which is found in the world.

The utility of entomological knowledge will farther appear from a circumstance which occurred in Sweden. The oak timber in the royal dock-yards had been perforated and greatly injured, when the king sent to Linnaeus, the father of natural history, to trace out the causes of the destruction of the timber. He detected the lurking culprit under the form of a beetle (*Lymexylon navale*) and by directing the timber to be immersed during the time of the metamorphosis of that insect furnished a remedy which secured it from its future attacks. Another instance which occurred among the elm trees in St. James' Park, London between the years 1820 and 1824 is recorded. These trees suddenly became affected in a very singular manner. The bark fell from the stem and whole rows died. There happened to be a company of soldiers stationed in the Park, and as the trees were barked too about the height of the soldier's bayonet, the suspicion fell on some unfortunate recruits as having occasioned the injury, and they were arrested; but nothing could be proved against them. Persons were now employed to watch the Park at night, but still in the morning the bark was lying in great quantities around the roots of the trees. At the same time the elms in a grove at Camberwell, near London, were also destroyed. This was ascribed to the effect of gas escaped from pipes used for lighting the road. Legal proceedings were commenced against the company for the removal of the nuisance. In this state of things, William Sharpe M'Lay, an eminent naturalist, profoundly acquainted with the history of insects was requested by Lord Sidney to draw up a report on the state of the elm trees, for the purpose of referring it to the Lords of the Treasury. He discovered it to be a beetle (*Hylensius dertructor*) belonging to the same genus as that which destroys the pines in Germany. By ascertaining its habits, he was enabled to point out a remedy, and the remainder of the trees were preserved.

Suffer me yet to call your attention to one other instance of the effect which ignorance on the subject of entomology is calculated to produce. A caterpillar of unusual size and singular form made its appearance on the trees of the Lombardy poplar in the State of New-York, some twenty years ago, as far as my recollection will now serve. The ignorant became alarmed; many idle reports were circulated; a dog was said to have been stung by one which occasioned swelling and death; rumour soon made it out to be a child; the newspapers circulated each idle tale. And now the work of destruction commenced. The axe was applied to the ornamental trees that shaded some of the finest streets of their villages. The same work of extermination was carried on at several farm-houses and gentleman's country seats. The stately poplars were levelled to the ground and burnt. The lover of nature remonstrated, but it was in vain to contend against the powerful current of prejudice. A little knowledge of the science of entomology might have satisfied the destroyers of those beautiful works of God, that the larva which they so much dreaded was harmless, that it would soon assume chrysalis form, and after lying inactive for a short time would put on wings of a brilliant hue, flit joyously on the air, and live on the nectar of flowers.

The celebrated Spanish fly (*Cantharis of Geoffry, and Lytta of Fabricius*), which is so invaluable in healing art, has often mixed with it in our shops, insects which so strongly resemble it, that the venders themselves are deceived, and none but the practiced eye of the entomologist could discover the deception, and yet some of these insects (and I have seen a considerable mixture in your own shops) belong to a different genus and are not only useless, but may be injurious.

To guard against the depredation of insects, we must first become acquainted with their genera and habits, and then by a course of scientific and practical experiments, we may be able to destroy them or avert their attacks. The larva (*Aegeria exitiosa*) that is found at the roots of peach trees has been carefully examined and correctly described by entomologists. Having ascertained that the worms enter the earth at the stem of the tree, about the beginning of August in this part of the country, a covering of cloth or skin tied around the stem about a foot above the ground extending three or 4 inches under the surface and retained there from the first of July to the middle of September has been found effectual in protecting our trees against the attacks of this enemy. There are three or four other species of insects that infest the peach itself, one of which only I consider as formidable in its attacks and most to be dreaded. A course of experiments on the character of these insects, and the best mode of guarding against their depredations is in progress by members of this Society. The results, together with careful drawings, it is believed, will be laid before the society in the course of the present summer.

Another insect which has not yet been satisfactorily described has, within a few years past, fastened itself upon the stems at our orange and lemon trees; and although it is so minute as to require the aid of good magnifying glasses to examine it, after being disengaged from the covering, which envelops it, yet it is so prolific that it now threatens (unless a remedy is soon discovered) to deprive us of the poor remnant of orange trees which the frost has left. Oily substances are known to destroy these insects, but in its application, the remedy prove worse than the disease; the pores of the tree are closed up, and perish in the course of the season. The *coccus* and *aphides* which are such pests to the green-house are better known and consequently may be more easily guarded against. I have invariably found that the immersion of the branches of plants infested with these insects in what is called by the apothecaries the "yellow wash," a composition of three drachms of corrosive subliment, mixed with a quart of lime-water, proves an effectual remedy.

Time will not permit me to dwell more minutely upon many other species of insects which infest our gardens and our orchards; we are every year subject to the ravages of others with which we are now unacquainted; for some of the most noxious insects in every country are not indigenous but have been imported. These few hints, I trust, will suffice to show the importance of a knowledge of entomology in successfully carrying on horticultural pursuits.

But an objection has been urged against this study, which the lovers of the science are anxious to combat, viz. that it requires us to inflict death upon its objects, and we are, therefore, charged with inhumanity. Cruelty consists in torturing or destroying any living thing from mere wantonness without any useful object in view. The entomologist is not one of these. His insects are, by processes which science has taught him, killed almost instantaneously. He abominates cruelty as much as those who condemn him, but he differs from them and his ideas of the amount of pain inflicted; he does not agree to the truth of the sentiment expressed by the great poet;

> The poor beetle that we tread upon,
> In corporal sufferance, feels a pang as great
> As when a giant dies.

His knowledge and experience convinced him that this contains more poetry than truth. It is a well-known fact that as we descend in the scale of animated being from the highest intelligence to the worm or polypus, there is a gradual diminution of sensibility. The pain of death must be more excruciating to man than to the animal since with the former, suffering is increased by mental reflections and by the dread of death. So also insects must suffer less than brute animals because they are differently constituted; they are cold blooded, destitute of the great sympathetic nerve, and breathe through orifices beneath their wings. But time will not permit me to treat this subject at large. Suffice it to say that whilst the Creator has formed insects as perfect as insects are required to be, yet an examination of their whole internal system must convince us that they possess less sensibility than even the tortoise, who is, notwithstanding, known to walk after his head has been separated from his body. Insects will leave their legs in the hand without experiencing any apparent pain. Say, informs us, that a butterfly, whose body has been perforated by an insect pin, flew off with it to the first flower and extracted it sweets without seeming to have been at all incommoded. Ants will walk when deprived of their heads. Bees will sting after their bodies have been cut in two. The silk worm and other of the lepidopterous family, after being deprived both of legs and wings will not only deposit their eggs, as if nothing had occurred, but will live their usual period. Now let us inquire, would a human being be as indifferent, and suffer apparently so little? Would he take his food and enjoy himself if his legs or his arms were amputated? Impossible! Besides, the period of an insect's life at the time when it is procured for the cabinet of the entomologist, is the last stage of its existence. It has already passed through various forms in several stages of its short life. The butterfly would have perished in a very few days, and the coleopterous insect would not have long survived; and let it be remembered, that the specimens which are treasured in the cabinet of the naturalist, which he values more than gold and on which he thus confers a kind of immortality, has, probably, been, by being thus collected, preserved from some rapacious bird, or fish, or insect, which would soon have devoured them. More have been destroyed in this manner, in a single day, than have been collected by all the entomologists in the world. But, if those who are so sensitive on these subjects should still declare that they cannot reconcile themselves to have any pain inflicted, even on the insect and for scientific purposes, I answer, in the language of the Rev. Mr. Kirby, one of the first naturalists of the age, and I would add, one among the most humane and excellent of men, "Pray Sir, or Madam, I would ask, should your green-house be infested by aphides, or your grapery by the semianimate coccus, would this extreme of tenderness induce you to restrict your gardener from destroying them? Are you willing to deny yourself these unnecessary gratifications, and to resign your favourite flowers and fruit at the call of your fine feelings? Or, will you give up the shrimps, which, by their relish, enable you to play a better part with your bread and butter at breakfast? If not, I shall only desire you to recollect, that for a mere personal indulgence, you cause the death of a greater number of animals than all the entomologists in the world destroy for the promotion of science."

But whatever objections may be urged against this branch of natural history, as connected with horticulture, none can be made against botany. Here no experiments are necessary that require the infliction of pain. Without a knowledge of systematic and physiological botany, we are unable to understand terms and observations and that must occur in every well-written work on horticulture. Botany has become a favourite study among the well-informed of both sexes in every civilized portion of the world. The attraction of flowers and fruit by their colours, taste, and smell—the delight of rearing a living thing, which grows under our eye, and developes itself from a shapeless mass to one of extreme beauty and loveliness—whose life is free from pain and whose death seals the promise of its reappearance will always interest us in favour of this study. And when it can be applied to useful purposes, when it can be made to add to the health and comfort of man, and when it makes him better and happier, surely it should find an advocate in every breast.

But it will be urged that however much true science might aid the cause of horticulture, yet that most of those who study the sciences have done little more than burden their memories with hard and unmeaning names. Suppose we admit it for the sake of argument, and allow that very few become proficients in the sciences. Is there then nothing gained? Why demand the study of mathematics, a science so generally recommended? Not certainly to make the bulk of those who attend to these studies either a strong nurse, or engineers, but simply to exercise and strengthen the intellect and to give the mind a habit of minute attention and investigation. If the natural sciences did no more than this, the pursuit of them would prove an advantage to man. But infinitely greater benefits may be derived from these studies; they may be applied to many useful purposes in life. They have enabled man to multiply the fruits of the earth, to bring from distant climes, plants and vegetables that will give subsistence to thousands. The introduction of the Irish potato alone into Europe, has been one of the greatest blessings conferred upon that land. The melon, the okra, the tomato, and the artichoke have all been brought from a distance and are now cultivated in almost all the temperate as well as tropical portions of the globe. The peach, the apple, the pear, and the plum, with their infinite varieties were originally confined to a small spot of earth and were of very little value till science and horticulture united in introducing them and improving their flavour. The pharmacopia of medicine is indebted to the botanist and horticulturalist for an immense number of ingredients that are calculated to avert the sufferings and to prolong the life of man. It was the will of heaven that man should be doomed to suffer pain and sorrow; but that same being, also in mercy, gave us a healing balm in many a vine clusters in the forest, in many a root, and the bark of many a shrub or tree. These have been discovered by the knowledge and labours of men of science, and now our gardens abound with remedies for many of the countless ills to which we are subject in this life. Some physicians of the present day have indeed gone so far as to assert that the healing art could be successfully carried on, without the use of vegetable medicines; but few, I believe, have carried this theory into practice. I am not aware that any thing has as yet been discovered among the minerals that can be substituted in the room of the Peruvian bark, the rhubarb, and an infinite number of vegetable medicines with whose names you are sufficiently familiar.

The garden and the orchard are calculated to afford the means of health and instruction to the man of science as well as the tenant of the cottage. A garden was the first habitation of man; it has ever since been a source of his purest pleasures, of his most healthful employments, and often the means of his sustenance. Multiply around the poor man's cottage the comforts of life and the means of enjoyment attach him to his garden and to his fruits, and you will save him from discontent in crime. These flowers of life will endear him to his home and his native land, and he will become a good citizen as well as a happy man. To the wealthy, these studies and employments should be equally dear. *Nihil est melius*, says Cicero, *nihil uberius, nihil homini libero dignius.* (Nothing is more profitable, nothing more suitable for a man of leisure.) Some of these studies by an unaccountable perversion of intellect have been so abused as to have been dragged into the surface of irreligion. It is well then, that the lover of nature who sees God in all things, who in the mirror of the Creation, beholds and adores the reflected glory of the Creator, should study these works in order to recommend that the great truths of religion as contained in the word of God, and make them subservient to the best interests of mankind!

The subjects to which our attention as horticulturalist should be directed are so numerous that time will only permit me to glance at a few that have struck me as most important. Our vegetable gardens might, particularly in the winter and spring, be made among the very finest in the world. Many of the vegetables which in Europe are raised with great care and expense in hot-houses, thrive and flourish with us in the open ground. Our turnips, carrots, ruta baga, kohl rabbi, green peas, spinach, salad, cellery, &c., can be had in perfection during the whole winter. Our melons can be sent to the New-York and Boston markets six weeks earlier than any that are produced in that region. Among the Musk-melons, several species they go under the name of Persian melons, and whose flavour is very delicious, ought to be more cultivated. There is another melon from Rio Janeiro, said to possess the flavour of a peach, which it would be well to introduce among us. The great secret in preserving these highly flavoured fruits from degenerating is to remove them entirely from all other plants of a similar genus and cultivate them carefully in a field by themselves. The Irish potato (*Solanum tuberosum*) instead of being planted, as is the case at present in our warmest sandy soils, ought for the sake of experiment to be planted on the coldest clay grounds. The finest flavoured potatoes are raised in cold climates—in Nova-Scotia, and in the coldest parts of England. Col. I'On, of our State, I am informed, has succeeded in preserving the flavour of the Lancashire potato by planting it in a part of his rice-field that is somewhat elevated. Among the turnips, big yellow Scotch, the yellow Maltese, and Hybride have succeeded remarkably well. Cabbages can be raised the whole year round, provided the different kinds suited to the seasons are sown in rotation and provided we choose a dry soil in winter and a moist one in summer. The impression is very general that this species of vegetable will prove inferior unless the seed is annually imported. This is certainly a mistake. The best fall winter cabbage brought to our market is produced by a lady who, for the last twenty years has been in the habit of preserving her own seed. The Cow-cabbage, from all that I have seen, I am inclined to believe will prove a failure. The heat and moisture of our climate during summer, together with the worms that prey on their roots, prove fatal to the great majority of these plants before they are a year old. The same fatality attends the most kinds of Cauliflower when planted in the spring. This, however, is not the case if the seed is sown in August since in many instances a fine crop of this delicious vegetable is produced in December and January, particularly if the winter does not prove too severe. The early French Cauliflower has recently been introduced as a spring vegetable, to be sown in February, and I have seen it in great perfection in, at least, one of the gardens of our city during the last spring. This species of broccoli, called the Russian sulfur, succeeded better with us during the last winter than any other that I have seen. The plants of the Sea-kale now in a state of culture seem fully to answer the expectation of those who have introduced them. The New Zealand spinach (*Tetragona expansa*) grows luxuriantly, with scarcely any cultivation. A new vegetable has very recently been introduced into Europe, through the zeal of David Douglass, an eminent botanist, under the patronage of the London Horticultural Society. It is called the *Oxalis crenata*, the flavour is said to partake of both the potato and chestnut. It is so productive that a bulb weighing one ounce, last year produced ninety bulbs weighing four pounds. This is said to be so great an addition to our culinary vegetables, that it is supposed by some, it may in time, supplant even the potato itself. The artichoke (*Cynara scolymus*) of which there are several very superior varieties, is a great delicacy and succeeds well in our mild climate. The Tart rhubarb has come to perfection in a few of our gardens and failed in others. This species called *Rheum undulatum* and *Rheum Hybridum* are deserving of a more careful cultivation. It is probable also that the Turkey rhubarb of commerce (*R. Palmatum*) may be found to succeed in some of the middle and upper districts of our State.

The fruit garden is deserving of no inconsiderable share of our attention. The apple does not generally succeed well in the maritime districts of our State. Still there are so many varieties of this delicious fruit, that the careful and zealous horticulturalist may yet discover some kinds that are adapted to our soil and climate. I tasted an apple about ten years ago, that was raised on Charlestown Neck, on a farm now belonging to Dr. Porcher, from a tree which had regularly produced apples of a very fine flavour for several years in succesion. There is an apple tree in Italy, called the Mala Cara, which is too delicate to survive the cold winter of our Northern States, that might, perhaps, be successfully cultivated here.

The Pear is another fruit which deserves our particular attention. Pear trees from the South of France appear to thrive better with us than those from the Northern States of our country. This is probably owing to their having been naturalized to the climate, which bears a considerable similarity to our own. The garden of the late Mr. Charles Florian May, of the city, has, for more than twenty years, produced regular an abundant crops of pears. The trees were imported from France. On the plantation of Col. Magwood, the fruit of some of our finest varieties, and among the rest the Sickle pear, came to great perfection last year. Although this is a tree, which in its natural state requires many years of patience before we can reap its fruits, yet inoculation and grafting will generally remedy this inconvenience. The Apricot (of which there are many varieties), on account of its blossoming early, is subject to be greatly injured by the late frosts of spring. I would recommend that an attempt be made to retard the blossoming of these trees by planting them in a northern instead of a southern exposure. The Peach and the Nectarine arrive at great perfection in some of the arts of our city and in some situations in the country. The great enemy against which we have to contend in this fruit is the insect (*Curculio*) which perforates the fruit sometimes at a very early stage and which has heretofore baffled all our skill in finding a remedy.* Our city will probably, however, be abundantly supplied in a few years from orchards planted in favourable situations in the country near the Rail-Road, the fruit of which may be conveyed to our market in a few hours.

There are scarcely any of our northern Plums that can be successfully cultivated along our sea-coast in the south. The variety called the Red Gage has succeeded best, and some other kinds have done well for a year or two in the very valuable fruit garden of Mr. Michel of the city. There is, however, a native plum very little known here, which in common with other varieties is called the Chickasaw plum, the flavour of which is nearly equal to that of the Green Gage, that can with great ease be cultivated here and is deserving of more of our attention.

Our situation on the sea-board of Carolina is remarkably well adapted to the cultivation of the Fig-tree. This is decidedly one of the most valuable of our fruit trees. It bears abundantly and frequently three crops in a season and is not subject to the ravages of insects or liable to be affected by our seasons. The fruit is always wholesome and is the more valuable on account of our being unable to eat it except when fully ripe. We have in our neglect of the cultivation of the fig acted up to the usual disposition of man in trying to procure that which is expensive and difficult and neglecting that which is cheap and can be easily obtained. Some of the finest varieties of this fruit have not yet been introduced into our gardens; a little attention from the members of this Society will

---

* See Note A [This footnote was omitted from Bachman 1833B2. See the end of this article for the only note.]

enable us to have a regular succession of the finest flavoured figs during five months in the year.

Orange and Lemon trees were formerly more extensively cultivated than present. The occasional severe frosts of our winters will probably ever remain an obstacle to our being able to cultivate these fruits with certainty. The Cherry, with the exception of the wild species, has seldom succeeded well; the variety called the Morella cherry, produces better fruit with us than any other that we have yet tried. Our native kinds however, admit a great improvement by cultivation. The Quince requires a colder climate than ours, and yet, I have occasionally known this species to bear good fruit for several years in secession. I remarked that they succeeded best on a clay soil and in a northern exposure. There are a great variety of the fruit bearing Pomegranates, all of which arrived at great perfection in this climate, and we should use our endeavors to introduce among us those varieties that are most valued in the east.

The Currant and Gooseberry we are unable to cultivate.

There is a variety of our native Mulberry (which has, however, become very rare with us) that is but little inferior to the much prized *Morus nigra* of Europe; the latter also I have seen bearing fine fruit in the gardens of Mr. Noisette and Mr. Howard of this city. It may not be foreign to our subject to notice here the *Morus multicaulis*, a new species of mulberry tree recently introduced, which for the raising of the silkworm will probably supplant the white mulberry now so generally cultivated. This tree puts out its leaves so much earlier than the other that they are three inches broad before those of the white Mulberry begin to unfold. The mulberry tree is easily cultivated. Our soil and climate are admirably adapted to its growth. Some of those that were planted by the first German Missionaries at Ebenezer, Georgia, during the time of governor Oglethorpe are still in a flourishing condition. The culture of the mulberry ought to be more attended to in our Southern country. The time may not be very far distant when the reduced prices of Cotton, in consequence of an extended cultivation, may render the raising of Silk, particularly the raw material, one of the staples of the South.

I regret to say, that my own experience, with regard to the successful cultivation of the Olive holds out but little encouragement. It is but seldom that amidst our varying seasons, the fruit of the olive has arrived at any degree of perfection. It is somewhat singular too, that whilst in the South of Europe, the olive is propagated with ease from cuttings from the largest stem to the smallest twig and even from the bark itself; yet, in our Southern States, it has been discovered that it is very difficult to propagate this tree in any other way than by the seed. Mr. Couper of St. Simon's Island, I am informed, has partially succeeded in obtaining fruit from trees imported from France.

With the Vine, we are likely to be far more successful. Whilst many varieties are not suited to our own climate, others and particularly native varieties have succeeded even beyond our expectations. To the perseverance and skill of Mr. Herbemont of Columbia and Mr. Abraham Geiger of Lexington, our southern country is indebted for much information on the subject of the culture of the grape and is now believed that many portions of the poorest pine barrens and our middle districts are admirably adapted to the growth of the vine. In the neighbourhood of our city, many varieties of the grape, for the use of the table, are produced, as our several exhibitions have abundantly testified.

The Camphor tree would, I am induced to believe, stand the severity of our coldest winters. There is one that was not long since growing in the garden of the late Mr. Young of Savannah that appeared never to have been in the least affected by frost. The Tea-plant (*Thea virides*, and *bohea*) has by successive planting of the seeds, from year to year been at last so acclimated in the garden of Mr. Noisette near this city that with very little protection, it may be cultivated and bring its seeds to maturity in the open ground; and we hope soon to be able to regale the Ladies who honour our exhibitions with a cup of their favourite beverage procured from plants raised on our own soil.

America has recently given to the world a new fruit called *Shephardia Argentea* or Buffalo berry. It is a bright red colour growing in clusters, and said to be very delicious; it may yet prove to us in the South a substitute for the Cherry. This tree, growing to a height of about fourteen feet was found on the banks of the River Platt and on the Missouri, and was brought to us by the indefatigable explorers of our western wilds; it has borne fruit near Boston, and two trees of this species have been for several years growing in the garden of the Editor of the Southern Agriculturalist and thus far seem well adapted to our climate.

The Strawberry, a fruit that has always been a favourite, is well deserving of the attention and encouragement of this Society. There are some varieties that are adapted to a dry, others to a moist soil. Some of them cannot endure the heat of our summers whilst others seem scarcely affected by heat or moisture. It is more than probable, therefore, that we may find varieties adapted to the soil of every garden. I am informed that Mr. James Gaillard, of Pineville, has different varieties of strawberries succeeding each other during the whole of the season.

The propagation of new varieties of fruits from the seed is highly deserving of the attention of the horticulturalist. Some of the finest varieties of fruits in the world have been produced in our own country without cultivation and were discovered as it were by mere accident. The original tree of the famed New-Town pippin is still growing on Col. Morris' farm, a few miles from New-York, and no one can tell whence its origin. The Sickle pear, one of the most delicious of the world, was found in the meadows of a gentleman, after whom it has been named, which had long been appropriated to the pasturing of cattle. The delightful Washington plum had a still more narrow escape. An inoculated tree was sent, I believe, from Long-Island to New York; it did not bear for many years. During this storm, the tree was shattered to pieces and broke off below the graft. A shoot from the natural tree sprung up, and here was the first origin of a plum, which, on account of its superiority was named after the greatest of men.

The subject of the naturalization of plants has but recently engaged with particular attention of horticulturalists in France, Holland, and England. The plan at present found the most effectual in acclimating plants from warm latitudes is to place them for a time in the hot-house; then into the green-house; after this, to remove them to the open ground surrounded by clumps of trees to shelter them from the severity of the weather, and when the plants have become sufficiently hearty, these protecting trees are removed. Although I have not been successful in my attempts according to this plan, yet it is certain that in England, they have succeeded in cultivating in the open air shrubs and plants, which heretofore were always considered as requiring the protection of the green-house. They have now, as they inform us, growing without any kind of protection many varieties of the Camelia, the Tea-plant (*Thea bohea* and *virides*), the fragrant Olive (*Olea fragrans*), the camphor tree (*Cinnamomum Camphora*), the ladies ear drop (*Fuchsia coccinea*), the *Oleander Splendens*, *Pittosporum*, the Myrtle and *Passiflora*; and if they can accomplish so much in the bleak and variable climate of England, how many trees and shrubs from Florida and the Indies, may we not be able in the course of time successfully to introduce into our soil and climate?

The cultivation of trees for shade and ornament should engage a portion of our attention, particularly in our city, where we can thus bring verdure into the air, produce an agreeable shade, and contribute to the health and comfort of our families. Among the shade trees which I would particularly recommend, as deserving of cultivation for their beauty and regularity of form, their evergreen leaves and quickness of growth is a species of Oak and properly called the water-oak. It is figured by Michaux, but so incorrectly, as to mislead us. It is, however, minutely described by Mr. Elliott as the *Quercus laurifolia*. A road these beautiful trees may be seen in front of the house and Vanderhorst-street, and Radcliffeborough, in the suburbs of the city. It is very desirable also to awaken a little more attention to the subject of respect to the dead by encouraging the planting of ornamental trees in our grave yards. We, to have such a veneration for those who have gone down to the dust that the poorest has his obituary and the obscurest his monument, are nevertheless very negligent of the burial place, where sleep the ashes of those whom we most esteemed and loved. The Indian mounds in our native wilds bear testimony to the veneration of the savage for the spot where the bones of his tribe repose. The Turk removes every thing from the mausoleum of the dead that might offend the eye and cultivates there the trees, shrubs, and flowers, that may invite the melody of the grove, and awaken and strengthen in the heart, the hopes of immortality; and yet from the grave-yards of our own Christian land present a picture of barrenness and desolation. Not a tree or shrub is placed there. No cap Cypress or Willow casts its shade, and no Rose or Narcissus decks that tomb. The Jerusalem oak (*Chenopodium anthelmendicum*), the exotic Melilotus, the Solidago, the Thistle and Nettle, which shelter the abodes of the adder and the toad, almost present a barrier to our entrance. And attention to this subject is beginning to be awakened in our country. The clean and ornamental grave-yards of the Moravians have long since been regarded with approbation, and the example set us at Boston in their new cemetery surrounded with every thing that can awaken the finer feelings of the heart and render such a place sacred to pleasing religious meditations ought to be imitated by other cities. Under any circumstances, we may render our grave-yards less offensive by giving them clean paths and shady trees. Our own evergreens, the Cedar, the Palm (*Chamerops palmetto*), and the Holly, will remind us of the sentiments of the early Christians who taught that they typified the mild and unfading lustre of Christianity so dear to the mourners' heart; and the Yew, the Cypress, and the Willow, dedicated to the silent solitude of the tomb might convey to us lessons of instruction and comfort.

With regard to the shrubs and flowers which we may easily cultivate in our gardens, our climate is admirably adapted to those which possess as great value on account of their beauty and fragrance, as any that the world can produce. Nearly all the plants of China and Japan, as well as a great number of those that are natives of the Cape of Good Hope, thrive with us in the open air as well as they do in their native climes. Some of the ladies who take an interest in our Society cultivate nearly all the variety of Roses that have yet been introduced into America. If the North excels us in the cultivation of the Crown imperial, the Peony and the Tulip, we can vie with it in the Myrtle, the Lagerstromia, the Oleander, the flowering Pomegranate, and our own fragrant Jessamine. Although the Ranunculus and the Hyacinth require some care and attention, yet they are so generally cultivated among us as to form the pride of our gardens in the spring. The fragrant *Lawsonia inermis* (the famed Henna of the East) has been blooming for the last seven years in the open ground of one of the gardens of our city. A few of the ladies of our city, this public spirit deserves the warmest thanks of this Society, have with much trouble and expense introduced among us many rare and valuable exotics, and their contributions at every exhibition of this Society convince us that they have not grown weary in the cause.

One of the most favourite exotics now cultivated is the Dahlia, which can be brought to very great perfection in our climate. In three of the gardens in and around our city are found many choice varieties of this beautiful Georgina; a flower so beautiful and so easy of culture ought to be found blooming in all our gardens. We should also pay more attention to the arrangement of the flower garden—to the planting of our bulbs and annuals and masses so as to insure a regular succession of flowers; for, with a little attention to this subject our Carolina gardens may be always in bloom.

The forests of Carolina abound in a vast variety of beautiful flowering trees and shrubs, which we ought to transplant into our walks and gardens and cherish and cultivate with care. Is there a tree in the world that is in every respect more worthy of admiration than our Magnolia Grandiflora, the majestic native of our woods? Our Calico tree (*Kalmia latifolia*) are white flowering Stuartias, sprinkled over, as it were, with flakes of snow. Our beautiful Robinias are found blooming abundantly in our mountains and along our water courses. Our Gordonais, our sweet scented shrub (*Calycanthus Florida*), one of our species of Similax, and vast numbers of other choice flowers perfume the air for many miles around. Our Azaleas, Phlox's, scarlet Lobelias, Bignonias, Honeysuckles, Jessamines and Pride of the Meadow (*Thyrsanthus frutescens*) give to the woods of Carolina a charm which not only fills the heart of the lover of nature with delight, but causes even dullness to pause to wonder and admire.

From this imperfect sketch it will be easily seen how much remains for this Society to accomplish. That it has already been productive of some good is evident from the increased attention which has of late been devoted to this subject and from the proofs which have been afforded at our several exhibitions. A lady in this city who has heretofore made a livelihood by the rearing and selling of plants has recently stated that such has been the demand for flowering shrubs and plants since the establishment of this Society, that she has been encouraged greatly to extend her establishment, and is now on a visit to the North in order to obtain choice and rare plants to accomplish this object. A few new gardens have lately been laid out in which great taste has been displayed; and at least, one scientific gardener, thoroughly skilled in his profession, has found employment among us. From these favourable beginnings, have we not some grounds for anticipating future success?

The subjects which are to engage the attention of this Society, are all of them innocent, if they are not otherwise profitable. The God of Nature has cast our lot on this teeming earth; be ours the task of doing all that in us lies to render that earth the abode of comfort and peace. If we do not give to man that which is profitable in a pecuniary view, we will remind him that many things may be kept rather for companionship than profit and that every little extra tie and enjoyment makes a man's home dearer to him. The vegetables which he has raised with his own hands in his own garden, the tree and the vine which his wife and his children have assisted him in planting, the fruits which they have admired and relished together, and the flowers which they have reared with mutual care, all serve to strengthen the bonds of conjugal, parental, and filial love.

And may we not hope, that at least, in the Floral department we may be aided by the fair hands and the approving smiles of the ladies of Charleston. They have done so much for us already that I am sure they require no solicitation to continue their exertions. I could wish that they might be encouraged to follow the examples of similar institutions in other cities and become members of our Society and aid us by their experience and encouragement. Under any circumstances, it is in the power of our fair friends not only to refine the taste of this community by their countenance, but to give a grace and a stimulus to this Society, which our own endeavours could never accomplish. Then may we expect that "this annual festival of flowers will be a banquet of delight where beauty will rule the hour," and pleasure and usefulness go hand in hand.

In endeavouring to advance the interests of this Society, we do not wish to elevate it above other institutions, whose objects are a more extended benefit to mankind, nor do not ask the community to make any costly sacrifices to support it. We desire to give a direction to the public taste, and by enlisting it in the various subject support a culture, we hope to improve the morals of the people and to add to the sum of human happiness.

We are told that the Arcadians were the most savage of all the Greeks till Pan taught them music. We admire the fable; let us profit by the moral. By directing the minds of our people to innocent and profitable employments, by creating a love for world pursuits and the beauties of nature, we may do much for our sunny South. We may form a population fitted to its beauty. We may be enabled to banish ignorance, vulgarity, and crime, and invite the angel of happiness and peace to make his everlasting abode with us.

Note. In preparing the above Address, the writer has referred to observations made in his common place book for years past. He finds that he has not always noted the names of authors from whom his information has been derived and is uncertain whether he quoted their language correctly. This will account for his not having been able to refer to authorities in all cases.

Experiments Made on the Habits of the Vultures

Audubon experimented to see if buzzards located their food by sight or smell, and in 1826 he published an article that noted, for example, that buzzards were attracted to a stuffed deer on the basis of sight alone. When controversy followed, Audubon asked Bachman to devise additional experiments. Bachman's experiments included attracting buzzards to a painting of a dead sheep. This experiment was repeated more than fifty times, and the buzzards never discovered putrid meat concealed nearby. Bachman noted that although buzzards have nostrils and some kind of olfactory organs, the species common in the American South clearly did not use them to locate their prey.

Bachman called for experiments on other species of vulture, and he urged Audubon's critics to repeat the same experiments or to conduct ones of their own: "It has always appeared to me an act of injustice to condemn any man for expressing an opinion on subjects of Natural History . . . particularly when the error could be so easily detected by instituting a similar course of experiments." At that point, the public criticism ceased. More recently, Bachman's conclusions have been questioned, but, as he stated, until "a similar course of experiments" disproves an experiment that was conducted fifty times, his conclusions deserve consideration.

Bachman's title for the article refers to a "carrion crow," which was a common name for the Black Vulture (*Coragyps atratus*). The Turkey Buzzard is often called the Turkey Vulture (*Cathartes aura*). Currently, the prevailing view among ornithologists is that the Turkey Buzzard relies to a large extent on smell to find its prey, but if so, it is one of the few birds in the world that does (Animal Diversity Web; University of Michigan Museum of Zoology). Since Bachman's often-repeated experiments seemed conclusive to him and to many witnesses, attempts should be made to replicate and to account for his results. Other experiments need to be designed to determine how far away an odor can be detected by the subspecies of Turkey Buzzard that inhabits the eastern United States. It is one of six subspecies in North and South America.

An Account of some experiments made on the habits of the Vultures inhabiting Carolina, the Turkey Buzzard,* and the Carrion Crow,** particularly as it regards the extraordinary powers of smelling, usually attributed to them. By J. Bachman [Charleston: John Bachman].

Although the Vultures inhabiting the Southern States are among the most common of our larger species of Birds, remaining with us during the whole year, building their nests in the hollows of fallen trees and stumps around our plantations, resting on our house tops and seeking their food around our markets and in the very streets of our cities, yet it appears that a difference of opinion exists with regard to some of their faculties and particularly whether they find their food by their sense of smell or of sight.

It has been the long established belief of all civilized nations since the time of the Romans that Vultures were possessed of extraordinary olfactory powers by which they were enabled to scent their food at the distance of many miles. Whether this opinion was founded on truth or whether it was a vulgar error, having its origin with the thousands of others which have been handed down from age to age originating in ignorance or superstition, cannot be fully ascertained until satisfactory experiments are made on the olfactory powers of the Vultures of Southern Europe, Asia and Africa.

All the writers on American Ornithology have ascribed to the Vultures of the U. States the same extraordinary powers of smell with the single exception of Mr. Audubon, who in a paper published in Jameson's Journal, Edinburgh, 1826, detailed a series of experiments made in America several years previous from which he came to the conclu-

* Cathartes Aura Ill.
** Cathartes Iota. Bon.

sion that these Birds were guided to their food altogether by the eye. He found by repeated experiments that Vultures were attracted by a derived deerskin, stuffed in imitation of that animal, and that in these instances when no effluvium could exist, they could not have been led to it by the scent. He next concealed a dead animal in the heat of summer in such a way that it could not be observed by the Vultures, although the scent was not abstracted; here it was suffered to decay without having been discovered by these Birds.

He next procured two young Carrion Crows, which he tamed and reared in a cage, the back part of which was so closed that objects approaching from that side could not be seen. Whenever he came with food to the front of the cage the Birds jumped at the bars, commenced hissing and putting their bills towards each other as if expecting to be fed mutually, as their parent had done; when, however, the cage was silently approached from behind with animals and flesh however putrid, no movement was made by the Birds to indicate their having observed by the effluvium that their favourite food was near.

The sentiments thus expressed by Audubon were at the time and are still treated with a good deal of severity, both in Europe and in his native country.

It has always appeared to me an act of injustice to condemn any man for expressing an opinion on subjects of Natural History merely because from his own investigation, he had arrived at different conclusions from those who had lived before him, particularly when the error could be so easily detected by instituting a similar course of experiments; such a course of conduct would be a bar to all improvements, and the sincere enquirer after truth would have to contend against a host of prejudices from those, who adopting the opinions of others, refused to make these enquiries which would satisfy their own minds that their opinions were fortified and confirmed by *experience*.

The details of the experiments made by Audubon are of such a character that either the conclusions at which he arrived are correct and our Vultures do not possess extraordinary olfactory powers or he has given to the world an unfair statement and is therefore not a man of veracity and is undeserving of the confidence of the community; should such an impression be unhappily made on the public mind, it will not only have a tendency to destroy his usefulness, but will deprive him of those pecuniary resources which are requisite to enable him to carry on successfully a very expensive publication. A publication which cannot fail to prove a very important acquisition to the natural history of our country and to establish an abiding monument to the fame of its Author.

The lovers of American Ornithology who feel under many obligations to the man who has devoted so many years of his life to this interesting and beautiful Department of Natural Science will not condemn him unheard; and those, particularly of our Southern States, would show themselves very careless observers of Nature, and very indifferent to the character and fame of Audubon, if possessing as they do so many favourable opportunities for investigation, they did not institute some inquiries, not only to do an act of justice to a distinguished Naturalist, but to ascertain an interesting fact in Natural History.

No one who will read Mr. Audubon's paper on the subject, containing a full detail of a number of experiments on the habits of our Vultures, can deny that if he intended to deceive the world, he certainly chose a subject where detection was easy and certain. In our Southern States, these Birds are so abundant as to have become a nuisance, particularly in our cities. It is but an act of justice to Mr. Audubon to state that in his frequent visits to Charleston, he has fearlessly invited investigation on this disputed subject.

During his absence he wrote to me on several occasions urging me to make further experiments; a number of engagements prevented me from devoting as much time to the subject as was necessary to investigate it in such a manner as to prove satisfactory to my mind, and I postponed it to a more leisure period. On a recent visit however of Mr. Audubon, I consented to institute these inquiries, in the prosecution of which, I was aided by the intelligence and experience of such disinterested Naturalists and men of Science as could be obtained.

It will be observed that our experiments were confined to our two species of Vulture, Cathartes Aura, and Cathartes Iota, which are so common in Carolina. There are three other species which have been described by Authors to frequent the Southern and Western portions of the United States. The Vulture Californianus (Lath.) undoubtedly exists West of the Rocky Mountains, as specimens were procured near the Columbia River by Mr. Douglass, but we have no evidence of its existing to the Eastward of that great chain. The existence in the United States of the great Condor* is only conjectured from a bill and a quill feather brought by Lewis and Clarke from the Columbia River and deposited in the Philadelphia Museum. The beautiful king of the Vultures** is said to appear occasionally in Florida, upon what authority I am unable to state. None of the gentlemen, however, who have resided at Key-West, and other portions of Florida for many years, have been able to find this Bird. Mr. Audubon, however, in his indefatigable exertions in that country found a different species which was not before known to exist in the U. States. The Caracara Eagle,† of which I have received several specimens through the kindness of Dr. B. B. Strobel and Dr. Leitner that had been procured between Tampa Bay and Key-West. Whatever powers of smelling these (with us rare species) may possess, I am unable to state from actual experiment. But it will probably be discovered that their olfactory powers have been as much overrated, as was the size of the famous Condor, whose quill feather even as late as 1830 was described as "twenty good paces long," and which on actual measurement has been found of less dimensions than that of several species existing on the Eastern Continent. But laying aside speculations on a subject which time and further observations only can decide, I proceed to a detail of facts that have come under my observation.

On the 16th Dec. 1833, I commenced a series of experiments on the habits of our Vultures, which continued till the end of the month, and these have been a renewed at intervals till the 15th of Jan. 1834. Written invitations were sent to all the Professors of the two Medical Colleges in the city. To the officers and some of the members of the Philosophical Society and such other individuals as we believed might take an interest in the subject. Although Mr. Audubon was present during most of this time and was willing to render any assistance required of him, yet he desired that we might make the experiments ourselves that we might adopt any mode that the ingenuity or experience of others could suggest as arriving at the most correct conclusions. The manner in which these experiments were made together with the results, I now proceed to detail.

---

* Cathartes Gryphus Temm
** Cathartes Papa Ill.
† Polyborus Vulgaris.

EXPERIMENTS ON THE HABITS OF VULTURES

There were two points in particular on which the veracity of Audubon had been assailed, 1st. Whether the Vultures feed on fresh or putrid flesh, and 2nd. Whether they are attracted to their food by the eye or scent.

On the first head it was unnecessary to make many experiments, it being a subject with which even the most casual observer amongst us is well acquainted. It is well known that the roof of our Market house is covered with these Birds every morning, waiting for any little scrap of fresh meat that may be thrown to them by the Butchers. At our slaughter pens the offal is quickly devoured by our Vultures whilst it is yet warm from the recent death of the slain animal. I have seen the Vultur Aura a hundred miles in the interior of this country, where he may be said to be altogether in a state of nature, regaling himself on the entrails of a Deer which had been killed and not an hour before. Two years ago, Mr. Henry Ward, who is now in London and who was in the employ of the Philosophical Society of the city, was in the habit of depositing at the foot of my garden in the suburbs of Charleston, the fresh carcasses of the birds he had skinned and in the course of half an hour, both species of Vulture and particularly the Turkey Buzzard came and devoured the whole. Nay, we discovered that the vultures fed on the bodies of those of their own species that had been thus exposed. A few days ago, a Vulture that had been killed by some boys in the neighbourhood and that had fallen near the place where we were performing our experiments, attracted on the following morning the sight of a Turkey Buzzard, who commenced pulling off its feathers and feeding upon it. This brought down too the black Vultures who joined him in the repast. In this instance the former chased away the two latter to some distance, an unusual occurrence, as the black Vulture is the strongest bird and generally keeps off the other species. We had the dead bird lightly covered with some rice chaff, where it still remains undiscovered by the Vultures.

2d. Whether is the Vulture attracted to its food by the sense of smell or of sight? A number of experiments were tried to satisfy us on this head, and all lead to the same result; a few of those I proceed to detail.

1st. A dead Hare,* a Pheasant,** a Kestrel† (a recent importation from

---

* Lepus timidus.
** Phasianus Colchicus.
† Falco Tinuncuins.

Europe) together with a wheelbarrow full of offal from the slaughter pens were deposited on the ground, at the foot of my garden. A frame was raised above it at a distance of 12 inches from the earth, this was covered with brushwood, allowing the air to pass freely beneath it, so as to convey the effluvium far and wide, and although 25 days have now gone by in the flesh has become offensive, not a single Vulture appears to have observed it, though hundreds have passed over it, and some very near it in search of their daily food. Although the Vulture's did not discover this dainty mess, the dogs in the vicinity, who appeared to have better olfactory nerves, frequently visited the place and gave us much trouble in the prosecution of our experiments.

2d. I now suggested an experiment which would enable us to test the inquiry whether the Vulture could be attracted to an object by the sight alone. A coarse painting on canvas was made representing a sheep skinned and cut open. This proved very amusing. No sooner was this picture placed on the ground that the Vulture's observed it, alighted near, walked over it, and some of them commenced tugging at the painting. They seemed much disappointed and surprised, and after having satisfied their curiosity, flew away. This experiment was repeated more than fifty times with the same result. The painting was then placed within 10 feet of the place where our offal was deposited. They came as usual, walked around it, but in no instance evinced the slightest symptoms of their having scented the offal which was so near them.

3d. The most offensive portions of the offal were now placed on the earth. These were covered over by a thin canvas cloth. On this which was strewed several pieces of fresh beef. The Vultures came, ate the flesh that was in sight, and although they were standing on a quantity beneath them, and although their bills were frequently within the eighth of an inch of this putrid matter, they did not discover it. We made a small rent in the canvass, and they at once discovered the flesh and began to devour it. We drove them away, replaced the canvass with a piece that was entire; again they commenced eating the fresh pieces exhibited to their view, without discovering the hidden food they were trampling upon.

4th. The Medical gentlemen who were present made a number of experiments to test the absurdity of a story, widely circulated in the United States through the newspapers, that the eye of a Vulture, when perforated and the sight extinguished, would in a few moments be restored in consequence of his placing his head under his wing; the down of which was said to renew the sight. The eyes were perforated; I need not add that although they were refilled it had the appearance of rotundity, yet the birds became blind and that it was beyond the power of the healing art to restore his lost sight. His life was, however, preserved by occasionally putting food in his mouth. In this situation they placed him in a small out-house, hung the flesh of the Hare (which had now become offensive) within his reach; nay, they frequently placed it within an inch of his nostrils, but the bird gave no evidence of any knowledge that his favourite food was so near him. This was repeated from time to time during an interval of twenty-four days (the period of his death) with the same results.

We were not aware that any other experiments could be made to enable us to arrive at more satisfactory conclusions, and as we feared if prolonged, they might become offensive to the neighbours, we abandoned them.

As my humble name can scarcely be known to many of those into whose hands this communication may fall, I have thought proper to obtain the signature of some of the gentlemen who aided me in, or witnessed these experiments, and I must also add that there was not an individual among the crowd of persons who came to judge for themselves, who did not coincide with those who have given their signatures to the certificate.

We, the subscribers, having witnessed the experiments made on the habits of the Vultures of Carolina (Cathartes Aura and Cathartes atratus) commonly called Turkey Buzzard and Carrion Crow, feel assured that they devour fresh as well as putrid food of any kind and that they are guided to their food altogether through their sense of sight and not that of smell.

  Robert Henry, A.M., *President of the College of South-Carolina.*
  John Wagner, M.D., *Pro. of Sur. at the Med. Col. State So. Ca.*
  Henry R. Frost, M.D., *Pro. Mat. Med. Med. Col. State So. Ca.*
  C. F. Leitner, *Lecturer on Bot. and Nat. His. So. Ca.*
  B. B. Strobel, M.D.
  Martin Strobel.

It now remains for Naturalists to account for the errors which have for so many ages existed with regard to the power of scent ascribed to our Vultures. Indeed, it is highly probable that facts elicited from the experiments of Audubon on our two species of Altar, strengthened by those instituted on this occasion, may apply to all the rest of the genera. Perhaps it may be discovered that the whole family of Vultures are altogether indebted to the eye in their search after food; indeed this may yet be found true in regard to the whole feathered tribe. It is, I believe, a common practice in England for persons who attend on the decoys for wild Ducks to carry a piece of burning peat which they hold near their mouths to prevent the birds from smelling them as it is also customary in America to burn gunpowder in various portions of the cornfields under an impression that the smell of the powder will frighten away the Crows. The Powder Manufacturers can have no great objection to this latter practice, and it must be confessed, that it is among the most innocent ways that powder can be wasted. But I fear that this will not benefit the crops of the farmer. These birds together with the Ravens[*] and even the wild Turkey[**] can be approached undercover of a bank or tree, and if they do not either see or hear you, it will, I apprehend, be a different matter for them to find you out by their olfactory powers. Indeed I am of opinion that whilst to Quadrupeds (particularly carnivorous ones) the faculty of scent is their peculiar province, this organ is but imperfectly developed in birds. As it does however exist (although in an inferior degree), I am not disposed to deny to birds the power of smell altogether, nor would I wish to advance the opinion that the Vulture does not possess the faculty of smelling in the slightest degree (although it has not been discovered by our experiments). All that I contend for is that he is not assisted by this faculty in procuring his food—that he cannot smell better for instance, than Hawks or Owls, who it is known are indebted altogether to their sight in discovering their prey. If our Vultures had to depend on their olfactory powers alone in procuring food, what would become of them in cold winters in Kentucky and other of our Western States, where they remained all the year and where the earth is bound up with frost for months at a time and where consequently, during that

[*] Corvus Corax.
[**] Meleagris Gallipave.

period, putridity does not take place? And if they depended alone on tainted meat for food, how soon would the whole race (at least in our temperate climates) die of hunger?

How easily error may be perpetuated from age to age, we may learn from a thousand other visionary notions, which the careful observations of recent travellers and naturalists have exploded. At this today, the belief is very general in this country that immediately after a Deer has been killed, the Vultures at the distance of many miles are seen coming a direct line against the wind, scenting the slaughtered animal. This may be accounted for with a little observation upon rational principles. When a Deer* is killed, the entrails are immediately taken out, these together with the blood which covers the earth to some extent, are seen by some passing bird. He directly commences sailing around the neighbourhood. He is observed by those at a distance; the peculiar motions of his wings, well known to those of his own species, communicates to them the intelligence that something good for them is perceived. These hastening to the place give information to those who are still further off, and in the course of an hour, a very great number are guided to the spot. But it will scarcely be argued that this great concourse of Vultures has been attracted by the effluvium of putrid flesh since for the animal has been killed but an hour before.

In the prosecution of our experiments, we discovered that the powers of sight in our Vultures, were not as great as those possessed by the Falcon tribe. A dead fowl was discovered by them at the distance of 70 or 80 yards; a sheep at 100 or 120 yards. These however were stationary objects lying on the ground. One of their own species, however, flying in the air is no doubt observed by them at a much greater distance. It may easily be conceived why the sight of the Vulture is less acute than that of Hawks and Eagles. The latter prey upon birds, quadrupeds, &c. for which they have to hunt, the former feed chiefly upon dead birds, quadrupeds or reptiles, and frequently those of large size, which it requires no extraordinary powers of vision to discover. An argument much relied upon by those who advocate the doctrine of the olfactory powers of Vultures is the circumstance of their usually flying against the wind as if

* Cervus Virginianus

to discover and follow some current of tainted air. This practice, it may easily be perceived, is not more common to the Vulture than to that of any other bird. It is a mistaken idea that birds in their migrations or on any other occasion prefer flying with the wind. This is inconvenient and uncomfortable to them, and the careful observer of the flight of birds is well convinced that all birds, the Vulture among the rest, prefer facing the wind, not to enable them to smell their food, but to render their flight more easy and pleasant.

It may next be enquired for what purpose are the wide nostrils and olfactory nerves given to the Vultures if they are not intended to assist them in procuring their food? To this, I answer, but the olfactory nerves of our Vultures are not larger than those of many other birds, and their nostrils are less even than those of the Hooping Crane,* which discovers its food (as I strongly suspect every other bird does) by the eye alone. The wide orifice in the beaks of Vultures and which is generally considered as the true nostril is probably a wise provision of Nature to enable a bird which from its filthy habits of feeding is continually exposed to have its nostrils closed up to blow out any substance calculated to obstruct them. The same may be said of the Hooping Crane, which from the manner of its digging for roots in the earth is liable to the same inconveniences. Several heads of the Vultures are now in the hands of individuals connected with our Medical Colleges for dissection. A satisfactory elucidation of the subject will require time, patience and an extensive knowledge of comparative anatomy in regard to the various species of birds. The results of these investigations will probably be communicated to the public in the course of a few months.

After having resorted to the means detailed above to satisfy myself of the accuracy of the statements of Audubon as regards the habits of the Turkey Buzzard detailed in Jameson's Journal, I once more carefully read over his remarks on the subject, and I now feel bound to declare that every statement contained in that communication is in accordance with my own experience, after a residence of twenty years in a country where the Vultures are more abundant than any other birds, and I have reason to hope from the character of the writers who have doubted his

---

* Grus Americana Temm.

veracity, that when they have read a detail of these experiments, they will either repeat them to the satisfaction of their own minds or place confidence in the statements of those who have taken this trouble; and that they with that generosity of feeling so distinctive of those who are engaged in liberal and kindred pursuits, will be gratified to assign to Audubon that meed of praise which he so undoubtedly merits.

John Bachman, *Member of the Philosophical Society of Charleston, S.C. and of the Academy of Nat. Sciences, Philadelphia.*
Charleston, 18*th Jan.* 1834.

Migration of North American Birds

Bachman produced an overview of migration that was uniquely comprehensive for its time and that included innumerable major insights. In "Natural Selection," the manuscript that was summarized for the *Origin of Species*, Charles Darwin praised Bachman's article as "excellent" (Darwin 1975: 491).

This article by Bachman represents another successful attempt to synthesize a large body of knowledge. In his essay on horticulture, he was writing more on the basis of widespread reading, but in this article on ornithology, he relied primarily on the analysis of observations that he had personally made. For several decades, Bachman had recorded extensive observations "in three very distinct portions of America" (New York, Pennsylvania, and South Carolina). In Charleston he was well located to make observations on many species of birds that passed through the area during their migrations. He bred numerous types of birds to observe their development and behavior. Having known Alexander Wilson and John James Audubon personally and having learned greatly from their work and from the writings of other ornithologists, Bachman was able to draw upon abundant evidence, and his generalizations were based on specific evidence and were derived using the inductive method of reasoning. His conclusions were the most reliable that had been published on the subject anywhere in the world.

Among Bachman's insights were that the reasons for migration were in many cases more closely related to what food was preferred than to the general availability of food in a specific locality at different times of the year. The same place could be occupied by one species of the same genus of birds at one time of the year and by another species at another time. He listed, for example, numerous birds that migrated in the winter from the north to the Carolinas while birds that had been residing in the Carolinas were migrating farther south, and in the summer the process was reversed. As long, though, as the preferred food continued to be available in any given locality, some types of birds (such as the Passenger Pigeon) would remain until the local supply was exhausted and would migrate gradually and sporadically regardless of the temperature unless snow covered their preferred foods.

In general, North American birds migrated south primarily for food and migrated north primarily to breed. It becomes easy to see how some types of birds had to migrate increasingly far to find enough of their preferred foods before returning in the spring to breed to the vicinity of their origin as a species.

Climate was important both for the availability of food and for its influence on breeding, but the influence of climate was much less direct than had been previously believed. Instinct caused some birds to migrate great distances quickly when they were nearly ready to breed and to migrate even to areas that did not yet have enough food for them. Shore birds must migrate before ponds freeze, but sea birds ordinarily do not migrate since the sea rarely freezes, and their preferred food is always available. Similarly, carnivorous birds are much less likely to migrate than birds that rely on insects and worms for food.

Bachman dealt with many other aspects of the subject. He explained the presence of a small part of the birds of North America in Europe as species that bred within the Arctic Circle. He cited instances of birds expanding their range when foods they preferred became more widely available through cultivation. He indicated that rails and swallows supposed by some renowned scientists to hibernate did not, but instead migrated to known locations. He gave examples of individual birds returning ten or more years to the same vicinity to breed. He noted that various birds were sufficiently reliable as indicators of changes in season to be used to determine when crops could be safely planted without fear of another frost taking place.

Bachman attempted to dispel the supposed mysteries of migration in order to demonstrate that "the laws by which the whole system of nature is governed are equally simple and majestic and are equally visible in the minutest as well as in the most stupendous of Gods works." What seems to have impressed Darwin most was Bachman's search for "the principles of nature" and the simplicity, but broadness of application of his explanations. Darwin's notebooks indicate that he was deeply impressed by the breadth of Bachman's knowledge about the distributions of species and by his ability to account for them.

On July 21, 1832, Bachman wrote Audubon that he was "preparing a piece on the migration of birds to read before the Philosophical Society" [1929: 179; cf. Stephens 2003: 164]. A manuscript version of this article is in the Thomas Cooper Library at the University of South Carolina. In 1836, the article was published in Benjamin Silliman's *American Journal of Science and Arts*.

Art. IX.—*On the Migration of the Birds of North America*. / Read before the Literary and Philosophical Society of Charleston, (S.C.) March 15th, 1833; / by Rev. J. Bachman.

For ages past, the migration of birds has been a subject of great interest to naturalists. The mysterious appearance and disappearance of many species, at different periods of the year, while many of them have never been seen in their migrations; the remote or unknown situations to which they retire; the sudden appearance of some birds in the spring, after one or two days of warm weather, and their equally sudden disappearance on the first cold day; all have conduced to create many vague and superstitious notions, in the minds of the uninformed, and have often left the intelligent student of nature in perplexity and doubt. Examples are seen in the accounts so often published of the swallows having been found in great numbers in caves and hollow trees, and in lakes and ponds; and of the common Rail or Sera (*Rallus Carolinus, L.*) having been discovered in gutters and hollow banks.

Some have supposed that birds, like some animals, are by their internal organization capable of becoming dormant during winter, and hence they readily listen to stories of birds having been found, concealed in great numbers in caves, the hollows of decayed trees, recesses of old buildings, and other secluded situations; whilst others have contended, that they were, during the winter, preserved under the water, beneath the mud.

Amidst such contradictory opinions on a subject concerning which the most intelligent naturalists are not yet agreed, there is a wide field open for inquiry and observation. The works of God amidst the wonders of nature are always worthy of investigation. If he has given to the birds of the air instincts which cannot be equaled by the boasted reason of man—if he has communicated to them some mysterious faculties, which have hitherto baffled the researches and wisdom of the wise—may it not be well for us at least to record the facts so that, although we may not be able to explain these hidden mysteries of nature, we may be humbled under a sense of our inferiority and thus be led to adore the wisdom of God.

Very little appears to have been written on the migration of North American birds; a topic probably regarded as of too little importance to merit the research necessary to a satisfactory result on such an intricate subject; for the elucidation of which, I have myself possessed some opportunities by witnessing the migration of birds in three very distinct portions of America.

That instinct is truly mysterious, which at particular seasons of the year teaches birds to take wing and leave their native haunts, pursuing their onward course, sometimes across arms of the sea, and in most cases over rivers, mountains, and forests, into far distant countries. It is equally surprising that many of them, commencing their migrations in summer, should thus anticipate the cold; while others returned from southern climes, before the snows of the north have disappeared, and whilst winter still "lingers in the lap of the spring."

Among animals and birds, we often discover a train of actions, all adapted to produce a certain effect by the agency of certain means without the exhibition of any part of a regular chain of thought, the essential characteristic of reason; this substitute for reason is called instinct, a term which has given rise to many unsatisfactory theories. I shall, therefore, pass them over with a few brief remarks on the difference between instinct and reason.

When certain species of birds at their first season of breeding, being without experience, build all their nests alike, those in form and materials, this may be called the result of instinct. On the other hand, when man guards against danger or makes provision for the wants of life or seeks relief from diseases by the application of medicines, he acts from reason, because he is instructed by the experience of the past. When birds at certain seasons of the year change the climate, in anticipation of cold or heat, they act from instinct, because to many of them, it is their first migration; and as they often migrate singly and not in flocks, in such cases no experience can aid them. On the other hand, when man makes provision for the changes of seasons and climates, he acts from reason and is instructed by his own experience or the experience of others.

Whatever difficulties there may be and accounting for that mysterious principle in birds called instinct and which induces them at certain seasons to change their abode, and again, after an interval of six months, to return to the neighborhood where the year before they reared their young; the facts of these migrations are an incontrovertible, and the reasons why they take place are becoming more and more apparent.

Those birds that migrate are from the very structure of their bodies admirably adapted to rapid and continued flight. Their feathers are so light that they float in the atmosphere for many hours with very little artificial support. The tubes of these feathers are hollow; the bones are specifically lighter than those of quadrupeds; the bones also are hollow, and instead of marrow are filled with air. They are furnished with lungs of an unusually large size, adhering to the ribs, and provided with aerial cells, insinuating themselves into the abdomen. These, added to the great length and strength of wing enable them, with ease and rapidity, to navigate the air, to elevate themselves above the clouds and pass from one country and climate to another.

We perceive, then, from the very structure of birds that they are admirably formed for rapid flight and migration. From a variety of accurate experiments, which have been made, at different periods, it appears that the Hawk, the Wild Pigeon (*Columba migratoria*), and several species of wild ducks fly at the rate of a mile in a minute and a half; this is at the rate of forty miles an hour, four hundred and eighty between the rising and setting of the sun, and nine hundred and sixty miles in twenty four hours. This would enable birds to pass from Charleston to our distant northern settlements in a single day, and this easily accounts for the circumstance that geese, ducks, and pigeons, have been taken in the northern and eastern states with undigested rice in their crops, which must have been picked up in the rice fields of Carolina or Georgia but the day before.* There is a well attested account of a falcon from the Ca-

---

* I had an opportunity, several years ago, in the state of New York, of examining the contents of the craws of several pigeons, taken from the same flock, which were pronounced by the country people to be rice. It proved, however, to be a different grain, the wild rice of the western lakes (*Zizania aquatic*).

nary Islands sent to the Duke of Lerma which returned from Andalusia to the Island of Teneriffe in sixteen hours, which is a passage of seven hundred and fifty miles. The story of the falcon of Henry the second is well known, which pursuing with eagerness one of the small species of bustards at Fontainebleau, was taken the following day at Malta and recognized by the ring which she bore. The swallows fly at the rate of a mile in a minute, which would be one thousand four hundred and forty miles in twenty four hours. That many birds continue their migrations by night as well as by day and are thus enabled to make an additional progress, may be easily ascertained from their notes, which in the autumn and spring, the seasons of their migration, we often hear by night. The cries of geese, cranes, and some species of land birds are distinctly heard, and others fly silently. Wild pigeons are frequently seen at early dawn in the higher atmosphere. They fly higher by night than by day and thus experience less inconvenience from darkness. The great Hooping Crane scarcely ever pauses in his migrations to rest in the middle states. I have heard his horse notes as he was passing over the highest mountains of the Alleghany, but he was always too high to be seen by the naked eye. This bird seems to take wing from his usual winter retreats in the south, ascends into the higher regions of the air, and scarcely halts, until he arrives at his breeding places in or near the polar regions.

There are very few birds that do not migrate either on account of food or climate. The observations of Captains Parry and Franklin, of Dr. Richardson and their associates, who wintered in the polar regions, prove that birds which never visit temperate climates in which naturalists formerly supposed were wholly confined to the arctic circle, believed to be intensely cold regions of the north in winter and migrate southerly to the distance of many hundred miles. These adventurous explorers of the polar regions speak of the dreariness and desolation of these countries in the winter and the almost total absence of animal life. During the whole winter, spent at Melville Island, a pair of ravens (*Corvus Corax*) alone were seen and these they state had frequently a white ring around their necks, "caused by the accumulated and crossed amounts of their own breath and giving them a very singular appearance." The snow Buntings (*Emberiza nivalis*), the Ptarmigan (*Tetrao Lagopus*), and two other species of Arctic Grouse (*Tetrao Salicti* and *T. Rupestris*) were their earliest visitors in the spring; and these birds are in Europe and in the farthest northern settlements of our continent found only in the coldest winters, and on the highest mountains; still we perceive that even they find the limits beyond which they cannot live in winter.

Birds migrate either to avoid the cold of winter or to find more congenial or more abundant food, and I am induced to believe that in general the latter is a stronger principle than the former. The small number that remained amidst the snows of the north are either carnivorous, such as a few of the Owls and Hawks, the Ravens (*Corvus Corax*), the Canada Jay, (*Corvus Canadensis*), and the northern Shrike (*Lanius borealis*). These pick up a scanty subsistence by feeding on a few of the smaller birds that remained or by following the hunters and the wolves and supporting life by picking the bones of the animals which they have left. Or they are composed of those birds that feed on the buds of trees, such as the Grouse, that live on the buds of the birch (*Betula*), Poplar (*Populus*), and several species of willow (*Salix*). Or those that feed on the seeds of the pine and spruce (*Abies*), as the Crossbills (*Curvirostra*), and pine Grosbeaks (*Phorhula enucliator*), or they are birds that are able to find subsistence on the seeds of plants that are protruded above the snow or on the seeds of grass found in the barn yards or haystacks of the farmers, such as a few species of the sparrow. But those immense numbers of birds that feed on insects and worms all migrate to those countries where they are abundantly supplied with this kind of food. These are the Swallows (*Hirundo*), the night Hawks and the Whippoorwill (*Caprimulgus*), the Tanagers, the fly catchers, and warblers. To them migration is essential to the support of life. The insects at that season disappear, the earth is bound in frost, or covered over with snow, and all the means of subsistence are removed; but long ere this these lively tenants of the air have updated the impulse of some mysterious instinct and have migrated to more congenial climes. To these may be added all the birds that obtain food from the muddy and moist places of the earth, such as the different species of Curlew (*Numenius*), the Snipes (*Scolopax*), and the sand birds (*Tringa*) as well as those ducks that obtain subsistence from freshwater ponds and rivers; these, finding their swamps, brooks, and shores frozen over, migrate from the north to milder regions where they can procure suitable food.

Those birds that migrate but partially and spend their winters in the northern states, though in a milder temperature than their places of summer retreat, such as the eagles, hawks, owls, and grouse, are enveloped in a warm, thick, and downy plumage, which in most of the species extends even over the legs and toes. Other birds are exposed to the water as well as the cold, such as some species of wild ducks (*Anas*), gulls (*Larus*), petrels (*Procellaria*), and puffins (*Puffinus*). These, gaining a subsistence from the sea, are not obliged to migrate on account of food. In addition to their warmth of covering, which shelter them from the cold, they are supplied with sacs, containing an oleagenous substance with which they regularly lubricate their feathers, thus rendering them imperious to moisture. Whilst floating on the surface of the water, they often draw up their feet beneath their warm covering of down, and thus every part of the body is protected against the influence of the cold. There is another circumstance with regard to the capacity of birds to endure cold which is not generally taken into consideration; it is the high degree of temperature. The temperature of the human body is generally placed at 97 or 98 of Fahrenheit, that of warm blooded, animals two or three degrees higher, and that of birds as high as 106 making a difference of 8 or 9 degrees between birds and men. A large mass of air penetrates the lungs and all the aerial sacs and canals of the bird, increasing the action of the heart and propelling the tide of circulation with great rapidity. The pulsation in birds follow each other in quick succession that they can scarcely be counted. The heat of their bodies being much greater than that of animals enables them to bear with ease the rigorous cold in the distant north and in the elevated regions of the air.

Some birds migrate only from one extreme of our union to the other. Thus the many species that go under the name of Sparrows that breed at the North, with the exception of three, the Snow Bunting (*Emberiza nivalis*), the three Sparrows (*Fringilla arborea*), and the white crowned bunting* (*Fringilla Leucaphrys*) spend their winters in tens of thousands in Carolina. When the meadow Lark (*Sturnus Ludovicianus*) and the brown Lark (*Anthus spinoletta*) find the snows of the north covering the

* It has been commonly believed, that this very rare species which breed at Labrador, does not migrate far to the south in winter. It however passes through Carolina early in autumn, and winters farther south.

earth, and hiding their favorite food, they retreat before it and seek sustenance in our Southern states. Other families of birds such as feed on ripe berries that abound in the winter also remain with us; these are the Robins (*Turdus migratorius*), the wax bird (*Bombycilla Americana*), and the blue bird (*Saxicola sialis*), which feed on the berries of the Tupelo (*Nyssa aquatic*), the Holly (*Ilex opaca*), the Cassena (*Ilex cassina*), and the small black and red berries of several species of Smilax and Prinos. The yellow crowned Warbler (*Sylvia coronata*) is the only Sylvia out of fifty species inhabiting the U. States that remain with us in the winter, and even this bird could not find a subsistence among us were it not that it almost changes its nature in winter and lives on the fruit of the wild myrtle (*Myrica cerifera*). This is also the case with the only fly catcher in Carolina, the Pewee (*Muscicapa fusca*), which sometimes fattens on the seeds of our imported tallow tree (*Styllingia cerifera*).

It is doubtful whether there are any birds that never migrate during the changes of the season. Hawks and Crows are infinitely more abundant in the north during summer, than in the winter; the greatest number of them retreat southerly; those of the south may at the same time proceeds still farther toward the Equinox. Our cardinal Grosbeaks (*Fringilla cardinalis*) are found in New Jersey during summer and are abundant in Virginia, hence the name of Virginia nightingale and yet during winter very few remain in those states. In the mean time as our number of birds of this species does not increase, it is very probable that those which have been raised among us remove still father to the south. As our summer birds such as the blue Grosbeak (*Fringilla cœrulea*), painted bunting (*Fringilla ciris*), and our warblers and fly catchers, abandoned us towards the close of autumn, we receive at the same time fresh supplies of feathered hordes from Canada and the northern portions of the United States. Many of these remain in our mild climate of Carolina, during the whole winter. Some of them such as the Fox colored sparrow (*Fringilla iliaca*), the Siskin (*F. Pinus*), the Purple Finch (*F. purpurea*), and the Woodcock (*Scolopax minor*) only approached our southern climates in proportion as they are pursued by the cold. These seem to beg their subsistence on their passage, and linger among us no longer than their necessities require.

When our winter birds return to their breeding places in the north, they are in the early period of spring replaced by analogous species from the tropics, which resort to South Carolina and principally along our maritime districts to engage in the affections and cares of reproduction. Of the many species of northern hawks, the red shouldered (*Falco lineatus*, Wilson) one of the most common species in the United States is the only one that remains in our low country during summer. In the mean time several interesting species from the south arrived among us of gentler and less destructive habits, feeding principally on insects and lizards. The beautiful swallowtail hawk (*Falco furcatus*), a Mexican species which seems to be ever on the wane, builds its nest on the highest trees of our forests. The Mississippi Kite (*Falco plumbeus*, Gmel.) with similar habits and also feeding whilst on the wing is found occasionally in groups of four or five soaring high in the air. This bird is so gentle when not on the wing that it generally suffers you to walk under the tree without being disturbed. The black winged Hawk (*Falco dispar*, Temm.) is another of our visitors. It bears so strong a resemblance to an Asiatic species (*Falco melanopterus*, Daud.) that, although it is described under another name, I have never been able to detect the slightest difference. It is occasionally met with as early as the beginning of February and breeds on a few of our islands along our sea-board. This species it has hitherto been supposed never migrated north of Florida. When the Gannets (*Sula bassana*, Lacep.) leave us for their northern rocks, we are visited by the two species of Pelicans (*Pelicanus onocrotalus*, and *P. fuscus*, L.) and by immense flocks of the wood Ibis (*Tantalus Loculator*, L.). The latter commence regular systematic attacks upon our rice fields and on the fish in our ponds, first muddying the water and then killing ten times as many as they can consume, leaving a rich repast for the alligator. As strange as it may appear in birds so large and numerous, their nests have never been found. No sooner do the Virginia Rail (*Rallus Virginianus*, L.) and the Sora leave us, then their place is supplied by two species of kindred genera, the purple and common Gallinules (*Gallinula martinica*, Gmel. *G. Chloropus*, Lath.). The latter is found breeding in nearly all the back waters of our rice fields; the former is seen but sparingly, and a large family of northern finches is succeeded by several interesting species, among the most beautiful of which are the Nonpareil or painted Bunting and the blue Grosbeak. Thus by a wise benevolent provision of providence, the varying seasons bring along with them a succession of the feathered tribe that either contribute to our sustenance or minister to our pleasures.

Whilst some of our northern birds make Carolina their southern limit of the winter, there are others that make it their northern boundary beyond which they dare not go at that season. Thus the Cat bird (*Turdus Felivox*), the white eyed flycatcher (*Muscicapa Cantabris*), the green Swallow (*Hirundo bicolor*), and several other species appear among us in small numbers after one or two warm days and winter. A few of these linger along our sea-board in sheltered situations during the winter, and if they are found in great abundance through the whole of that season in Florida and Mexico. The whole crane and heron family (the latter composed of twelve American species) will spend their winters in South Carolina, with the exception of a few stragglers, from among the great blue heron, a very small number of the white heron, and a few of the young of the hooping crane; yet all of these species of birds are numerous and Florida in winter.*

* The following herons breed in Carolina, and all of them in communities with the exception of the least bittern (*Ardea exilis*), a rare species, which conceals its nest among the rushes in fresh water ponds, where it deposits three nearly white eggs: Great heron (*Ardea Herodias*), Great white heron (*A. lieu*, Tem.), Snowy heron (*A. candidissima*), Louisiana heron (*A. Ludoviciana*), Yellow crowned heron (*A. violacea*), Night heron (*A. nocticorax*), Blue crane or heron (*A. coerulea*; the young of this species are white until they are two years old), Green heron (*A. virascens*), and least bittern (*A. exili*). The American bittern (*A. minor*) remains in our marshes during the spring, until about the twelfth of May, when it retires to its breeding places in the farthest north. The *Ardea Pealii* of Bonaparte, or which as has been ascertained by Audubon is the young of the *Ardea rufescens* of Buffon. Having had living specimens in my possession for some time, I am enabled to state that the dowdy feathers of the young, whilst in the nest, are Brown. The birds then continue white until the second year, when they assume a rufescent color. They are found breeding in great numbers on the islands of the southern extremity of Florida. In the same places are found also, the newly discovered heron (the largest of all our American species), which Audubon describes under the name of *Ardea occidentalis*. The brown crane (*Grus Canadensis*, of Temm. and Bonaparte) is undoubtedly the young of the great hooping crane, as I have ascertained in a pair in confinement, which either in the second or third year of their age assumed the form and plumage of the adult bird, the *Grus Americana*.

Many birds who make occasional and partial migrations only [travel] to procure a supply of food. Thus the common partridge (*Perdix Virginiana*) in seasons when there is a scarcity of grain in New Jersey crosses the Delaware River, into Pennsylvania. The same has been observed along the Susquehanna and Hudson. The flight of these birds is so heavy that they are seldom able to reach the opposite shores on the wane, but drop into the water when they are weary and swim across. This is also the case with that most delicious of all birds, the wild turkey. Along the Ohio, Missouri, and Mississippi rivers, numbers of these in the seasons of a scarcity of their accustomed food cross those rivers, partly by flying and then by swimming, and in their wet and exhausted state are taken in great numbers either in the rivers or as they arrive on the opposite shores. The wild pigeon is another of those birds that are supposed to be driven among us only by the extreme cold of the north. This is a mistake. These birds appear in Carolina only at very long and uncertain intervals. Sometimes they visit us in cold, but frequently in mild winters. I have seen wild pigeons in immense flocks in Canada in the coldest winters, when the thermometer was below zero. It is to be remarked, that the previous autumn had produced an abundance of beech nuts and buckwheat, their favorite food, and that the ground was not covered with snow. It is only when the forests of the west have failed in their usual supply of mast and berries, that the wild pigeons come among us to claim a share of the acorns and berries of our woods, and they refuse grains scattered over our rice fields.

It is, perhaps, not improbable that the occasional changes in the migrations of the birds of our continent may, in the course of time, introduce among us some species of birds from the south and west that are not now found here. A large number of the feathered race follow the improvements of civilized man. No sooner does cultivation commence than many birds, which were unknown to the forest around them, are seen in his fields and orchards. A new species of grain attracts the graminivorous bird—a particular plant or tree, on which certain caterpillars or insects feed, invites the Sylvias, Vireas, and Muscicapas; and the tubular flowered plants of the West Indies, transplanted into the soil of Florida, are already beginning to attract some of the many species of hummingbirds of the south. In the days of Wilson (one of the most observing of our American ornithologists), the great Carolina Wren (*Troglodytes Ludovicians*) and the pine creeping warbler (*Sylvia Pinus*) together with several other species that are now common in the northern states (where I sought for them for many years in vain) were there unknown. They have now extended their summer migrations as far north, at least, as Boston. The cliff swallow (*Hirundo Lunifrons*, Say), a Mexican species, was first seen on the banks of the Ohio in 1815. These birds excited much interest from the peculiar structure of their nests, built of mud, and clustered together, resembling a bunch of the gourds. From year to year, they continued to increase and advance eastwardly in their migrations until they have now extended across the continent, as far as Canada and Maine. The olive-sided flycatcher, (*Muscicapa Cooperii*, Nutt.) has but recently made its appearance in the north, and on the mountains of Virginia; and in the latter situations, the newly described Bewick Wren of Audubon (*Troglodytes Bewickii*) has supplanted all the other species of that genus. The fork-tailed flycatcher (*Muscicapa savanna*, Bonap.) has, only within a few years, commenced leaving the tropical wilds of Guiana, and a few stray birds of that species are almost annually seen in the middle states. The solitary flycatcher (*Vireo solitarius*, Vieill.), which was so rare with us ten or twelve years ago that scarcely a bird of that species could be found in a year has of late become so abundant that in the month of February five or six can be counted in particular situations near our city in a single day, and their sweet notes form a considerable addition to the concerts of our feathered choir. The orange crowned warbler (*Sylvia celata*, Say), so long confined to the far west and the orange groves of Florida, has become equally common in our immediate neighborhood. The pectoral sandpiper (*Pelidna pectoralis*, Say) and the long-related sandpiper* (*Tringa himantopus*, Bonap.), which were formerly so exceedingly rare that Wilson knew nothing of their existence, are now found every summer in small numbers along our sea coast. It may not be unworthy of remark in this place and in

---

* From specimens in various stages of plumage which I possess of the long legged sandpiper, I am disposed to believe that Swainson and Richardson in their Fauna Boreali-Americana have been deceived by the variations in the plumage and size to which this bird is subject and have described it three times under the names of *Tringa himantopus*, *T. Audubonii*, and *T. Douglassii*.

confirmation of the views now advanced that no less than eight or nine species of birds either wholly undescribed or not previously known to exist in the United States have recently been discovered in the neighborhood of the city. A few of these may have long existed in the country, and escaped the researches of former naturalists, but I am under an impression that some may have but recently come among us.* From these facts, we may easily perceive, that after all the additions that have been made to our American ornithology by Wilson, Bonaparte, Cooper, Nuttall, Richardson, and especially the indefatigable Audubon, the field still remains open to the investigation of the student of nature, and promises a rich reward.

There is one singularity in the migration of American birds that is as yet involved in some obscurity. A vast number of northern warblers and fly catchers do not pass over the low countries of Carolina in their migrations and the closest observers have not been able to find a single specimen of many species that are abundant in the north, and that all migrate southerly in autumn. It is possible that migratory birds pass southerly in two immense channels, one leading from Hatteras, or some of our capes a little farther south, and then across the gulf of Mexico to the West India Islands where they spend the period of our winter in immense numbers. They are often met at sea during the period of their migration and are frequently known to alight on the rigging of vessels, where they rest for an hour or two, and then again pursue their onward course. The other path of migration and probably the most common, to which I refer, is along the Allegheny mountains which extend through the whole interior of our country. They vary occasionally in their flight so as to follow not only the range of mountains, but the courses of rivers. In these views I am supported by Audubon and Nuttall, and they are strengthened by the fact that the Rose-breasted Grosbeak, the Baltimore Oriole, the Scarlet Tanager and a number of species of warblers that seldom visit the maritime districts of Carolina are found to pass along our mountains and from thence through the states of Louisiana, Mississippi,

---

* Some of these birds have since been figured and described by Audubon under the following names: *Muscicapa dominicensis*, Briss.; *Parus Caroliniensis*, Aud.; *Fringilla Bachmanii*, Aud.; *Fringilla Macgillivraii*, Aud.; *Sylvia Swainsonii*, Aud.; *Sylvia Bachmanii*, Aud.; *Rallus elegans*, Aud.

and Arkansas. Some of these birds remain in Mexico; some enter within the Tropics, and others possibly pass beyond them in order to find a climate similar to that which they have left.

It has recently been ascertained that some of the birds that are found in the north of Europe and have hitherto not been known to exist in America, migrate from the polar regions, along the Rocky Mountains sometimes as far south as Mexico and in their spring migrations return by the same route. The Magpie (*Corvus Pica*) and the Bohemian waxwing (*Bombycilla garrula*) and a few others are of that number. Several other birds peculiar to the American continent never visit the cultivated districts of the United States, but take the same course in their annual migrations; among these are the black water-ousel (*Cinclus Pallassii*, Tem.), the evening Grosbeak (*Fringilla Cooperi*), Clarke's Crow (*Corvus columbarius*), and the Columbia Jay, a most splendid bird, figured by Anderson, rivalling in beauty the bird of Paradise. The spotted thrush of Latham (*Turdus nœvius*), the arctic Blue-bird (*Erythaca arctica*, Swain.), the *Emberiza picta*, Swain., and of the saffron headed Troophial (*Icterus xanthocephalus*, Bon.) are also of this number. Those birds only that breed in the arctic circle visit both Continents. It is computed that out of almost four hundred and fifty species already known in North America, only twenty seven land, and eighty one[*] water birds are natives of both continents; consequently three hundred and forty two species are peculiar to our own continent. The land birds that visit both continents are composed of Eagles, Hawks, Owls, and most of the genus corvus, and a few other species possessing great strength of wing and warmth of covering, enabling them to migrate with ease and to bear the rigors of the polar regions. The water birds are either composed of Ducks, which breeding the far north, are enabled to reach the regions of Norway and Russia, and visit the shores of Europe, or they are of the Goal, Tern, and Petrel, species which finds sustenance every where on the bosom of the ocean, and may therefore with great facility navigate the widest seas. Still, it will be observed that the number of birds that migrate from one continent to another is very small, and I am under an impression, that

---

[*] This number has been considerably increased since the publication of Richardson and Swainson's Fauna Boreali-Americana, in which it is to be feared the mania for adding new species of birds have been indulged to a considerable extent.

these migrations take place but seldom. Such is also the case with our animals of which very few are found on the eastern continent. In fact, our whole kingdom of nature not even excepting the insects and plants presents peculiarities which well entitle it to the name given in it by its first discoverers of "the new World."

Whether many of our migratory birds that leave us early in the season, may not pass beyond the tropics and retire to latitudes in the southern hemisphere of the same temperature with that which they left is a subject that remains for the investigation of future naturalists. Why may they not take advantage of the reversion of the seasons, and rear a second brood in South America? The purple martin which is found in our whole country during summer as far as the 60th degree North latitude is known to breed in South America during our season of winter, and this is also the case with several of our rarest sylvias. Even admitting that our birds do not migrate to the southern hemisphere, it is probable that some of the species may breed in two distinct portions of North America. The stork, after it leaves Europe, is known to raise in other brood in Africa. Audubon found the white headed Eagles (*Falco leucocephalus*) and the Fish Hawk (*F. Haliœtos*) having nests with their young full-fledged and able to fly in the month of November in Florida. The Barn Owl (*Strix flammea*) sometimes lays its eggs in the unoccupied buildings of the city in November, and I last year had a young bird of the great horned Owl (*Strix virginiana*) sent me on the 3d of December, which had been taken from a nest in this vicinity. Now this is a season when our northern countries are blocked up in frost and snow, and it is not improbable that many of these birds, following the opening of spring, may raise a second brood in the more northern climates.

The Rail (or Soree as it is called in Virginia) and the swallows have occasioned more speculations and created more superstitious ideas with regard to their winter residents than any other of our American birds. The erroneous opinions with regard to the Rail have probably arisen from the sudden manner in which it appears and disappears in the middle states, and the un-philosophical notions with regard to the swallows have originated in Europe and from thence and transmitted to our country.

Rails, after having been absent during the whole summer from the middle States, suddenly make their appearance early in August in an immense numbers along the Delaware, Schuylkill, and James rivers. In a single night, their clamorous voices are heard in tens of thousands on those reedy shores, where but the day before not one could be found. Here they remain till about the middle of October, when suddenly their well known cackle ceases and in the places where the day before many hundreds were seen, not a solitary one remains. They seem so heavy of flight that they are sometimes taken by hand and hence the oft repeated inquiry whence come and whither goes the Soree. Many believe that these birds are scarcely capable of flight and must find some retreat in hollow banks or perhaps under the ice or mud. The truth is they migrate altogether by night and like the Woodcock and other kindred species fly admirably in twilight or in the dark. They breed very far north. An intelligent Indian trader informed me that he had found great numbers of their nests whilst hunting for the egg of the wild goose (*Anser Canadensis*) along the reedy marshes of the most northern lakes. It is not generally known that when they leave the middle states, they appear in the rice fields and marshes of Carolina, where they remain a short time before they migrate still farther south and in the spring again visit us as they are passing on to their northern breeding places. There is then nothing in the migrations of the Rail that cannot be accounted for on the principles of nature.

All the absurd theories with regard to the hibernation of Swallows have originated from the habits of a few species common to our country and to Europe. The chimney swallow of Europe (*Hirundo rustica*), resembling our barn swallow (*Hirundo rufa*), in every thing but its habits of building in chimnies so perfectly that they cannot be distinguished from each other, and the bank swallow (*Hirundo riparia*), which is a native of both continents, and our chimney swallow (*Cypselus pelasgius*, Temm.) have occasionally been found in holes on the banks of rivers, in the hollows of decayed trees, or in the recesses of old buildings, clinging together sometimes in great numbers, nearly in a torpid state. Hence it was asserted that these were their winter retreats and that here they remained in a state of torpidity from the cold of autumn to the sunny days of spring. This doctrine has been espoused by a number of intelligent naturalists of Europe from the amiable and observing White of

Selborne even to the great Cuvier, who makes use of the following language: "Some birds retire into remote places, to some desart cave, some savage rock, or ancient fortress." He evidently had no opportunities for a satisfactory examination. Dr. Good has also asserted of the chaffinch of Sweden (*Fringilla coelebs*) that many of the males indulge in a profound sleep in Sweden whilst females migrate to Holland towards the winter and duly returned to them in the spring.

From dissection (the details of which it is unnecessary to give here), it has been ascertained that from the internal structure of swallows, and the same may be said of all birds, it is impossible for them to live beyond a day or two in a torpid state. In this declaration I am supported by the investigations of the celebrated John Hunter. I have seen the American chimney swallow as well as the rail placed under the water to try the experiment whether they could exist in that element, and they have invariably been drowned in a few minutes and no warmth or even electricity could afterwards revive them. The habits of swallows drinking from brooks and rivers, while they are on the wing and of their picking up flies and insects, whilst skimming the surface of the water, has no doubt given rise to the deceptions in persons supposing that they had seen them going under the water as a winter retreat. When birds of this species have been found in nearly a dormant state, it was either in the autumn or early in the spring, generally the latter. These are the seasons of their migration. At night they sought those retreats as usual to sleep; here they were overtaken by a cold change in the atmosphere, and here they would have died in a very short time if the weather had not become milder. These birds have, I apprehend, never been found in this situation in winter. Besides, our senses can satisfy us where the swallows spend their winters. Of the six species of swallows that inhabit the United States, all of them but the cliff swallow which has but recently made its appearance in the country are seen in thousands performing their annual migrations, along the environs, and even the very streets of our city. The green swallow (*Hirundo bicolor*) is found in Florida during the coldest weather of that country and was during the last winter (1832) seen every day with the exception of about two weeks in considerable numbers in the neighborhood of Charleston. The barn swallow and purple martin (*H. purpurea*) leave us earlier and return later, the chimney swallow follows last in the train on its return from the south as it is the first to leave us in autumn. Thus we perceive that there is nothing mysterious, nothing unnatural in the migration of the swallows.

When the period of migration arrives, birds evince an uncontrollable restlessness of disposition, as if conscious that an important undertaking was at hand. The Snow and Canada Geese (*Anser hyperborean* and *A. Canadensis*), which I have had for some years in a state of domestication, (although in other respects perfectly tame) make constant efforts on the return of every spring to obey the impulse of nature and take their departure for the north. Although a joint from a wing of each has been removed, yet they make attempts at flying, and when at this season they are enabled to escape from their enclosure, they hurry off in a northern direction as if determined to make their long journey on foot. Wilson gives a well authenticated anecdote of a female wild goose having been domesticated by Mr. Platt of Long Island, which, after flying off on the following spring, returned to the autumn with three of its comrades or young and the birds were all living several years afterwards. I have preserved in an aviary, robins, finches and orioles that had been procured when young at the north, and no sooner did the spring (the time of migration) arrive than they exhibited by their constant fluttering a disposition to escape and the moment this was effected they flew off not to the south or west but as directly in the line of migration as if guided by a compass. These are facts of which the humblest individual may inform himself, but which neither our wisdom or philosophy can explain.

The manner in which birds perform their migrations is also deserving of notice. At the approach of autumn, when the cold is beginning to drive the insects to their winter retreats, when the earth begins every where to present the image of desolation and death, when many terrestrial animals are preparing for themselves a shelter from the cold, it is then and sometimes a few weeks earlier (as if in anticipation of this season) that birds assemble, in troops, to set out on their annual aerial voyage to southern climes. The young in most species instinctively flock together as if disdaining to enquire the path of migration from the old. Some taught by an instinct of nature, which way to bend their course departs singly, and make their long and weary journey alone, others go into straggling flocks, sometimes you see the air almost darkened with the swallows and night hawks (*Caprimulgus Virginianus*), as other species crowd into close columns during their flight. This is particularly the case with the wild pigeons, wax birds, (*Bombycilla Carolinensis*), black birds (*Icterus Phœnicus*), the cow bunting, (*I. Pecoris*), the wild geese, ducks and several species of Tringas or sand birds. Some species move slowly and seemed only urged along either by the cold or by a scarcity of their accustomed food. Others pass rapidly and effect their migrations in a very few days. Some flit along the Earth's surface and rest, here and there, as if to take a glance at the fields, gardens and habitations of man, whilst others mount high in the air and soar almost among the clouds as if scarcely deigning to cast an eye on the cities and villages and the puny efforts of their inhabitants and on the mountains and vallies beneath them. These aerial voyagers by an admirable instinct seize upon a favorable moment in which winds and the weather are fitted for these migrations; they are not carried along by the wind, but are obliged from the construction of their feathers to fly against it. They have a foreknowledge of frosts and snows for weeks before they arrive and they have a mysterious but sure monitor within them to tell them of the coming of spring. They require no chart and no compass to enable them to navigate the air and pass through the region of clouds, the thunder and the storm. They arrive at the end of their destined to voyage, and there in the Grove, the forest, the mountain, the field or the garden, they find food, shelter and a home prepared by the hand of providence; there in all probability, they revisit the very neighbourhood and probably build their nests in or near the same tree, or bush, or tufts of grass in which the year before they reared their young. This too may have been the scene of their infancy and here they may have carroled their earliest song. The disposition of birds to revisit, annually, the place where they have once bred is remarkable. A blue bird that was marked so as to be known built its nest for ten successive years in a box that had been prepared for the purple martin. A pewee (*Muscicapa fusca*) has been known to revisit the same cave for nine successive years. A robin, bred for a still longer time in the same apple tree, and a red tailed hawk (*Falco borealis*), which is distinguished from all others of the species on account of its plumage having accidentally become white, has for the last twelve winters kept possession of a dead pine in an old field in Colleton district (South Carolina).

Whilst many species of birds perform their migrations during the day a great number traveled by night. The lover of nature who in the seasons of the migrations of birds, sees the flock after flock passing it over his head, all day long, or witnesses the wrens, blue birds and creepers just stopping for a few moments to seize a worm or insect and then as if impelled by destiny, rising again on the wing and urging onwards has also the evidence that many pass over him at night. He hears unusual sounds in the air. The single sharp note of the rice bird repeated all around him is succeeded by the crake of the snipe resembling the grading of the wheel repeated at long intervals, and the woodcock (*Sclopax minor*) wheels around him uttering notes like the loud ticking of a watch so rapidly repeated that they cannot be counted. He ascends higher and still higher in the air like the lark of Europe till he seems to have risen above the clouds, when suddenly his voice is hushed and in zigzag lines he descends rapidly to the earth and alights near the same spot from whence he arose. This is repeated for several successive evenings and at early dawn till suddenly, he commences his annual migration and is seen no more. The yellow crowned and the night herons utter their hoarse croak as they pass high and rapidly on, and at a still greater distance like unearthly sounds is heard the not unmusical cry of the Canada goose. In the mean time the rails, owls, thrushes, warblers, and many other birds glide silently by him like spirits of the air; and without being superstitious, there comes over him a sensation of mingled admiration and fear and he feels the truth of the language of inspiration. "Great and marvelous are thy works Lord God Almighty."

The arrival and departure of birds affords a pretty sure indication of the state of the weather and the advance of the seasons. Living constantly in the air and exposed to all its variations, they become either from instinct or habit acquainted with the changes of the atmosphere, with the winds, the weather and the seasons. Captain Parry and Dr. Richardson inform us of the anxiety with which the Northern Indians watch to the approach of the first bird, the harbinger of spring. On the 12th of April, says Dr. Richardson, "the arrival of the swans, geese and ducks gave certain indications of the return of spring. On the 14th a robin appeared;

this bird is considered by the natives as an infallible precursor of warm weather;" and Capt. Parry says "the snow bunting was the first precursor of spring that appeared. When the well known notes of the whippoorwill (*Caprimulgus vociferous* and *C. Carolinensis*) are heard, the farmer is reminded that the time for the planting of corn is at hand. The fish hawk's return to the rivers of the north is regarded by the fisherman as a proof of the season for the taking of shad has arrived. When the swallow appears, the danger of frost is believed to be over; and if the Cuckoo of Europe is hailed by the old and the young as an evidence of the return of spring, and if we have in common with them admired the beautiful sentiment of the poet, "Sweet bird thy bower is ever green, / Thy sky is ever clear; / Thou hast no sorrow and thy song, / No winter in thy year," the inhabitants of the middle and northern states of our country feel equally interested and pleased when they hear the soft and melodious note of the blue bird, the robin, and the wood thrush (*Turdus mustelinus*) reminding them that "the winter is past and gone and at the time of the singing of birds has come."

Previous to a storm, the birds give indications of its approach. Our vultures in great numbers rise in circles till they are almost lost in the region of the clouds; the stormy petrels (*Thlassidroma Wilsonii*, Bon.) crowd in great numbers around the vessels and follow in their wake as if seeking the protection of man; the seagulls and terns make the shores reecho with their hoarse clamorous notes; the loon (*Colymbus glacialis*) is excessively restless and his screams are heard at the distance of more than a mile; and the barred owl (*Strix nebulosa*) utters his funeral cries even in the day. But when fine weather is about to return, the whole scene is changed, and every hedge and copse and grove is rendered vocal and the whole feathered tribe seem to rejoice at the prospect of the cessation of the storm and the anticipation of bright skies and sunny days.

But although our subject is far from being exhausted, I am admonished that it is time to bring these desultory remarks to a close. If I shall have fortunately succeeded in throwing even a ray of light on that which has hitherto appeared mysterious in nature; or if I have been enabled to awaken in a single mind a sentiment of admiration and gratitude to that superintending providence who teaches "the stork in the heavens to know her appointed time, and the turtle and the crane and the swallow to know the time of their coming." I shall be doubly recompensed for those pleasing studies of nature which have enabled me to offer these remarks.

The farther we pursue this subject of the more we shall be convinced that there is a wise arrangement in nature which governs instinct and action and creates being in beauty and happiness. The laws by which the whole system of nature is governed are equally simple and majestic and are equally visible in the minutest as well as in the most stupendous of Gods works. From the beauty and harmony of that system of nature by which we are surrounded, the mind is insensibly led to admire and adore that mighty cause, the fountain of wisdom and perfection, who though unseen, is ever present, who is "the source of all matter and mind and modes of existence."

The temple of nature, wide and wonderful as it is stands ever open, inviting the ignorant as well as the wise to enter and learn those lessons which are calculated not only to enlighten the mind but to improve the heart, and the chief object of science and Philosophy should be to lead us to the Altar of the benevolent Author of all things, and to make our experience and knowledge subservient to his grand designs.

Species of Squirrel Inhabiting North America

On August 14, 1838, Bachman made a presentation to the Zoological Society of London in which he described fourteen species of North American squirrels, six of which were described scientifically for the first time. Charles Darwin was present, and after the meeting, he asked Bachman numerous questions and recorded the answers in detail in his notebooks for research (Darwin's Notebooks D31-34 and C251e-255 [Barrett et al., 1987: 340, fn. 29-1]).

This article on squirrels is one of a series that Bachman prepared on various genera of small mammals, and it illustrates his comprehensive approach to dealing with all known examples of the most closely related species. Most naturalists described one or a few new species at a time and made no attempt to distinguish all closely related species from one another. Bachman attempted to reconcile previous descriptions; he observed live squirrels and studied museum specimens; and he compared all known types in terms of the number of teeth, size, color, and other characteristics. He regularly requested that skulls be sent along with skins to provide evidence that was crucial for accurate identification.

In this article, Bachman summarized information about eight adequately described types and distinguished six more species and several permanent varieties (some of which had been described previously as separate species). Bachman's new species were the Texian Squirrel (*Sciurus texianus*), Golden-bellied Squirrel (*Sciurus subauratus* Audubon and Bachman), larger Louisiana Black Squirrel (*Sciurus auduboni*), Sooty Squirrel (*Sciurus fulignosus*), Columbia Pine Squirrel (*Sciurus richardsoni*), and Downy Squirrel (*Sciurus langinosus*).

Darwin met Bachman shortly before his breakthrough in working out the principles for his concept of evolution. He had recently returned from his voyage around the world on the *Beagle*, during which he had made many important observations, but in 1838 he had published little and was mainly known as a promising geologist. He knew so little about zoology that he asked George Robert Waterhouse, the Society's curator, and others to identify the animals he had brought back. Bachman was nineteen years older than Darwin and had already published major articles on the migration of birds and on all known American species of hares and shrews. While in London, he was asked to identify all smaller mammals in the collection of the Zoological Society, and he did so with the full agreement of John Richardson, who had written the most important monograph on North American zoology, and of Waterhouse, who at the end of the same meeting of the Society named a species of hare in honor of Bachman (*Lepus bachmani*).

The ways in which Bachman distinguished between species and varieties and his comments on how one species replaced another in similar habitats was of great interest to Darwin. In his lecture, for example, Bachman distinguished four permanent varieties of the Fox Squirrel, which had been called the Black Squirrel by Catesby and Bartram, but since the name Black Squirrel and *Sciurus niger* had been used previously for another species by Linnaeus, the Fox Squirrel had been renamed *Sciurus capistratus* by Bosc. "There is great difficulty in finding suitable characters by which the majority of our species of Squirrel can be designated, but in none greater than in the present. All our naturalists seem to insist that we have a *Sciurus niger*, although they have applied the name to the black varieties of several species." This species varies greatly in color, but can readily be distinguished by its white nose and ears. Bachman noted that the varieties interbreed, and ordinarily a nest of young squirrels will be found to contain offspring resembling the color of each parent in equal numbers rather than having the colors of the parents blended. By contrast, he found that the Cat Squirrel could be any shade from light gray to almost black and occasionally white (without being albino). The Northern Gray Squirrel also has a black variety.

On December 12, 1837, Bachman wrote Edward Harris, the naturalist who had become his friend as well as Audubon's, that "this Genus I have found intolerably troublesome, and when I have finished it, I fear that others will have to correct my blunders" (Bachman 1888: 159). Some species have been subsequently shifted to another genus, and some are now subspecies, but most have retained the full names Bachman gave them (Wilson and Ruff 1999).

Bachman's article on squirrels was originally published in the *Proceedings of the Zoological Society of London* (Part 4 [1838]: 85–130). Parts of his presentation were summarized, and its title was given only in the table of contents: "Monograph of the Species of Squirrel Inhabiting North America." The article was reprinted in the *Magazine of Natural History, and Journal of Zoology, Botany, Mineralogy, Geology, and Meteorology*, and long abstracts were published in Silliman's *American Journal of Science and Arts* and in *Isis, oder Encyclopädische Zeitung; von Oken*.

August 14, 1838
William Yarrell, Esq., in the Chair.

A series of skins belonging to species of the genus *Sciurus* including, with one or two exceptions, all which are known to inhabit North America were upon the table; and the Rev. Dr. Bachman, of S. Carolina brought them severally before the notice of the Members. Six of the species exhibited were new, and for these he proposed the specific names of *Texianus, lanuginosus, fuliginosus, subauratus, Auduboni,* and *Richardsoni*. Dr. Bachman's manuscript notes upon the habits and characters of the North American Squirrels, with descriptions of the newly characterized species were also laid before the Meeting.

The first species noticed by Dr. Bachman is the *Sciurus capistratus* of Bosc or Fox Squirrel; *Vulpinus* of Gmel.; *niger*, Catesby; *variegates*, Desm.; the Black Squirrel of Bartram. Its essential characters consist in its large size, and having the tail longer than the body, the hair coarse, and the ears and nose white. The dental formula is *inc. 2/2, can. 0-0/0-0, mol. 4-4/4-4*. In a very young individual, supposed to have quitted the nest only a day or two, Dr. Bachman found an additional anterior grinder on each side in the upper jaw, but very minute. The additional molar teeth, he concludes, are shed at a very early period as they were not present in two other specimens subsequently examined, and which were some days older than the former one. The Fox Squirrel is the largest found in the United States and is subject to great differences of colour, but it still exhibits such striking and uniform markings, that this species may always be distinguished. Three principal varieties are noticed; in the first, which is the gray variety and the most common, the white of the nose extends to within four or five lines of the eyes; the ears, feet, and belly, are white; forehead and cheeks, brownish black; the hairs on the back are dark, plumbeous near the roots; then a broad line of cinereous; and then black, and broadly tipped with white, with an occasional black hair interspersed, especially on the neck and fore-shoulder, giving the animal a light gray appearance; the hairs in the tail are for three-fourths of their length white from the roots, then a ring of black, with the tips white. This is the variety given by Bosc and other authors as *Sc. capistratus*.

The second variety (the Black Fox Squirrel) has the nose and ears white, a few light-coloured hairs on the feet, the rest of the body and tail black; there are, occasionally, a few white hairs in the tail. This is the original black squirrel of Catesby and Bartram, (*Sc. niger*).

In the third variety, the nose, mouth, under-jaw, and ears, are white; head, thighs, and belly, black; the back and tail, dark-gray. This is the variety as alluded to by Desmarest, *Ency. Méthod. Mammalogie*, p. 333.

A fourth variety, very common in Alabama, and also occasionally seen in the upper districts of South Carolina and which has on several occasions been sent to Dr. Bachman as a distinct species has the ears and nose white, a prominent mark in all the varieties, and by which the species may be easily distinguished. The head and neck are black; back, rusty-blackish brown; and neck, thighs, and belly, bright rust colour; tail annulated with black and red. This is the variety erroneously considered by the author of the notes on MacMurtius' translation of Cuvier (Append. vol. i p. 433.) as the *Sciurus rufiventer*.

The three first varieties noted above, Dr. Bachman describes as being common in the lower and middle districts of South Carolina; and although they are known to breed together, yet it is very rare to find any specimens indicating an intermediate variety. Where the parents are both black, the young are invariably of the same colour; the same may be said of the other varieties; where, on the other hand, there is one parent of each colour, and almost equal proportion of the young are of the colour of the male, the other of the female. On three occasions he had opportunities of examining the young produced by progenitors of different colours. The first nest contained four, two black and two gray; the second, one black and two gray; and of the third, three black and two gray. The colour of the young did not in a majority of instances correspond with that of the parent of the same sex. Although the male parent was black, and the young males were frequently gray and *vice versa*.

Dimensions of the Fox Squirrel.

	in.	lines.
Length of head and body	14	5
Tail (to end of vertebræ)	12	4
Tail to the tip	15	2
Palm and middle fore-claw		19
Sole and middle hind-claw	2	11
Length of fur on the back		8
Height of ear posteriorly		7

This species is said to exist sparingly in New Jersey: Dr. Bachman has not observed it further north than Virginia, nor could he find it in the mountainous districts of that state. In the pine forests of North Carolina it becomes more common; in the middle and maritime districts of South Carolina it is almost daily met with although it cannot be said to be an abundant species anywhere.

Sciurus Texianus. Texian Squirrel. This name is proposed by Dr. Bachman for an apparently undescribed species which he saw in the Museum at Paris. It was said to have been received from Mexico. In the Museums of Berlin and Zürich, he also found what he conceives to be the same species; and in the British Museum there is a specimen obtained at Texas by Mr. Douglas agreeing with the others in almost every particular. Dr. Bachman also states that, among his notes there is a description of a specimen received by a friend from the south-western parts of Louisiana, which, on a comparison with memoranda taken from the other specimens, does not appear to differ in any important particular. Hence, he thinks it probable that this species has a tolerably extensive range extending perhaps from the south-western portions of Louisiana, through Texas, into Mexico.

The Texian Squirrel is about the size of the Fox Squirrel. On the upper surface there is a mixture of black and yellow and on the under parts deep yellow. The under sides of the limbs and also the parts of the body contiguous are whitish. Fore-legs externally and the feet rich yellow: ears on both surfaces yellow with interspersed white hairs: nose and lips, brownish white: hairs of tail rich rusty yellow at base with a broad black space near the extremity and finally tipt with yellow.

Dimensions.

	in.	lines.
Length of body	13	6
Tail to end of hair	15	0
Tarsus	3	0
Height of ears to end of fur	0	6½

The Texian Squirrel bears some resemblance to the *Sciurus capistratus*. The latter species, however, in all the varieties hitherto examined by Dr. Bachman, has uniformly the white ears and nose.

This species would appear to replace the *Capistratus* in the south-western parts of America.

SCIURUS SUBAURATUS. Sci. corpora suprà cinereo, flavor lavato, infrà austere aureo, caudâ corpora longiore. Dentes, inc. 2/2, mol. 4-4/4-4.

The designation "Golden-bellied Squirrel," and the specific term *subauratus*, are given by Dr. Bachman to a species, of which two dead specimens were procured in the markets of New Orleans by Mr. Audubon. Their size was between that of the Northern Gray and the Little Carolina Squirrel. There was no trace of the small anterior upper molar generally found in the species of the genus *Sciurus*. The upper incisors are the deep orange brown colour and of moderate size: under incisors a little paler than the upper; the head is of moderate size; whiskers longer than the head; the ears are short and pointed and clothed with hair on both surfaces. The body seems better formed for agility than that of the small Carolina and in this respect approaching nearer to the Northern Gray Squirrel. The tail is broad and nearly as long as that of the last-named species.

The colour of the whole upper surface is gray with a distinct yellow tint. The hairs, which give this outward appearance, are grayish slate colour at their base, then very broadly annulated with yellow; then black, and near the apex annulated with the yellowish white. The sides of the face and neck, the whole of the inner side of the limbs, feet, and the whole of the under parts of a deep golden yellow; on the cheeks and sides of the neck, however, the hairs are obscurely annulated with black and whitish; the ears are well clothed on both surfaces with tolerably long hairs of the same deep golden hue as the sides of the face; hairs of the feet are mostly blackish at the root, and some are obscurely tipped with black; hairs of the tail black at the roots, and the remaining portion of a bright rusty yellow; each hair three times in its length annulated with black; the under surface of the tail is chiefly bright rusty yellow; whiskers longer than the head, black.

Dimensions.

	in.	lines.
Length of head and body	10	6
Tail (vertebræ)	9	2
Tail including fur	12	0
Palm to end of middle fore-claw	1	7
Length of heel to point of middle nail	2	6
Height of ear posteriorly	0	5
Length of fur on the back	0	7
Breadth of tail with hair extended	8	6

Weight, one pound and a quarter.

Sciurus magnicaudatus, Harlan's Fauna, p. 170. *S. macrourus*, Say. Long's Expedition, vol. I p. 115.

This species Dr. Bachman remarks, that although he has seen many specimens labelled under the above name, yet the only true *S. macrourus* which has come under his own observation is one in the Philadelphia Museum. Not being in possession of his own memoranda upon the species, he quotes the description published by Say.

Sciurus aureogaster, F. Cuv. et Geoff. Mamm. Californian Squirrel.

Dr. Bachman's acquaintance with this species rests upon the examination of some specimens in the Museum of the Zoological Society from which he draws up the following description.

The general hue above is deep gray grizzled with yellow: the under parts and inner side of the limbs are deep rusty red; chin, throat, and cheeks, pale gray. Limbs externally, and feet, coloured as the body above. Hairs on the toes chiefly dirty white. Tail large and very bushy. Hairs on the tail black, twice annulated with dirty yellow, and broadly tipped with white—the white very conspicuous where the hairs are in their natural position. Ears thickly clothed, chiefly with blackish hairs, the hinder basal part externally with long white hairs extending slightly on the neck. All the hairs of the body are gray at the base, those of the upper parts annulated first with yellow, then black, and then white. Whiskers black, the hairs very long and bristly. The under incisors almost as deep and orange colour as the upper.

Habitat Mexico and California.

Dimensions.

	in.	lines.
From nose to root of tail	12	0
Tail to end of hair	10	6
Heel to end of claws	2	5½
Nose to ear	2	1½
Height of ear posteriorly	0	7½

A second specimen, the locality of which was not given, differed from the above in having a much richer colouring. The belly was a very bright rust colour the hairs on the tail were black at the roots, then broadly annulated with rusty yellow, then a considerable space occupied by black, the apical portion white, but when viewed from beneath, a bright rust colour like that of the belly was very conspicuous, occupying the basal half of the hair. The upper parts of the body were grizzled with black and white, and many of the hairs were annulated with rust colour. Over the haunches and rump, the hairs are annulated with rusty yellow and black. The hairs of the feet were chiefly black.

The original specimen on which this species was founded is in the Museum in Paris, and Dr. Bachman quotes the following description from Mr. Waterhouse's manuscript notes.

"General colour, grizzled black and white. Throat, chest, belly, inner-side of legs, nearly the whole of the fore-legs, and the forepart of the hind-legs, rusty red. Tail very broad; the hairs black; red at the base, and white at the apex; lips white; feet black with a few white hairs intermixed; forepart of head also black with scattered white hairs. Chin blackish in front shading towards the throat into gray."

	in.	lines.
Nose to root of tail	11	6
Tail to end of hair	11	0
Tarsus	2	4 ⅓

Sciurus cinereus. Gmel. Cat Squirrel, Pen. Arct. Zool. i. 137.

A little smaller than the Fox Squirrel; larger than the Northern Gray Squirrel; body Stout; legs rather short; nose and ears not white; tail longer than the body. Dental formula, *incis.* 2/2, *can.* 0-0/0-0, *mol.* 4-4/4-4, = 20.

Of this species Dr. Bachman remarks, "It has sometimes been confounded with the Fox Squirrel and at other times with the Northern Great Squirrel. It is, however, in size intermediate between the two and has some distinctive marks by which it may always be known from either. The Northern Great Squirrel has, as far as I have been able to ascertain from an examination of many specimens, permanently five grinders in each upper jaw, and the present species has but four. Whether at a very early age the Cat Squirrel may not, like the young Fox Squirrel, have a small deciduous tooth, I have had no means of ascertaining; all the specimens before me, having been obtained in autumn or winter and being adults, present the dental formula as given above. The Fox Squirrel is permanently marked with white ears and nose, which is not the case with the Cat Squirrel; the former is a southern species, the latter is found in the middle and northern states.

"The head is less elongated than that of the Fox Squirrel; nose more obtuse; incisors rather narrower, shorter, and the less prominent; the molars, with the exception of their being a little smaller, bear a strong resemblance to and are arranged in a similar manner to those of the former species. The neck is short; legs short and stout; nails narrower at base than those of the Fox Squirrel; shorter and less arched; the tail also is shorter and less distichous; the body is shorter and thicker, and the whole animal has a heavy, clumsy appearance. The fur is not as soft as that of the Northern Gray Squirrel, but finer than that of the Fox Squirrel.

"This species as well as the last is subject to great varieties of colour. I have observed in Peale's Museum specimens of every shade of colour from light gray to nearly black. I have also seen too in cages which were nearly white, but without the red eyes, which is a characteristic mark in the Albino. There appears, however, to be this difference between the varieties of the present species and those of the Fox Squirrel; the latter are permanent varieties, scarcely any specimens being found in intermediate colours; in the present there is every shade of colour, scarcely to be found precisely alike.

"The most common variety, however, is the Gray Cat Squirrel, which I shall describe from a specimen now before me.

"Teeth orange; nails dark brown near the base, lighter at the extremities. On the cheeks there is a slight tinge of yellowish brown, and this colour is extended to the neck; the inner surface of the ears is also of the same colour; the fur on the outer surface of the ear, which extends a little beyond the outer edge and is of a soft woolly appearance, is light cinereous, and on the edge of the ear, rusty brown. Whiskers black and white, the former colour predominating. Under the throat, the inner surface of the legs and thighs, and the whole undersurface, white. On the back the hairs are dark cinereous near the roots, then the light ash, then annulated with black and at the tip white, giving to the fur an iron-gray appearance. The tail, which does not present the flat distichous appearance of the majority of the other species, but is more rounded and narrower, is composed of hairs which, separately examined, are of a soiled white tint near the roots, then a narrow marking of black, then white, then a broad line of black, and finally broadly edged with white.

"Another specimen is dark gray on the back and head, and a mixture of black and cinereous on the feet, thighs, and under surface. Whiskers nearly all white. The markings on the tail are similar to those of the other specimen.

Dimensions.

	in.	lines.
Length of head and body	11	3
Tail (vertebræ)	9	6
Tail to the end of the hair	12	6
Height of ear posteriorly	0	6
Palm and middle fore-claw	1	6
Heel and middle hind-claw	2	9
Length of fur on the back	0	7

"This has been to me a rare species. It is said to be common in the oak and hickory woods of Pennsylvania, and I have occasionally met with it near Easton and York. I also observed one in the hands of a gunner near Fredericksburg, Virginia. In the northern part of New York it is exceedingly rare, as I only saw two pair during fifteen years of close observation. In the lower part of that state, however, it appears to be more common, as I recently received several specimens procured in the county of Orange.

"This squirrel has many habits in common with other species, residing in the hollows of trees, building in summer its nest of leaves in some convenient crutch, and subsisting on the same variety of food. It is, however, the most inactive of all our known species. It mounts a tree, not with the lightness and agility of the Northern Gray Squirrel, but with the slowness and apparent reluctance of the little Striped Squirrel (*Tamias Lysteri*). After ascending, it does not amount to the top, as is the case with other species, but clings to the body of the tree on the side opposite to you, or tries to conceal itself behind the first convenient limb. I have never observed it escaping from branch to branch when it is induced in search of food. To proceed to the extremity of the limb, it moves cautiously and heavily, and returns the same way. On the ground it runs clumsily and makes slower progress than the Gray Squirrel. It is usually fat, especially in autumn, and the flesh is said to be preferable to that of any other of our species.

"The Cat Squirrel does not appear to be migratory in its habits. The same pair, if undisturbed, may be found taking up their residence in a particular vicinity for a number of years in secession, and the sexes seem mated for life."

Sciurus leucotis. Northern Gray Squirrel.

    Gray Squirrel. Pen. Arct. Zool. vol. i p. 135. Hist. Quad. No. 272.

    Sci. Carolinensis. Godman non Gmel.

    Sci leucotis. Gapper, Zoological Journal, vol. v. p. 206, published in 1830.

Larger than the Carolina Gray Squirrel; tail much longer than the body; smaller than the Cat Squirrel; subject to many varieties of colour.

    Dental formula, *incis.* 2/2, *mol.* 5-5/4-4, 22.

Dr. Bachman states, that this species, which is very common in the Northern and middle states, has hitherto been improperly confounded with the Carolina Gray Squirrel. It appears to have the additional anterior *molares* permanent, in this particular agreeing with several other American Squirrels. The fact, that many of them have only 4-4/4-4, he alludes to as indicating the necessity for modifying the dental formula hitherto assigned to the genus *Sciurus*.

The incisors are strong and compressed, a little smaller than those of the Cat Squirrel, convex, and of a deep orange colour anteriorly; the upper ones have a sharp cutting edge, and are chisel-shaped; the lower are much longer and thinner. The anterior grinder, although round and small, is as long as the second; the remaining four grinders are considerably more excavated than those of the Cat Squirrel, presenting two transverse ridges of enamel. The lower grinders corresponding to those above have also elevated crowns. The hair is a little softer than that of the Cat Squirrel, and is most harsh on the forehead.

The nose is rather obtuse; forehead arched; whiskers as long as the head; ears somewhat rounded, concave; both sides of the ear covered with hair, that which clothes the outside being much the longest. In winter the fur projects upwards, about three lines beyond the margin.

Dr. Bachman observes that although this species exists under many varieties, there appear to be two very permanent ones. These are,

    1. Gray variety. The nose, cheek, around the eyes, extending to the insertion of the neck, the upper surface of the fore and hind feet, and

a strike along the sides yellowish brown. The ears on their posterior surface are dirty white, edge to brown. On the back from the shoulder there is an obscure stripe of brown, broadest at its commencement, and running down to a point at the insertion of the tail. In a few specimens the stripe is wanting. On the neck, sides of the body, and hips, the colour is light gray; the hairs separately are for one half their length dark cinereous, then light umber, then a narrow mark of black and tipped with white; a considerable number of black hairs are interspersed, giving it above a gray colour; the hairs in the tail are light yellowish brown from the roots, with three stripes of black, the outer one being widest, and broadly tipped with white; the whole under surface is white.

"There are other specimens where the yellowish markings on the sides and feet are altogether wanting. Dr. Godman (vol. ii. p. 133) asserts that the golden colour on the hind feet is a very permanent mark. The specimens from Pennsylvania in my possession have generally this peculiarity, but many of those from New York and New England have gray feet, without the slightest mixture of yellow."

2. Black variety. This variety, on several occasions, Dr. Bachman has seen taken from the same nest with the Gray Squirrel. It is of the size and form of the gray variety. It is dark brownish black on the whole of the upper surface, a little lighter beneath. In summer its colour is less black than in winter. The hairs of the back and sides of the body and tail are obscurely grizzled with yellow.

Dimensions of the Northern Gray Squirrel.

	in.	lines.
Length of head and body	11	9
Tail (vertebræ)	10	0
Tail to the tip	13	0
Height of ear	0	7
Height to the end of fur	0	9
Palm to end of middle claw	1	10
Heel to end of middle nail	2	6
Length of fur on the back	0	7
Breadth of tail with hairs extended	4	2

As regards its geographical distribution, the northern limit of this species is not determined; it however exists as far as Hudson's Bay; was formerly very common in the New England States, and in the less cultivated portions is still frequently met with. It is abundant in New York and the mountainous portions of Pennsylvania. Dr. Bachman has observed it on the northern mountains of Virginia; it probably extended still further south: in the lower parts of North and South Carolina, however, it is replaced by a smaller species. The black variety is more abundant in Upper Canada, in the western part of New York, and in the States of Ohio and Indiana. The Northern Gray Squirrel does not exist in Georgia, Florida, or Alabama; and among specimens of Squirrels sent from Louisiana, stated to be all the species existing in that State, he did not discover the present species.

In its habits Dr. Bachman describes the *Sc. leucotis* as one of the most active species of Squirrel existing in the United States. It rises with the sun, and continues industriously engaged in search of food during four or five hours in the morning. In the middle of the day it retires for a few hours to its nest and then resumes its labours till sunset. In the warm weather of spring and summer it builds a temporary residence in the crutch of some tree or in the fork of some large branch. A pair of squirrels are employed on this nest, which is formed of dry sticks and twigs, and lined with moss. In the winter months the squirrels reside together in the hollows of trees, and there the female brings forth her progeny. No instance has come under Dr. Bachman's observation of their breeding in a state of domestication.

During the rutting season the males engage in frequent contests, and often wound each other severely. The very current notion that they emasculate one another in these encounters, is supposed by Dr. Bachman to have originated in the circumstances of the *testes* diminishing in bulk at a certain period of the year, or in these organs being retracted within the *pelvis*.

The food of the Northern Gray Squirrel is like that of the species in general, nuts, seed, and grain; it gives, however, the preference to the several kinds of hickory. Its fondness for the green corn and young wheat renders it very obnoxious to the farmer, and various inducements are consequently held out for their destruction. In Pennsylvania an ancient law existed, offering three pence a head for every one destroyed; and in this way, in the year 1749, the sum of eight thousand pounds was paid out of the treasury in premiums.

It is this species of Squirrel which occasionally migrates in such a vast bodies, but instances of this nature are of much rarer occurrence now than formerly. Autumn is the season of the year at which the migration takes place, and they instinctively direct their course in an eastward direction. Dr. Bachman states that he once witnessed a body of them in the act of migrating, and saw them cross the Hudson in various places between Waterford and Saratoga. They swam the deep and awkwardly with the body and tail entirely submerged. Many were drowned in the passage, and those which reached the opposite bank were so exhausted that the boy stationed there had no difficulty in killing them or taking them alive.

Sciurus Carolinensis, Gmel. Little Carolina Gray Squirrel.

This species is smaller than the Northern Gray Squirrel, and has the tail, which is the same length as its body, narrower than in that species. The colour above is rusty gray, beneath white, and not subject to variation.

The head is shorter, and the space between the ears proportionally broader than those of the Northern Gray Squirrel; the nose also is sharper; the small anterior molar in the upper jaw is permanent, being invariably found in all the specimens examined by Dr. Bachman; and is considerably larger than in the other species. All his specimens, which give evidence of the animals having been more than a year old, instead of having the small threadlike single two as in the northern species, have a distinct double tooth with a double crown; the other molars are not unlike those of the other species in form, but are shorter and smaller; the upper incisors are nearly a third shorter. The body is shorter, less elegant in shape, and has not the appearance of sprightliness and agility for which the other species is so eminently distinguished. The ears, which are nearly triangular in shape, are so slightly clothed with hair internally that they may be said to be nearly naked; externally, they are sparely clothed with short woolly hair, which does not, however, extend beyond the margins, as in the other species; the nails are shorter and less hooked; the tail is shorter, and does not present the broad distichous appearance of the other. Teeth light orange colour; nails Brown, lighter at the extremities; whiskers black; nose, cheeks, and around the eyes with a slight tinge of rufous gray. The fur on the back is for three-fourths of its length dark plumbeous, then a slight marking of black, edged with brown in some hairs, and black in others, giving it on the whole upper surface an uniform dark ochreous colour. In a few specimens there is an obscure line of lighter brown along the sides, where the ochreous colour prevails, and a tinge of the same colour on the upper surface of the fore-legs above the knees. The feet are light gray; the hairs of the tail are, for three-fourths of their length from the roots, yellowish brown; then black, edged with white; the throat, inner surface of the legs and the belly, white.

Dimensions.

	in.	lines.
Length of head and body	9	6
Tail (vertebræ)	7	4
Tail to point of hair	9	6
Height of ear	0	6
Palm to end of middle claw	1	3
Heel to end of middle nail	2	6
Length of fur on the back	0	5
Breadth of tail with hairs extended	3	0

Dr. Buckland remarks that the present species has long been confounded with the Northern Great Squirrel, but that any naturalist who has had an opportunity of comparing many specimens of both, and of witnessing their natural habits, cannot fail to regard them as a distinct species. Specimens of the former, which he had received from North Carolina, Alabama, Florida, and Louisiana, scarcely presented a shade of difference when placed beside those of South Carolina; whilst in the Northern Great Squirrel the great variations in colour form a prominent characteristic distinction.

As regards the geographical range of the Carolina Squirrel, Dr. Bachman states it to be abundant in South Carolina, Alabama, Mississippi, and Georgia, especially in low grounds or swamp localities; it is the only known species in the southern peninsula of East Florida, and it also occurs, though not abundantly, in Louisiana. Dr. Bachman has received it from North Carolina and believes that he has seen the species in the southern part of New Jersey. Its habits he describes as very different from those of the Northern Gray Squirrel: its bark is less full, but much shriller and more querulous. Instead of mounting high on the trees when alarmed, it clings round the trunk on the opposite side, and hides itself under the Spanish mosses which are trailing around the trees. It is much less wild, and consequently more readily captured than the northern species. Its favourite haunts are low swampy situations and amongst the trees which overhang the streams and borders of the rivers: its nest is composed of leaves and Spanish moss and is generally placed in the hollow of some cypress. In one respect, it differs from all the other species of the genus in being, to a certain extent, nocturnal in its habits. Dr. Bachman has frequently observed it by moonlight as actively engaged as the Flying Squirrel; and the traveller, after sunset in riding through the woods is often startled by its noise.

Sciurus Collicei. For a description of this species, of which the original specimen is in the Collection of the Zoological Society, Dr. Bachman refers to Dr. Richard Simmons' Appendix to Capt. Beechey's Voyage.

Sciurus nigrescens. A species described by Mr. Bennett, in the Proceedings of the Zool. Soc. for 1833, p. 41.

Sciurus niger, Linn. non Catesby. The Black Squirrel.
A little larger than the Northern Gray Squirrel; fur soft and glossy. Ears, nose, and the whole body, pure black; a few white tufts of hair interspersed. Incis. 2/2, canines 0-0/0-0, molars 4-4/4-4, = 20.

Of this species Dr. Bachman remarks, "Much confusion has existed with regard to this species. The original *Sciurus niger* of Catesby is the black variety of the Fox Squirrel. It is difficult to decide from the descriptions of Drs. Harlan and Godman whether they refer to specimens of the black variety of the Northern Gray Squirrel or to this species which I am about to describe. Indeed, there is so strong a similarity that I have admitted it as a species with some doubt and hesitation. Dr. Richardson has, under the head of *Sciurus niger* (see Fauna Boreali-Americana, p. 191), described a specimen from Lake Superior, of what I conceive to be the black variety of the Gray Squirrel; but at the close of the same article (p. 192.), he has described another specimen from Fort William, which answers to the description of the specimens now before me. There is great difficulty in finding suitable characters by which the majority of our species of Squirrel can be designated, but in none greater than in the present. All our naturalists seem to insist that we have a *Sciurus niger*, although they have applied the name to the black varieties of several species. As the name, however, is likely to continue on our books, and as the specimens before me, if they do not establish a true species, will show a very permanent variety, I shall describe them under the above name.

"Dr. Godman states (Nat. Hist. vol. ii. p. 133.) that the Black Squirrel has only twenty teeth; the specimens before me have no greater number, with the exception of one, evidently a young animal a few months old, which has an additional tooth on one side, so small that it appears like a white thread, the opposite and corresponding one having already been shed. If further examinations will go to establish the fact that this additional molar in the Northern Gray Squirrel is persistent and that of the present deciduous, there can be no doubt of their being distinct species. Its head appears to be a little shorter and more arched than that of the Gray Squirrel, although it is often found that these differences exist among different individuals of the same species. The incisors are compressed, strong, and of a deep orange colour anterioraly. Ears, elliptical and slightly rounded at tip, thickly clothed with fur on both surfaces, that on the outer surfaces, in a winter specimen, extending three lines beyond the margins; there are, however, no distinct tufts. Whiskers a little longer than the head. Tail long and distichous, thickly clothed with moderately coarse hair.

"The fur is softer to the touch than that of the Northern Gray Squirrel. The whole of the upper and lower surface as well as the tail are bright glossy black; at the roots the hairs are a little lighter. The summer fur does not differ materially from that of the winter, it is however not quite so intensely black. In all the specimens I have had an opportunity of examining, there are small tufts of white hairs irregularly situated on the under surface, resembling those on the body of the Mink (*Mustela vison*). There are also a few scattered white hairs on the back and tail.

Dimensions.

	in.	lines.
Length of head and body	13	0
Tail (vertebræ)	9	1
Tail including the fur	13	0
Palm to end of middle fore-claw	1	7
Length of heel to the point of middle claw	2	7
Length of fur on the back	0	8
Breadth of tail with hair extended	5	0

"The specimens from which this description has been taken were procured, through the kindness of friends, in the counties of Rensellaer and Queens, New York. I have seen it on the borders of Lake Champlain, at Ogdensburg, and on the eastern shores of Lake Erie; also near Niagara on the Canada side. The individual described by Dr. Richardson and which may be clearly referred to this species was obtained by Capt. Bayfield at Fort William on Lake Superior. Black squirrels exist through all our western wilds and to the northward of the great lakes, but whether they are of this species or of the black variety of the Gray Squirrel, I have not had the means of deciding."

Dr. Bachman had for several successive summers an opportunity of studying the habits of this species in the northern parts of the United States. It seems to prefer valleys and swamps to dryer and more elevated situations and to possess all the sprightliness of the Northern Gray Squirrel. A colony of them had taken up their abode by the side of a retired rivulet, where they were closely and frequently watched by Dr. Bachman. He remarked that when drinking they did not lap, but protruded the mouth a considerable way under the surface of the water: supported upon the tail and *tarsi*, they would remain for a quarter of an hour wiping their faces with their paws; when alarmed, their favourite place of retreat was a large white pine tree (*Pinus strobus*): their bark and general habits did not differ much from those of the Northern Gray Squirrel.

SCIURUS AUDUBONI. Larger Louisiana Black Squirrel.

Sciurus corpora suprà nigro, subtùs fuscescente; caudâ corpus longitudine œquante.

A new species, for which Dr. Bachman is indebted to Mr. Audubon. It has the fur very harsh to the touch, and is rather less in size than the *Sciurus niger.*

SCIURUS FULIGINOSUS. Sooty Squirrel.

Sciurus corpora suprà nigro et fuscescenti-flavo irrorato, subtùs fuscescente; caudâ corpora valdè longiore: dentes inc. 2/2, mol. 5-5/4-4.

Dr. Bachman remarks of this species, "I am indebted to J. W. Audubon, Esq., for a specimen of an interesting little Squirrel obtained at New Orleans on the 24th March, 1837, which I find agreeing in most particulars with the specimen in the Philadelphia Museum, referred by American authors to *Sciurus rufiventer.*

"Dr. Harlan's description does not apply very closely to the specimen in question, but seems to be with slight variations that of Desmarest's description of *Sciurus rufiventer.*

"The following description is taken from the specimen procured by Mr. Audubon. It was that of an old female, containing several young, and I am enabled to state with certainty that it was an adult animal.

"I have given to this species of the character of 22 teeth, from the circumstance of my having found that number in the specimen from which I described. The animal could not have been less than a year old. The anterior molars in the upper jaw are small; the inner surface of the upper grinders is obtuse, and the two outer points on each tooth are elevated and sharper than those of most other species. In the lower jaw of the molars regularly increase in size from the first, which is the smallest, to the fourth, which is the largest. Head short and broad; nose very obtuse; ears short and rounded, slightly clothed with hair; feet and claws rather short and strong; tail short and flattened, but not broad, resembling that of the *Sc. Hudsonius.* The form of the body, like that of the little Carolina Squirrel, is more indicative of the strength than of agility.

"The hairs on the upper part of the body, the limbs externally and feet, are black, obscurely grizzled with brownish yellow. On the under parts, with the exception of the chin and throat, which are grayish, the hairs are annulated with brownish orange and black, and a grayish white at the roots. The prevailing colour of the tail above is black, the hairs however are brown at base and some of them are obscurely annulated with brown, and at the apex pale brown. On the underside of the tail the hairs exhibit pale yellowish brown annulations.

Dimensions.

	in.	lines.
Length of head and body	10	0
Tail (vertebræ)	6	9
Do. including the fur	8	6
Fore foot to point of middle fore-claw	1	8
Hinder foot to point of longest nail	2	1
Height of ear posteriorly	0	4
Length of fur on the back	0	7
Weight without intestines, ¾ lb		

"I am under an impression that this little species is subject to some variations in colour, the present specimen and that in the Philadelphia Museum having a shade of difference, the latter appearing a little lighter. In Louisiana it is so dark in colour as to be familiarly called by the French inhabitants, 'Le petit noir.' This Little Black Squirrel is an inhabitant of low swampy situations along the Mississippi and is said to be abundant in its favourite localities.

"As yet I am unacquainted with any species of Squirrel fully agreeing with *Sc. rufiventer.*"

Sciurus Douglassii, Gray. *Oppoce-poce*, Indian name.

A species about one-fourth larger than the Hudson's Bay Squirrel; tail shorter than the body. Colour: dark brown above and bright buff beneath. Dental formula; *incis. 2/2, can. 0-0/0-0, mol. 4-4/4-4, = 20.*

The incisors are a little smaller than those of *Sc. Hudsonius*. In the upper jaw the anterior molar, which is the smallest, has a single rounded eminence on the inner side; on the outer edge of the two there are two acute points, and one in front; the next two grinders, which are of equal size, have each a similar eminence on the inner side, with a pair of points externally; the posterior grinder, although larger, is not unlike the anterior one. In the lower jaw the bounding ridge of the enamel in each tooth forms an anterior and posterior pair of points. The molars increase gradually in size from the first, which is the smallest to the posterior one, which is the largest.

This species in the form of its body is not very unlike the *Sc. Hudsonius*; its ears and tail, however, are much shorter in proportion. In other respects also, as well as in size, it differs widely.

Head considerably broader than that of *Sc. Hudsonius*; nose less elongated and blunter; body long and slender; ears rather small, nearly rounded, slightly tufted posteriorly; as usual in this genus, the third inner toe is the longest, and not the second, as in the *Spermophiles*. The whiskers, which are longer than the head, are black. The fur, which is soft and lustrous, is on the back, from the roots to near the points, plumbeous, and at the tip with brownish gray; a few lighter coloured hairs interspersed, and gives it a dark brown tint: when closely examined it has the appearance of being thickly sprinkled with minute points of rust colour on a black ground. The tail, which is distichous but not broad, is for three-fourths of its length of the colour of the back; in the middle of the hairs are plumbeous at the roots, then a regular markings of brown and black and tipped with soiled white, giving it a hoary appearance; on the extremity of the tail the hairs are black from the roots tipped with light brown. The inner sides of the extremities and the outer surfaces of the feet, together with the throat and mouth, and a line above and under the eye, are bright buff.

The colours on the upper and under parts are separated by a line of black, commencing at the shoulders and running along the flanks to the thighs. It is widest in the middle, where it is about three lines in width, and the hairs, which projected beyond the outer margins of the ears, and form a slight tuff, are dark brown and in some specimens black.

Dimensions.

	in.	lines.
Length from point of nose to the insertion of the tail	8	4
Tail (vertebræ)	4	6
Tail including fur	6	4

Height of ear posteriorly	0	6
Palm to end of middle fore-claw	1	4
Heel and middle hind-claw	1	10

Sciurus Hudsonius, (Pennant). The Chickaree Hudson's Bay Squirrel. Red Squirrel.
    Common Squirrel. Foster, Phil. Trans., vol. 62, p. 378, an. 1772.
    Sciurus vulgaris, var. E. Erxleben Syst., an. 1777.
    Hudson's Bay Squirrel. Penn. Arct. Zool., vol. 1. p. 116.
    Common Squirrel. Hearne's Journey, p. 385.
    Red Barking Squirrel. Schoolcraft's Journal, p. 273.
    Red Squirrel. Warden's United States, vol. i. p. 330.
    Ecureuil de la Baie d'Hudson. F. Cuvier, Hist. Nat. de Mam.
    Sc. Hudsonicus. Harlan. Godman.

The Hudson's Bay Squirrel, a well-known species, is a third smaller than the Northern Gray Squirrel; tail shorter than the body; ears slightly tufted. Colour, reddish above, white beneath.
    Dental formula: *incis.* 2/2, *can.* 0-0/0-0, *mol.* 4-4/4-4, = 20.

Sciurus Richardsoni. Columbia Pine Squirrel.
    Small Brown Squirrel. Lewis and Clarke, vol. iii. p. 37.
    Sciurus Hudsonius, var. β. Columbia Pine Squirrel. Richardson, Fauna Boreali-Americana, p. 190.

Smaller than *Sc. Hudsonius*; tail shorter than the body; rusty gray above, whitish beneath; extremity of the tail black.

This small species was first noticed by Lewis and Clarke, who deposited a specimen in the Philadelphia Museum, where it still exists. I have compared it with a specimen brought by Dr. Townsend and find them identical. Dr. Richardson, who appears not to have seen it, supposes it to be a mere variety of the *Sciurus Hudsonius*. On the contrary, Dr. Townsend says in his Notes, "It is evidently a distinct species; its habits being very different from those of the *Sciurus Hudsonius*. It frequents the pine-trees in the high range of the rocky mountains west of the great chain, feeding upon the seeds contained in the cones. These seeds are large and white, and contain much nutriment. The Indians eat a great quantity of them and esteem them good. The note of this squirrel is a loud jarring chatter, very different from the noise of *Sc. Hudsonius*. It is not at all shy, frequently coming down to the foot of the tree to reconnoitre the passenger and scolding at him vociferously. It is, I think, a scarce species."

The difference between these two species can be detected at a glance by comparing the specimens. The present species, in addition to its being a fourth smaller and about the size of the *Tamias Lysteri*, has less of the reddish-brown on the upper surface, and it may always be distinguished from the other by the blackness of its tail at the extremity, as also by the colour of the incisors, which are nearly white instead of the deep orange of the *Hudsonius*.

The upper incisors are small and of a light yellow colour; the lower are very thin and slender, and nearly white. The first, or deciduous, grinder, as in all the smaller species of Pine Squirrels that I have examined is wanting; the remaining grinders, both in the upper and lower jaw, do not differ very materially from those of Douglas' Squirrel.

"Dental formula: *incis.* 2/2, *can.* 0-0/0-0, *mol.* 4-4/4-4, = 20.

"The body of this most diminutive of all the known species of genuine squirrel in North America, is short, and does not present that appearance of lightness and agility which distinguishes the *Sc. Hudsonius*. Head large, less elongated, forehead more arched, and nose a little blunter than *Sc. Hudsonius*; ears short; feet of moderate size. The third toe on the fore-foot but slightly longer than the second; the claws are compressed, hooked and acute; tail shorter than the body; the thumb-nail is broad, flat and blunt.

"The fur on the back is dark plumbeous from the roots, tipped with rusty brown and black, giving it a rusty gray appearance. It is less rufous than the *Sc. Hudsonius*, and lighter coloured than the *Sc. Douglasii*. With the feet on their upper surface are rufous: on the shoulders, forehead, ears, and along the thighs, there is a slight tinge of the same colour. The whiskers, which are a little longer than the head, are black. The whole of the under surface, as well as a line around the eyes, and a small patch above the nostrils, smoke-gray. The tail for about one half its length presents on the upper surface a dark rufous appearance, many of the hairs being nearly black, pointed with light rufous: at the extremity of the tail for about an inch and three-fourths in length, the hairs are black, a few of them slightly tipped with rufous. The hind-feet from the heel to the palms are thickly clothed with short adpressed light-coloured hairs; the palms are naked. The sides of the body are marked by a line of black commencing at the shoulder and terminating abruptly on the flanks: this line is about two inches in length and four lines wide.

Dimensions.

	in.	lines.
Length of head and body	6	2
Tail (vertebræ)	3	6
Do. including fur	5	0
Ears posteriorly	0	3
Do. including fur	0	5
Palm and middle fore-claw	1	3
Sole and middle hind-claw	1	9

SCIURUS LANUGINOUS. Downy Squirrel.

Sciurus corpore suprà flavescenti-griseo, lateribus argenteo-cinereis, abdomine albo: pilis mollibus et lanuginosis: aurbus brevibus: palmis pilis sericeis crebrè instructis; caudâ corpore breviore.

"A singular and beautiful quadruped, to which I have conceived of the above name appropriate, was sent to me with the collection of Dr. Townsend. He states in his letter, 'Of this animal I have no further knowledge than that it was killed on the North-west coast, near Sitka, where it is said to be common: it was given to me by my friend W. F. Tolmie, Esq., surgeon of the Hon. Hudson's Bay Company. I saw three other specimens from Paget's Sound, in the possession of Capt. Brotchie, and understood him to say that it was a burrowing animal.' Sitka is, I believe, the principal settlement of the Russians on Norfolk Sound and Paget's Sound, a few degrees North of the Columbia River.

"If the head is broader than that of the *Sc. Hudsonius*, and the forehead much arched. The ears, which are situated far back on the head, are short, oval, and thickly clothed with fur; they are not tufted as in the *Sc. Hudsonius* and *Sc. vulgaris* of Europe, but a quantity of longer fur situated on the outer base of the ear and the rising two or three lines above the margins give the ears and the appearance of being somewhat tufted. In the Squirrels generally, the posterior margin of the ear doubles forward to form a valve over the auditory opening, and if the anterior one curves to form a helix; in the present species the margins are less folded than those of any other species I have examined. The whiskers are longer than the head; feet and toes short; rudimental thumb armed with a broad flat nail; it may also be slender, compressed, arched and acute; the third on the fore-feet is a little the longest as in the Squirrels. The tail bears some resemblance to that of the Flying Squirrel and is thickly clothed with hair, which is a little coarser than those on the back. On the fore-feet the palms are only partially covered with hair; but on the hind feet, the under surface, from the heel even to the extremity of the nails, is thickly clothed with short soft hairs.

"The fur is softer and more downy than that of any other North American species, and of the whole covering of the animal indicates it to be a native of a cold region.

"Dental formula: *incis*. 2/2, *can*. 0-0/0-0, *mol*. 4-4/4-4, = 20.

"The upper incisors are smaller and more compressed than those of *Sc. Hudsonius*; the lower ones are a little longer and sharper than the upper: the upper grinders on their inner surface have each an elevated ridge of enamel; on the outer crust or edge of the two, there are three sharp points instead of two obtuse elevations, as in the Squirrels generally, and in this particular it approaches the *Spermophiles*. In the lower jaw, the grinders, which are quadrangular in shape, present each four sharp points.

"The incisors are of an orange colour; and the lower incisors are nearly as dark as the upper. Whiskers pale brown. Nails white. The fur on the back from the roots to near the extremity is a whitish gray; some hairs are annulated near the tips with deep yellow and at the tip black: on the sides of the body the hairs are annulated with cream colour. There is a broad line of white around the eyes; a spot of white on the hind-part of the head, a little in advance of the anterior portions of the ears. The nose is white, and this colour extends along the forehead and terminates above the eyes, where it is gradually blended with the colours on the back. The cheeks are white, a little grayish beneath the eyes. The whole of the under surface is white, as are also the feet and inner surface of the legs, the hairs being uniformed to the roots. The hairs above the tail are for the most part of a light ash colour at the roots; above the ash colour on each hair there is a broad but not well-defined ring of light rufous; this is followed by dark brown, and at the tips of the hairs are rufous and gray. Many of the hairs of the tail, however, are white, some of them are black, and others almost uniform rusty yellow.

Dimensions.

	in.	lines.
Length of head and body	7	11
Tail (vertebræ)	4	8
Tail including fur	6	0
Palm and middle fore-claw	1	0
Sole and middle hind-claw	1	9
Length of fur on the back	0	7
At the tip of tail	1	10
Height of ears, including fur, measured posteriorly	1	5

"On the back and tail there are so many white hairs interspersed, the white spot on the head being merely occasioned by a greater number of hairs nearly or wholly white, that there is great reason to believe that this species becomes much lighter, if not wholly white, during winter.

"In the shape of the head and ears, and in the pointed projections of the teeth, this species approaches the Marmots and Spermophiles; that in the shape of its body, its soft fur, its curved and acute nails, constructed more for climbing than digging in the earth, and in the third toe being longer than the second, it must be placed among the Squirrels."

Mr. Waterhouse exhibited a new species of Hare from the collection made for the Society by the late Mr. Douglas and proposed to characterize it under the name of *Lepus Bachmani*: he thought it probable that the species had been brought from California. It was thus described:

LEPUS BACHMANI. *Lep. intensè fuscus, pilis fuscescenti-flavo nigroque annulatis; abdomine sordid albo: pedibus suprâ pallidis, subtùs pilis densis sordid fuscis indutis: caudâ brevi, alba, suprà nigricante, flavido ad-*

spersâ: auribus externè pilis brevissimis cinerescenti-fuscis, interne albidis, ad marginem externum, et ad apicem flavescentibus obsitis: nachâ pallidè fuscescenti-flavâ.

"Fur long and soft, of a deep gray colour at the base; each hair annulated near the apex with pale brown and black at the points; on the belly the hairs are whitish externally; on the chest and fore-part of the neck the hairs are coloured as those of the sides of the body; the visible portion is pale brown, each hair being dusky at the tip; chin and throat gray white. The hairs of the head coloured like those of the body; and in distinct pale longitudinal dash on the flanks just above the haunches: the anal region white. The general colour of the *tarsus* above is white; the hairs, however, are grayish-white at the base, and then annulated with very pale buff colour (almost white), and pure white at the points; the sides of the *tarsus* are brown; the long hairs which cover the under part of the *tarsus*, as well as that of the fore-feet, deep brown. The fore-feet above very pale brown, approaching to white; the hairs covering the toes principally white: the claws are slender and pointed, that of the longest toe very slender. Ears longer than the head, sparingly furnished with hair, the hairs minute and closely adpressed; externally, on the forepart, grizzled with black and yellowish white, on the hinder part grayish-white; the apical portion is obscurely margined with black; at the base the hairs are of a woolly nature, and of a very pale buff colour; the hairs on the occipital part of the head and extending slightly onto the neck are of the same colour and of the same woolly character; the ears internally are white, towards the posterior margin of obscurely grizzled with blackish, at the margin yellowish.

Dimensions.

	in.	lines.
Length	10	0
Tarsus	3	0
Tail and fur	1	3
Ears externally	2	8
Nose to ear	2	5½

Habitat S.W. coast of N. America, probably California.

"This animal may possibly not be adult; but neither in the teeth, so far as can be ascertained from a stuffed specimen, nor in the character of the fur, can I see any reason for believing it young, excepting that it is much under the ordinary size of this species of the genus to which it belongs; and although it may not be an adult, it certainly is not a very young animal. Compared with *Lep. palustris*, with which species it was sent over by Mr. Douglas, it presents the following points of distinction. Although the present animal is not above one-third of the size of that species, the ears measure nearly a quarter of an inch more in length: in fact, they are here longer than the head, whereas in *Lep. palustris* they are much shorter. The next most important difference is in the feet, which instead of having comparatively short and adpressed hairs which do not conceal the claws are in *Lep. Bachmani* long and woolly, especially on the under part, and not only conceal the claws, but extend upwards of a quarter of an inch beyond their tips. The claws are more slender and pointed, especially those of the fore-feet. Besides these differences there are some others, which perhaps may be considered of minor importance: the fur is much softer and more dense; the longer hairs are extremely delicate, whilst in *Lep. palustris* they are harsh. As regards the colour, *Lep. palustris* has a very distinct rich yellow tint, which is not observed in the present species, the pale annulations of the hairs which produced the yellow tint, being replaced by brownish white or pale brown."

Changes of Colour in Birds and Quadrupeds

For its methodology, this is one of Bachman's most significant contributions to natural history. His topic is the "mutations to which some birds and quadrupeds are subject, from the young to the adult state, and at different periods of the year." Few systematic observations had previously been made over sufficiently long periods to determine all regular variations in color for any species. His conclusions about when and how changes in color took place were based partly on observations of species in captivity, and since he was aware that in confinement, maturity can be delayed, he also made extensive observations of animals in the wild. Using inductive reasoning, he was able to discern patterns that had not previously been described.

It was in this article that Bachman noted a number of writers "have indulged in various speculations, and adopted a number of contradictory theories.... In our investigations of nature, we are perhaps too prone to build our theories first, and afterwards seek for the facts which are to support them." Not infrequently, the young or females of one species had been described as separate species from adult males, and even adult males that changed colors seasonally were also sometimes described as separate species.

One group of naturalists was certain that all birds molted only once a year, and another group was equally certain that all birds molted twice a year. Bachman positioned himself thus: "Although it has been more my intention in this paper to record the facts which I have been able to collect on this subject than to frame a new theory or endeavour to restore an exploded one; yet it has often occurred to me, that the advocates of the two opposite theories have run into extremes, in consequence of the adherents of each, with too much pertinacity to a preconceived opinion.... The observations which I have made thus far have led me to conceive that the truth lies between these two extremes." He found the same divergence of opinion on the changing coloration of mammals, and he added, "how far other quadrupeds in a state of nature are subject to this biennial shedding of the hair is an inquiry which we must make not from books, but from a careful examination of the different species." As he emphasized, scholarship in natural history must consist of observation and experiment rather than summaries of assertions or deductions based on assumptions and inadequate evidence. He defended and justified his methodology, saying, "If some apology be requisite for these somewhat desultory remarks and for having entered so much into detail, it may be offered in defence that the subject on which it treats, although apparently trifling and unimportant, leads to many inquiries in philosophy and in natural science."

Bachman organized his information to answer four questions, and although he called for further research, he provided sufficient information to disprove many previous assertions and to establish basic patterns that were later confirmed. His questions were:

1. "Does the change of plumage in some birds arise from a change of feathers, or from the feathers themselves assuming at one period a different colour from that which they have at another?"
2. "Whether old birds, whose plumage has arrived at full maturity, are subject to the same general law with those who advance from immature to perfect plumage."
3. "Do birds which are subject to these semi-annual changes in plumage receive their new colours in the spring in consequence of a fresh moult, as they do in summer, or are they produced by a gradual fading or brightening of the feathers without a fresh moult?"
4. "... whether the same laws may be applied to those variations in colour to which the hair of quadrupeds is subject."

"Observations on the Changes of Colour in Birds and Quadrupeds." *Transactions of the American Philosophical Society*, n.s., 6 (1838), article 4: 197–239. Read 19 May 1837.

No attentive observer of nature can have failed to remark the striking and wonderful mutations to which songbirds and quadrupeds are subject from the young to the adult state and at different periods of the year. The young of our Indigo bird and Blue Grosbeak are clothed with brown, but in the adult state are brilliantly blue. The young of the Painted Bunting is of an humble ash colour; and, after undergoing a succession of changes, puts on a livery of bright purplish lilac, vermilion and glossy green. The young of the White Ibis and the Whooping Crane wear a homely brown garb, whilst the adult is pure white; on the other hand, the young of our Blue Heron and Reddish Egret are white; and the adult in the one case bright blue, and in the other rufous. There are other birds, such as the Reed bird, the American Goldfinch, the Yellow crowned Warbler (*S. coronate*), and various species of Gulls and Sandpipers, that disguise themselves so during six months of the year whilst on their migrations as to be with difficulty recognisable. Some of our quadrupeds also as if seeking for notariety, at one season are clothed in white; and as if courting obscurity in the other, assume the humbler dress of the earth and the dried leaves around them. To account for these mutations or even to describe the process by which they are effected has been a source of much perplexity to naturalists and philosophers. In Europe, Cuvier, Temminck, Yarrell, Drs. Flemming and Whitear, Mr. Montagu, and several writers in the recent numbers of the journals of Paris, Dresden and Halle have indulged in various speculations and adopted a number of contradictory theories. Some have contended that birds moult but once, others twice a year; some that the annual mutations are produced by a gradual fading or brightening of the feathers; whilst others have contended that it is produced by a sudden moult. Dr. Flemming, who adopts the theory that the feathers of birds do not arrive at maturity and drop off till they are a year old supposes that those feathers which are received in the spring, remain till the following spring, and that those added in autumn, moult on the succeeding autumn, thus making two irregular moultings in a year. One school of naturalists have supposed that the change in those quadrupeds which become white in winter is effected by the gradual lengthening and blanching of the summer fur; whilst another presumed that this wonderful mutation can only occur by a shedding of the summer hair, which is replaced by the snowy pelage of winter. In our country, few observations have been made on this subject. Our celebrated ornithologists, Wilson, Bonaparte, and Audubon, appear to have dwelt very sparingly on the changes of plumage in birds, which, however overlooked, seems to belong to their department; and with the exception of a paper by Mr. Ord, in the Transactions of the American Philosophical Society, I do not recollect having read any article on this department of the physiology of birds and quadrupeds. In our investigations of nature, we are perhaps too prone to build our theories first and afterwards seek for the facts which are to support them. Hence, naturalists, having the same field of inquiry before them, and reading from the same book of nature, which is open to all, are very apt to be swayed by their preconceived notions and thus retard the progress of science by unprofitable disputes.

The time perhaps has not yet arrived, when any certain theory can be built on this part of physiology. A sufficient number of experiments and observations have not been made with care and judgment; nor have such a body of facts been collected as will enable us to judge which theory is founded on nature and truth. Systems which we adopt in haste from an examination of a few isolated facts, without awaiting the slow progress of time, are frequently obliged to be abandoned in after years. When Bonaparte wrote his long, spirited and interesting article on Peale's Egret Heron, and framed his theory that "the Egret Herons are entirely of a snowy whiteness, without any coloured markings on the plumage whatever," he could not have conceived that the next student of nature who visited the spot where his specimen was obtained should discover that this identical species became brown and rufous; and thus compelling nature herself to scatter his fine spun theory to the winds.

It is proposed in this article to give such facts relating to the moulting of birds and quadrupeds, as some experience has enabled me to collect. Should this have a tendency to elicit further inquiry on a subject so interesting to naturalists, and lead to the study of the causes of these wonderful changes, my object will be attained.

1. Does the change of plumage in some birds arise from a change of feathers or from the feathers themselves assuming at one period a different colour from that which they have at another?

On this head the following observations were made.

Falco leucocephalus. A pair of birds were sent to me in the spring of 1830 taken from a nest in the neighbourhood of Beaufort, S.C. two years before. The old birds having appeared unusually large and destitute of the white heads and tails of the Bald Eagle, the individual who had preserved the young sent them to me under an impression that they were the long disputed Sea Eagle (*F. albicilla*) of Europe. Perceiving that they were the young of our *F. leucocephalus*, and being aware that some of these birds, when in confinement, required five and six years before they attained their full plumage, I did not much value these troublesome and expensive pets. On a closer examination, however, I observed that the male had some of the feathers of his head streaked with white whilst the outer edges were bordered with brown. I discovered also, when he spread out his tail, that some of the inner portions of his feathers were broadly and irregularly patched with white. I concluded that the important change I was desirous of witnessing was in progress. The birds were therefore carefully fed and noticed. The female had evidenced no appearance of change or progress towards a change until the moulting season arrived. This commenced in June: both birds moulted freely, the female three weeks later than the male. In the latter, the feathers of the head came out nearly pure white; those of the tail continued still irregularly barred with brown and ash, but became gradually whiter till about the latter end of October, when he had all the markings of an adult bird. The feathers of the female assumed nearly the colours of the male at the time I first received him. In the beginning of November, all further change in the plumage of the female appeared to have been checked. The birds remained in my possession till the following January, when they were sent to Europe.

The above observations, in addition to the facts they present with regard to the process by which changes in colour are effected in this species, enable us to form some idea of the time it may require to produce the full plumage. Birds in confinement, deprived of the exercise, air and food to which they are accustomed in a state of nature, often moult irregularly, and the time they attain full plumage is considerably protracted. An individual of this species which was for some time in my possession did not moult for eighteen months, although apparently in good health; and when at last this process took place, it was in January, the coldest month of the year in Carolina. In the present instance the male received the white feathers in his head and tail a year earlier than the female, which, I have reason to believe, would not have been the case if they had been left at liberty. Here, however, is one instance established with a tolerable degree of certainty of this species arriving at full plumage in three years.

Strix asio. The young of this species was first described under the above name and the old as *strix nævia*. Its general colour above is reddish brown, but when it receives its mature plumage it is mottled with white, ash and pale brown. I have possessed very few opportunities of examining the change of plumage in this species. Having, however, once found an individual sitting on its nest while yet in the red stage, I conclude that they did not arrive at full plumage till they are more than a year old.

In the month of February (the year was not noted, and is unnecessary to our present inquiry) I saw in the possession of some lads, who were dragging it through the streets by a string an individual of this species so singularly marked that I was induced to ask them for it. They stated that they had found it with a broken wing in the woods. The wing was nearly severed, but the wound apparently healed; a proof that the injury was of long standing. The bird, though lean, was in moult. The lesser wing coverts, a portion of the head, and in a regular spot on the breast had nearly attained the markings of the mature bird; the rest of the feathers that had not been moulted remained the colour of the young. It survived but a few days. Whether this was a spring moult or that of summer retarded in consequence of the wound, I had no means of ascertaining.

Psittacus Carolinensis, Carolina parrot. This bird has become so rare in Carolina that I only once noticed a small flock of five or six among the cypress trees of the Salt Katcher swamps. In the autumn, however, of 1831 a friend received from St. Augustine five young birds of this species. They were never in my possession, but I visited them occasionally, and scarcely ever without the expense of a wound, which they were at all times ready to inflict upon any strange visiter. They continued in the uniformed plumage of the young of this species till the beginning of February, when they all about the same time commenced moulting, more perceptibly and more extensively in some of the individuals than in others. A striking increase of brightness was visible in all the new feathers. Those on the neck, which first came out a yellowish green, gradually and irregularly became bright yellow; but in all cases, as far as

I had an opportunity of judging, the change of plumage was in the new feather. Absence from the city prevented me from seeing these birds after they had arrived at full plumage.

Icterus Baltimore, Baltimore Oriole. The only opportunity afforded me of observing the changes of colour in this beautiful species was in the state of New York in May 1815. A young male had been obtained and confined in a cage, where it was for some time fed by its parents. In the month of October of the same year it moulted; the young feathers were much brighter than those which were dropping out; and in two months afterwards the bird was in full plumage. Our ornithologists, Wilson and Audubon, who state that this bird requires three years before it arrives at full plumage, may have been led into this mistake from having seen birds of the late brood of July, not in perfect plumage in the following spring. Mr. Audubon, indeed, recently informed me that he had satisfied himself of this error in examining an aviary kept by some gentleman in Baltimore, and I acknowledge myself indebted to him for the hint which induced me to refer to the long neglected notes I had made on this species.

Icterus spurious, Orchard Oriole. This sprightly species was for several years preserved in an aviary, where I employed my leisure moments in studying the habits of this beautiful race. There, among various others of the feathered tribe, it built its pensile nest and reared annually two broods of young. Its curious *gestures*, its varied and often melodious notes, as well as its continually varying colours rendered it one of the most interesting and admired species. The young were yellow on the under surface, and the males could scarcely be distinguished from the females. They moulted in autumn, the last brood not until January, being two months later than the first. There was no perceptible change in colour for two weeks, when the birds of the first brood assumed the black patch under the throat; the later birds still retained the yellow in those parts. They retained these colours during spring, whilst they were engaged in the cares and duties of reproduction. The moulting season now commenced in August, when the elder males became mottled on the back with irregular streaks, and the younger assumed the black patch under the throat. In the month of January the first assumed the bright chestnut colour on the breast, and were in full plumage, and the others but one stage removed from it; thus passing through all their variations of colour to their full plumage in less than two years. Our naturalists have extended the time to four years; a pretty long time to be concealed under a mask.

Fringilla cœrulea, Blue Grosbeak. A pair of young birds of this species was taken from the nest in May 1836 and raised by a friend, who since presented them to me. They were then of an humble drab colour, and commenced moulting about the beginning of December. The female died in the moult. The new feathers in the male came out blue, edged outwardly with brown; the moulting has proceeded rather slowly, owing, no doubt, to the cold weather. On 21st of March young feathers were still making their appearance on various parts of the body, whilst the first feathers that appeared after the moult were bright blue, even to their extremities; the latter ones are still along their outer edges near their points tipped with brown and drab, which continue to brighten as the feathers become more matured. The male undoubtedly attains his bright and perfect plumage before he is a year old although I have occasionally seen a young male in early spring with a few brown feathers (the result, no doubt, of an imperfect moult), yet out of many hundreds that I have noticed in the breeding season, I have never seen any males that were not in full plumage. They are not subject to a change of colour during winter, as supposed by Wilson, nor do they require three years to attain their full plumage, as stated by Audubon.

Fringilla ciris, Painted Bunting. This beautiful and social species is very common among the caged birds of the south. Having long preserved it in an aviary, where it raised two or three broods in a season, I have it in my power to state from personal observation the changes through which it passes from immature to perfect plumage. For the period of a year, the male strongly resembles the female, being of an olive green colour above. The birds moulted late in autumn, without any perceptible change of colour. In this homely dress they in the following spring commenced building their nests, rearing their young, and, with their sprightly song, cheering the females engaged in incubation. In August and September they began to moult; the new feathers on the head came out bright blue; those on the breast of a light ash colour, tinged with carmine; the colours on the head appeared to have received in their full perfection, immediately after the moult; those on the breast and neck continued to brighten gradually for some weeks. The perfect plumage, as far as it is acquired in this species in confinement, was obtained in less than two years. Our ornithologists have assigned to the period of four or five years for this process. To the objections which may be urged against experiments made on birds in confinement, as affording no certain guide in ascertaining the time at which they arrive at full plumage, it may be observed that in caged birds, excluded from the full influence of the sun, air and suitable food, the period may be extended, but not accelerated.

The above instance may suffice to show the process which nature pursues in effecting the changes of plumage from the young to the adult state in land birds. I shall, therefore, omit adding the notes I have prepared on several others: particularly on those of *Polyborus vulgaris*, Carracara Eagle, from Florida, a pair of which I have now in confinement; *Falco borealis*, Red Tailed Hawk; *Picus erythrocephalus*, Red Headed Woodpecker; *Fringilla cardinalis*, Red Bird; *Tanagra rubra*, and *æstiva*, Scarlet Tanager, and Summer Red Bird; *Columba leucocephala*, White Crowned Pigeon, &c. In all of them, however, there is a striking uniformity, the changes being effected during the space of a year, and in most of them the feathers become gradually brighter immediately after the moult.

I shall now proceed to note a few observations on those water birds that are subject to striking variations in plumage from the young to the adult state, which will prove that nature, although varying in some particulars in some of the species is still subject to the same general laws.

Grus Americana, Whooping Crane. The young of this bird is of a brownish ash colour, and at a year old is still one fourth smaller than the adult bird. In this state it has been considered by all our writers on American ornithology with the exception of Audubon as a distinct species.

Grus Canadensis, the Canada Crane. Dr. Richardson, who found its eggs (which were smaller than those of the Whooping Crane) in the polar regions, is also under the impression that it is a distinct species. It will be recollected, however, that all our birds, as far as we are acquainted with their histories, breed at a year old, even before they have attained their full plumage; and at that young birds lay smaller eggs than old. How long a time the Whooping Crane requires to arrive at full maturity, I have had no means of ascertaining. I however had an opportunity of witnessing the change of colour in a pair of this species, which convinced me that the Canada Crane was the young of the Whooping Crane. The birds were obtained, it was said, from Florida, in what manner I was not informed, and were represented as two years old. They were very tame, eating from the hand, and evidencing no vicious disposition; preferring, as food, the sweet potatoe (*Convolvulus batata*) and Indian corn (*Zea mays*). They had not quite completed their moult. They appeared a shade lighter than the Canada Crane and considerably larger. The feathers were white nearly to their extremities, where the ash colour prevailed the brown edges of the feathers continued to become narrower. At each successive visit the change was rendered more visible. In a few weeks some of the feathers became pure white, while others were slightly tinged with a cinereous colour. The change was not perfect, though nearly so, when the birds were removed from the city, and all who saw them unhesitatingly pronounced them Whooping Cranes.

Ardea cærulea, Blue Heron. Our writers on American ornithology were not, until recently, aware that the young of this species was white. In this state of plumage it so much resembles the Snowy Heron (*Ardea candidissima*) that it can scarcely be distinguished from that species except by its black legs and toes. I had many opportunities of witnessing the changes of plumage to which this bird is subject and discovered that, both in captivity and in a state of nature, the time and process do not materially vary. The young birds continue pure white till late in autumn, when they moult. No perceptible change in colour takes place except in a very few instances; a few feathers have a slight tinge of blue near their inner webs. All the old birds retire south of Carolina in winter; a few of the young, and white plumage remain. When this species return from their winter retreat in Florida and Mexico, they possess their beautiful trains, some white, others white and blue. On some the breasts are spotted; some of the feathers having become blue, others still remaining white: every where, however, a tendency to the change which is approaching is visible in each new feather, its inner vaine being more or less marked with blue. A few still continue white, probably the late brood of the former year. It is amusing to witness the breeding place of these birds. A thousand nests may sometimes be seen on some small islands among the reserve dams of the rice fields of Carolina. Here is seen an indiscriminate mixture of all colours, engaged in incubation and providing for their families. A blue male may be seen mating with a white female, a blue female choosing for her companion a white male. Birds strangely spotted, some with white, others with blue trains sweeping through the air, rising in the hundreds over your head, and presenting so many variations in plumage, that the young ornithologist is tempted to believe that he has a half dozen new species to describe. In August a regular moult commences, and before the old birds leave Carolina, in autumn, they have acquired their blue colour; thus attaining their full plumage in less than two years.

Ardea rufescens, Reddish Egret. This Heron, which I have had in a state of domestication in its changes of plumage, and the period in which these changes are effected (as far as an imperfect experiment was made) seems to partake strongly of the character of the last mentioned species.

The various and opposite changes which are species of *Grus* and *Ardea* undergo from the young to the adult state is striking and wonderful. In all of them there is no perceptible difference in the plumage of the sexes of each species, except in the least Bittern (*Ardea exilis*), where it is very striking. In the *Ardea herodias, A. minor, A. Ludoviciana, A. virescens* and *A. exilis*, no very striking mutations take place from the young to the adult. The *Ardea occidentalis* (great White Heron of Audubon), *A. alba* and *A. candidissima* are white in all stages of growth, and at all seasons. The *Grus Americana* and *Ardea nycticorax* are ash or brown, when in the first year of their existence, and then the former becomes white, and the latter greenish black and white; whilst on the other hand, the *Ardea cærulea* and *Ardea rufescens* are white when young, and blue or rufous when in full plumage.

Plotus anhinga, Black Bellied Darter. I discovered several nests of this rare and singular species in the immediate vicinity of each other, in one of the dark and gloomy morasses of Carolina in June last. No naturalist has heretofore spoken as having seen its nest except Mr. Abbot, in his letter to Mr. Ord, who describes the eggs as blue, and the nest containing six young and two eggs. I have sometimes thought that my excellent old friend mistook some heron's nest for that of this species as the nests I examined contained only three eggs, and in one instance four, on which the females had for some time been sitting, and the eggs were white, covered over with a calcareous matter like those of Cormorants. The young were brought home, fed on fish, and became very familiar and amusing. The sexes when young are alike of a brown colour. The male commenced moulting in November. The young feathers came out black and seemed to assume nearly all their bright colours immediately. When it had nearly arrived at full plumage, in January, it was unfortunately killed by a dog, in revenge for the severe bites it was wont to inflict on all intruders upon it at meal time. In this species the full plumage is received before it is a year old. In all the instances I have enumerated, it will be perceived that in the changes which occur in the plumage of birds from the young to the adult state, nature is nearly uniform in her operations. That these changes occur immediately after the moulting of the bird and that many feathers may, in process of time, become of a different colour; in short that the colour may change without a change of plumage.

This leads to a further inquiry.

2. Whether old birds, whose plumage has arrived at full maturity, are subject to the same general law with those who advance from immature to perfect plumage. In other words, does their plumage become perfect at once or is it subject to gradual changes, as in the case of young birds?

It is admitted that in a great majority of birds, as in the Crow, Blackbirds, Blue Jays, &c., these colours are permanent, and there is no perceptible change in colour from the formation of the feathers till they have matured and drop off. But in many species this is not the case. Many instances have fallen under my observation to satisfy me fully that feathers change their colours in adult as well as in young birds.

A female wild Turkey was sent to me eighteen months ago by my friend Dr. Tidyman, in order to enable me to ascertain whether this species when taken wild and full-grown could be domesticated. It had been just caught in a trap and was excessively wild. By subjecting it to confinement with the tame variety and excluding the light in part, it lost all its wild habits, and eventually became so gentle as to approach my hand to be fed. In the month of October last it moulted like the tame turkeys, with which it now associated. But I was surprised to find many of the tail and wing feathers and some of the feathers on the back coming out of a light ash colour, whilst others were nearly white; insomuch, that some of my friends were induced to pronounce it only one of our varieties of the tame turkey. Shortly afterwards these feathers began to change in colour and gradually become darker and brighter till after the expiration of a month, it had again all the rich plumage of the wild female turkey.

A male of the Summer Duck (*Anas sponsa*) has been in my possession for several years. In the moulting season, which occurred in August last, he lost all his fine plumage, and the new feathers were at first so much of the colour of the female that the sexes could scarcely be distinguished by the plumage. Shortly afterwards, however, a change commenced; from day to day the beautiful tones of the male were returning; his rich colours were gradually restored; and in the course of six weeks, this, the most elegant of all the species was again in full plumage.

These observations are calculated to strengthen the opinion advanced by the Rev. Mr. Whitear, contained in the twelfth volume of the Transactions of the Linnæan Society of London, that "a change in the colour of the plumage of birds does not always arise from a change of feathers, but sometimes proceeds from the feathers themselves assuming at one season of the year a different colour from that which they have at another."

To this theory I am disposed to subscribe to a certain extent, and under some limitations. That feathers are changed in colour without a change of the plumage is admitted and, I think, satisfactorily proved. But, as far as I have been able to ascertain, this change is always preceded by a recent moult.

At this stage of our inquiries we arrive at an interesting point. There are many birds that are subject to a semi-annual change of colour, after they have arrived at full maturity. The male Rice Bird (*Emberiza oryzivora*) is of a brownish yellow colour during six months of the year and returns to its breeding place in the northern and middle states, clothed in a livery of white and black. The male American Goldfinch (*Fringilla tristis*) is of a homely olive brown colour during winter whilst in the summer it is a bright yellow. The Yellow Crowned Warbler is ash coloured during half of the year; of a beautiful blue in the other. Our Plovers, Tringas, Gulls, &c., are subject to the same changes. Our ornithologists have widely deferred in investigating this part of the physiology of birds. The subject of inquiry here seems to present itself under a new aspect, and the inquiry is:

3. Do birds which are subject to these semi-annual changes in plumage receive their new colours in the spring in consequence of a fresh moult, as they do in summer, or are they produced by a gradual fading or brightening of the feathers without a fresh moult?

Mr. Ord, in a well written article published in the third volume, new series, of the Transactions of the American Philosophical Society, has advanced the following theory. "It's being now satisfactorily proved that a change of colour obtains, in some birds, in the winter and spring, without a change of plumage, I am disposed to conclude that the state of moulting, properly so called, takes place in all birds but once a year."

To this theory the following difficulties seem to present themselves. The colour in the plumage of birds, especially in the small feathers most subject to a change, appears to be advanced in the extent it is intended to arrive at in a few weeks, or, at furthest, in a few months. At that point it seems to become stationary and to remain so for a considerable length of time. If the same feathers are afterwards to receive a fresh set of colours, there must be some secretions in the body of the bird, and a fresh impulse given to the feathers already advanced to maturity, imparting to them properties which they did not possess before. When it is necessary for a bird in summer to receive a new dress, differing in colour from the old, there are no secretions by which it can impart fresh colouring matter to its old feathers, which have long become stationary in their growth and colour. It is, then, essential that these feathers should be thrown off and those substituted, which, in the progress of their growth and their advance to maturity may receive those hues destined for them by nature. This law of nature seems to be so simple, that it can be easily comprehended; but if a different and opposite process is observed in giving new and entirely different colours to the old feathers of the bird in spring, the law of nature, so uniform in other respects, cannot be traced. The feathers and other appendages of birds may (without adopting the nice distinctions and scientific terms he used in botanical science) be compared to the leaves and other appendages of plants. In early spring, the juices, together with the influence of sun and air, impart that nourishment which causes the leaf to expand, and assume its beautiful colours; but when the leaf has arrived at maturity, no fresh growth of the tree will give a new colouring to its leaves. New ones may be formed, and these may continue to grow and flourish, but the old ones fade and drop off. Even among our evergreens of the south, whose leaves are persistent, as the orange tree, &c., which may be said to have two periods of growth (April and September), the leaves when once matured seem to have lost the power of further circulation; they ceased to grow; their rich colours fade; and they only await the time appointed them by nature to return to the ground. If the feathers of birds, then, which have been long stationary in their growth are capable of receiving a new set of secretions and of assuming opposite colours, we must seek for some new law of nature not hitherto understood.

The origin of the feathers is the matrix which is placed in the skin or under it. The structure of the matrix is a pulpy substance called bulb, and a capsule, which is composed of several layers. The bulb furnishes the material of the stem and vane, which, when complete, disappears, leaving no residue but the almost imperceptible ligament, connecting the quill to the bottom of the cavity, which receives and embraces it. Whenever a new feather is to be formed, a new matrix is necessary to the process. The early connection of this matrix with the body is by means of vessels. From these the pulp or bulb derives its nourishment. The feather is whilst young enclosed in a sheath, and this as well as the quill itself is filled with a coloured fluid. In a few weeks the secretions have been imparted to the feathers, and the sheath by a process of absorption becomes dry and is rubbed off. In the tube of the feather now remains a jointed membranous body, which every one has observed in the barrel of a common quill. This shaft is now filled with air. After this process, the object of nature in providing covering and the means of flight to the bird seems to have been accomplished, and the feather ceases to grow.

In order to renew the nourishment of the quill, it would be necessary to renew the vascular connection. It is doubtful whether any reviviscence of colour can take place without this. As long as this lasts, the assimilative powers within the feather may continue to act. When this ceases, these are extinguished with it. As the vascular fluid, then, which fills the sheath and the quill gradually disappears as also the bulb which nourishes it, after the feather has arrived at full maturity and becomes stationary in its growth, it would appear that after this period no further change of colour can take place. This point, however, has not been so satisfactorily investigated by physiologists as to enable us to express a positive opinion, and, until this is done, we are obliged to resort to an examination of individual species in order to ascertain how far there is a uniformity in these changes of colour in all birds.

The fact that birds in winter are by a wise Creator furnished with a thicker covering than they possess in summer does not appear to have received that attention from naturalists who have made inquiries on the subject, that its important bearing on this point seems to require. Several of the birds subject to the semi-annual changes in colour spend their winter in climates comparatively cold. The *Sylvia coronata* is seen by thousands in Carolina, where the thermometer during winter stands for several days in succession below the freezing point. The *Fringilla tristis* also braves our coldest winters; and the *Saxicola sialis* and *Sylvia petechia* linger among our copse woods and orange groves. These birds are closed with a covering suited to the rigorous season. In the summer their dress is lighter as well as more gaudy. If, then, there is no dropping off of feathers, it would be reasonable to inquire what becomes of their winter clothing? This would have to be borne through all the heat of summer till the moulting season; and their thick blanket of down would damp their ardour and silence many a joyous song.

I am well aware, however, that we may build theories and indulge in speculations, which, however plausible they may appear, and however satisfactory to our own minds may be easily overturned by a single glance at the manner in which nature performs her operations. Admonished, therefore, to resort to a better book than philosophers can write, the book of nature itself, I was determined in some hour not devoted to any higher pursuits to make the inquiry by a simple examination of facts, not difficult to collect. The point in dispute seemed susceptible of an easy solution, by a careful examination of those birds that are subject to semi-annual changes of colour and to ascertain if they moulted once or twice in a year.

The following was the result of my investigations. I copy from notes taken on a succession of visits into the country.

February 23, 1837. Visited a country residence about two miles from

Charleston. Slight indications of early spring. Weather windy and unsettled; a slight frost during the night. A few of the trees in the city in blossom, especially the peach and plum. In the country the *Laurus geniculatus*, *Prunus chicasa*, *Acer rubrum* and *Salix nigra*, the only trees in bloom. Procured a number of specimens in ornithology.

The following had no change from winter plumage, and no appearance of a moult. Mocking Bird (*Turdus polyglottus*), one specimen. Pine Creeping Warbler (*Sylvia pinus*), two specimens. *Sylvia coronate*, two specimens. Solitary Thrush (*Turdus minor*), one specimen. White Eyed Fly-Catcher (*Vireo noveboracensis*), one specimen. *Fringilla tristis*, ten specimens, of both sexes. The males of the latter species differed considerably in colour in the different specimens. Whilst some strongly resembled the females in their plumage, others were considerably brighter in colour, especially on the breast; and in two or three specimens there were irregular markings of black on the frontlets, as if nature during the last moulting season had been making an effort at advancing the birds to the bright plumage of which they had just been deprived.

The following had commenced moulting. Wax Bird (*Bombycilla Carolinensis*), two specimens, males. Both birds were considerably advanced in the moult; new feathers in great quantities were coming out on every part of their bodies; these were much more brilliant than the old, and seem to receive their bright colours at once. Crested Titmouse (*Parus bicolor*), one specimen, moulting along the sides and on the head. Solitary Fly-Catcher (*Vireo solitarius*), one specimen. (This bird, contrary to the assertions of ornithologists, has a soft and melodious song. It is becoming more common every spring.) A great many young feathers were appearing on the head, around the lores of the eyes, and on the breast and back. Savannah Finch (*Fringilla Savanna*), one male specimen; far advanced in the moult; about one half of the feathers on the breast young, and still sheathed. Bay Wing Bunting (*Fringilla graminea*); a row of young feathers appearing on each side of the neck. Chipping Sparrow (*Fringilla socialis*); moulting commenced on the breast, and a few young feathers on the head.

The same day I examined some living birds, in a state of confinement. The male of the Caraccara Eagle (*Polyborus vulgaris*) flew fiercely at my face whilst I was inspecting the female. Neither of them was in moult.

The English Pheasant (*Phasianus colchicus*), the Wild Turkey (*Meleagris gallipavo*) and Virginian Partridge (*Perdrix Virginiana*) showed no indication of moulting. This was also the case with the different species of pigeon, and among all the *Emberizæ* and *Fringillæ*. The Song Sparrow (*Fringilla melodia*) was only shedding its feathers.

March 3. Visited the country. No very perceptible advance in vegetation. Night cool, with slight frosts. The only additional plants and flower were *Leontodon taraxicum*, *Laurus sassafras* and *Oxalis stricta*. These facts with regard to the advance of spring are noted in order to enable us to form some idea of its probable effects on the time of the moulting of birds.

Obtained the following birds.

Bombycilla Carolinensis, fifteen specimens. About one half of this number were moulting on the breast, neck, and under the throat; some of them extensively; and in one of them especially, it appeared as if every feather on the body, except those in the tail and scapulars, was already replaced by new ones, half formed. In some of the specimens moulting had not yet commenced or had already been completed. Bonaparte states in regard to the genus *Bombycilla*, "that they moult once a year." *Fringilla graminea*, two specimens moulting extensively on every part of their bodies. I noticed that these two birds, which were males, were in full song. *Fringilla socialis*, two specimens. In one a large quantity of new feathers were coming out from the sides; and in the other the whole head was covered over with young feathers still sheathed. *Troglodytes ædon*, one specimen. Far advanced in the moult along the breast.

The following had no appearance of any change in colour or of moulting.

Fringilla hyemalis, two specimens; *Fringilla cardinalis*, three specimens; *Regulus cristatus*, two specimens; *Troglodytes Carolinensis*, two specimens; *Sylvia coronata*, eleven specimens; *Turdus migratorius*, six specimens; *Fringilla tristis*, twenty-two specimens.

The *Fringilla tristis* and *Sylvia coronata* seemed wild and restless as if preparing to migrate. Their numbers had also considerably diminished, and it is possible that they may not remain long enough to enable me to ascertain the process of their change of colour in spring.

March 6. Since I last visited the country there has been a fall of snow about 3 inches in depth, remaining on the ground for twenty-four hours; an unusual occurrence in S. Carolina. This cold change brought back a number of birds that appear to have left us for the season.

Obtained and examined the following species of birds. *Sylvia coronata*, twelve specimens. About one half of these (both males and females) had commenced moulting. The new feathers were of the bright colour which this bird assumes in summer plumage; the old remain stationary. *Fringilla Savanna*, seven specimens. These birds were receiving new feathers on various parts of their bodies; those on the head were all changing; under the chin, the ash coloured feathers of winter were replaced by those of pure white, the colour of summer. *Turdus migratorius*, two specimens. In one of these specimens a considerable number of young feathers were appearing along the breast, under the chin and on the head. *Troglodytes Carolinensis*. A row of new feathers along the sides.

No change was apparent in the following species. *Corvus ossifragus*, six specimens; *Corvus cristatus*, two specimens; *Fringilla tristis*, sixteen specimens; *Picus pubescens*, three specimens; *P. villosus*, two specimens; *P. auratus*, two specimens; *Icterus Phœniceus*, eleven specimens, females; the males of the last species are found in separate flocks at this season, a few of the young males occasionally associate with the females.

After the above period, I was for a short time in Georgia, where, through the kindness of my friends, who furnished me with specimens, I was enabled to examine a number of species of birds. On my return to Charleston I resumed my inquiries into this subject, and through my own exertions and the aid of some of my young friends had an opportunity of inspecting a considerable number of specimens of various species subject to a semi-annual change of colour. The result of examinations made on the 13th, 16th, 17th, 21st, 23d and 28th of March; on the 3d, 10th, 17th and 20th of April; and on the 4th and 5th of May, I will condense under a notice of each species, as the notes taken on these different occasions would swell this article to an unreasonable length.

Fringilla Pensylvanica, White Throated Sparrow. Of this species I was enabled to examine about thirty specimens, in addition to those in an aviary. The light ash colour under the chin is in every instance replaced by a fresh moult; the new feathers coming out as pure white. This is the case in both sexes. The birds too appear to moult on every part of the body, some having the whole head covered with young feathers still sheathed; the same process is going on with others, on the breast; and in others, moulting first commences in spots on the back. Thus nature seems to leave one portion of the clothing as a covering to the bird, whilst it is renewing the rest; and the feathers of the one part of the body are fully formed before moulting commences on other parts.

Fringilla Savanna, Savannah Finch; *Fringilla palustris*, Swath Sparrow; *Fringilla melodia*, Song Sparrow; *Fringilla pusilla*, Field Sparrow. The same process was observed in the species. They all molted extensively, but irregularly; and in many specimens every feather except to those in the wings and tail appeared to be renewing.

Fringilla erythropthalma, Towhee Bunting. Of this species I received seven specimens. Two males have the whole head covered with new feathers; one other was moulting extensively on the back; not an old feather was remaining on that part; and in another this process was going on among the feathers on the breast and sides. In the remainder there were a few young feathers, but no regular moult.

Sylvia petechia, Yellow Red-Poll Warbler. This species I was extremely anxious to examine, in reference to this subject. The difference of its plumage in winter and summer is as striking as that of *Sylvia coronata*. It however becomes yellow olive in spring, whilst the other becomes blue, strieked with black, with a yellow spot on its rump and crown. The *Sylvia petechia* has been described under two names, that of the above and *Sylvia palmarum*, by Bonaparte and Audubon; the latter recently acknowledged his error. I obtained of this species twenty-four specimens. Every individual was moulting extensively. The new feathers came out at once in summer dress, leaving not a shadow of doubt in my mind that this bird receives its bright colours from a change of feathers and not from a change of colours in the old feathers.

Sylvia pensilis, Yellow Throated Warbler. This is one of the small number of the species of this large genus that breeds in Carolina. Its note was first heard on the 16th of March. Obtained, at various times, thirteen specimens. In every instance the whole of the yellow on the throat was replaced by young feathers just pushing forward, and in several individuals moulting was general over other parts of the body. The above observations apply equally to *Sylvia trichas*, Maryland Yellow Throat, of which I inspected about five specimens.

Fringilla tristis, American Goldfinch. This was one of the birds referred to by Mr. Ord, as a proof of the correctness of his theory that birds moult but once a year. Although this species is common in Carolina in winter, feeding on the seeds of the long moss (*Tillandsia usnoides*) that hangs in festoons from the limbs of our venerable live oaks (*Quercus virens*), and at a later period on the imported, but very common chick weed (*Alsina media*) it generally leaves us for the north early in March, and, no doubt, undergoes its changes of plumage in the spring in the immediate vicinity of Philadelphia. This bird had left us for about two weeks, but was driven back to this vicinity in consequence of a succession of cold days; and from the 16th to the 22d of March I had an opportunity of obtaining forty specimens. On the former day two or three individuals were brought to me in which moulting had commenced. From this period until every straggler appeared to have left us, on the 23d, I discovered that the process of moulting was advancing rapidly. It seemed to commence in the old males, extending itself to the females; and on my last examination not an individual of either sex was brought to me that was not in extensive moult. The young feathers on the head came out black; those on the rest of the body of a bright yellow. In the old feathers no change whatever had taken place. Mr. Ord is correct when he states that this species is in full song in March; this was the case in all the individuals I observed; but he appears to have overlooked the opportunity of ascertaining that it was also in full moult. The theory that the song of birds is silenced in consequence of the exhausting process of moulting is not calculated to bear the test of close examination. The *Fringilla tristis*, *Troglodytes ædon*, *Fringilla Pensylvanica*, *F. graminea*, *Sylvia pensilis*, and many others, were in full song, although they were moulting very extensively. This is also the case at the moment I am writing in several species now in my aviary. It will probably be found that the vocal powers of birds are called forth by an increased development of the sexual organs, in the vivifying season of spring, and that their song is suspended in autumn, when these organs are sensibly diminished.

Psittacus Leucocephalus, White Fronted Parrot. A bird of this species, now in the possession of my friend Dr. Wilson of this city, was carefully examined on the 20th of March. It was in fine health, repeated several words, and made attempts at a song. We found it in extensive moult, and all who examined it were satisfied that it was receiving a full set of new feathers on every part of the body. This species exhibits a variety of colours, green, white, yellow, pink and red. The feathers came out apparently in the bright hues which the bird assumes when in perfect plumage. In pursuing my investigations on the change of plumage in birds, I have observed that old birds, moulting in the spring, usually receive their bright colour at once, whilst many species that moult in autumn have these colours imparted to them by a gradual process.

Psittacus coccinocephalus, Scarlet Headed Parrot; *Psittacus purpureus*, Purple Bellied Parrot. I have had in my possession for several years past a living bird of the first named species and frequently seen one of the latter in the cage of a friend. Both species moult in summer and again in the months of February and March. The moult, with the exception of the larger feathers in the tail and wings, was as general in spring as in autumn.

Emberiza oryzivora, Reed Bird. This species, which is so singular in its habits and changes, passes under different names in various parts of the country, referring either to its colour, food or song. In the eastern states it is usually called Bob-Link, from its notes bearing a fancied resemblance to those syllables. In New York I have heard the farmers call it Skunk Bird in consequence of its black breast and sides and broad white streak on the back of bearing a resemblance in colour to that animal (*Mephitis Americana*). In the middle states it is called Reed Bird in consequence of its feeding on the seeds of that plant (*Zizania*); and in the southern states the Rice Bird from the extensive depredations it commits on our rice fields; hence its specific name (*Oryzivora*).

In autumn, the males of this bird, as is well known, lays aside its black and white summer dress, not by a change of colour in the old feathers, but by a thorough moulting, extending even to the large feathers in the wings and tail. It now becomes of the plain colour of the female. In this dress it continues till March, when it gradually changes again, and in May is once more in full summer plumage. Mr. Ord states, from personal observation, on birds confined in cages, that "during the time the male is undergoing this metamorphosis, there is no change of feathers, their colours being altogether the results of their organic secretions." I have no positive evidence to prove that there was any inaccuracy in his observations. These birds generally make their appearance in Carolina from the 1st to the 10th of May, when they have already attained full plumage. Having never kept them during winter and spring, I am obliged to consider the examinations of Mr. Ord in reference to this species as conclusive. This will, then, show that the Rice Bird is an exception to the general rule, and that, whilst all our other known species that assumed to distinct colours in a year moult in the spring as well as in autumn, the Rice Bird is an instance where a different process takes place. This, however, affords no proof of the truth of the assertion that all birds moult but once a year, since it is now certain that the majority of birds moult twice a year. I do not however consider it impossible that even in regard to this species some mistake may have occurred. These changes progress silently and rapidly, and unless watched with great care are effected without our observing the process. The males made their first appearance this year in our rice fields in scattered flocks on the 5th of May; and on that and the following days I had an opportunity through the politeness and attention of several planters of examining fifty specimens. Nine-tenths of the birds were in full plumage, and there was no appearance of their having moulted; but in five or six specimens, where the black colour on the breast had not been fully restored, I perceived several places where rows of young feathers still sheathed existed in spots of an inch or two in extent. These feathers, instead of coming out yellowish brown, as is the case with this species when they moult in autumn, were black, narrowly edged on the points with brown. Whether the moult was accidental in the species or whether it extends to every individual in the species, I had no means this spring of ascertaining; certain it is, that the young feathers perceptible on the specimens sent with this communication, do not undergo the process of the change from brown to black, which must be the case if the colours of spring are imparted without a fresh moult.

Sylvia æstiva, Blue-Eyed Yellow Warbler. This species, in autumn, is of a pale yellow colour, changing in spring to bright yellow, streeked on the breast with orange. It probably undergoes this change within the tropics. On the 27th of April I had an opportunity of examining eleven specimens; two of these were not in full plumage, and I observed that in one instance the whole head was covered with new feathers, and in another the young feathers on the back had recently been produced and were still in closed in a tube.

Fringilla cyanea, Indigo Bird. This is a rather rare species in the maritime districts of Carolina and breeds but sparingly in these parts. On the 3d of May I obtained three male Indigo Birds. They were in full song, and I was enabled to discover them by their notes. One of the birds was in perfect plumage, and I could observe no evidences of a moult; the other two were mottled, and the points of the feathers were edged with brown. They were receiving new feathers and extensively on every part of their bodies; one of the birds would probably have been in full plumage in a week, the other was less advanced and would have required a longer time. Whether these were young birds receiving their bright colours for the first time or old birds renewing their colours by a fresh moult, I could not ascertain.

The arguments of Dr. Flemming (vol. II., pp. 26, 27) in favour of his theory that a spring moult is unnecessary to a change of colour in feathers, as confirmed by the examination of Captain Cartwright on the Ptarmagin (*Tetrao lagopus*) of Labrador, is very far from being satisfactory to my mind; on the contrary, it appears to me that Cartwright's own notes, taken on the spot, ought to have caused the Doctor to have hesitated before he adopted so confidently a theory which may yet be discovered on a more careful examination to be wholly founded in error. The following is an extract from the Captain's Journal (see Transactions, On the Coast of Labrador, vol. I., p. 278): "when I was in England, Mr. Banks [Sir Joseph Banks], Dr. Solander, and several other naturalists, having inquired of me respecting the manner of these birds changing colour, I took particular notice of those I killed and can aver for a fact that they get at this time of the year (September 28) a very large addition of feathers, all of which are white, and that the coloured feathers at the same time change to white. In spring, most of the white feathers drop off and are succeeded by coloured ones or, I believe, all the white feathers drop off, and they get an entire new set. At the two seasons they change very differently; in the spring beginning at the neck and spreading from thence; now they began at the belly and end at the neck." Captain Cartwright here asserts that the Ptarmagin, in autumn, receives a very large addition of white feathers, but that the coloured feathers are changed to white. I perceive no difficulty in explaining this autumnal change. Presuming that the birds moulted in the middle of August, as is the case in Labrador; some of the feathers would not come out pure white, but would gradually become so, as is the case with other birds I have mentioned, and they would thus for a time retain a mottled appearance. The summer moult in birds extends over the whole body even to the wing and tail feathers in every species I have examined, and I presume that the Ptarmagin does not form an exception. The only point of difficulty lies in the spring moulting, and in this Captain Cartwright is pretty explicit. He states with some hesitation his belief from an examination of many specimens that "all the feathers, in spring, drop off." There could have been no difficulty in ascertaining this fact from the manner in which young feathers are sheathed during the process of moulting. Here, then, we have another well attested proof of a double moult in those species of birds that are subject to a semi-annual change of colour.

Before concluding my remarks on land birds, it may be necessary to state how far I have found my observations to apply, in regard to a second moult in birds in general. I have examined no species in which an individual was not occasionally found that was not moulting sparingly during spring. This may, however, in some species have been accidental. The Mocking Bird, Blue Bird, Cardinal Grosbeak, Loggerhead Shrike (*Lanius ludovicianus*), and some others, I have reason to believe do not moult to any extent in spring, although a considerable number of feathers drop off which are not replaced; and yet in many other species of the genera *Turdus* and *Fringilla*, even where no very great change of colour occurred, the moult extended to every part of the body; whilst the Mocking Bird and the Brown Thrush (*Turdus ferrungineus*) did not appear to moult. The Solitary Thrush (*Turdus minor*), which loses nearly all its spots on the breast in spring, seemed on the 29th of March to have acquired a full set of young feathers on the breast.

Although in many species the moult appeared to be complete in other respects, yet in but two species, the Savannah Finch and Swamp Sparrow, did I discover that the scapulars and tail feathers were moulted; these are stronger and firmer, and as far as my observations have gone, seem in most of the species to be shed but once a year; this will not surprise us when we take into consideration the diversity in the operations of nature in other animals. The horse, for instance, sheds his hair on every part of the body at least once a year, and yet the hairs in the mane and tail continued to grow during the life of the animal. On the other hand, the thinner and lighter hairs of animals and feathers of birds are frequently removed and easily replaced. The feathers on the breast of female Canary Birds and most other species that I have noticed drop off during incubation and are successively replaced during the season at every time they recommence building their nests.

On the spring moulting of water birds the following information has been collected.

In the spring of 1825 a specimen of our beautiful Fresh Water Rail (*Rallus elegans*) was sent to me from the country. It was so completely in moult that it could not be preserved as a specimen. The young feathers came out in their full brightness.

On the 7th of April 1835 I examined several Shearwaters (*Rhincops nigra*). About one half of the birds were in the moult; the remainder still retained their winter dress and colours.

On the 20th of April 1836 I procured three males of the Semi-palmated Snipe (*Totanus semipalmatus*). They were changing from winter to summer plumage and were moulting very extensively.

Early in the month of May 1833 I saw a large string of Black Bellied Plovers (*Charadrius helveticus*) in the markets of New York, and being desirous of obtaining some specimens for the cabinet, as this bird has not arrived at full plumage in its spring passage along the sea shore of Carolina, I found so many of them in fresh moult, and their young feathers still sheathed that it was difficult to find specimens suited to my purpose.

In September 1835 a Turnstone (*Strepsilas interpres*) was sent to me by a friend. It had been wounded and captured by a son whilst in its plain autumnal plumage. I entrusted it to a lady who fed it on moistened cornmeal and on bread soaked in milk. It survived the winter and was in fine health on the following April, when it shed its winter dress, and the new feathers, although not perfectly bright at first became so in three weeks.

March 20th, 1837, saw about two dozen of the Ruddy Duck (*Fuligula rubida*) in the Charleston market. All that I examined were males. They were receiving their ruddy colour of spring, and were moulting very extensively.

March 21st, examined a Black Headed Gull (*Larus atricilla*). On every part of the head the ash coloured feathers were becoming replaced by young ones, still sheathed; these were of a dark lead colour, appearing in some lights quite black.

March 1834, received from Boston specimens of the Herring Gull (*Larus argentatus*), together with a number of northern species, which are now lying before me. In the majority of them, moulting had progressed to a considerable extent. Mr. Ord conceives Montagu, who had made similar observations on the species, as "labouring under the influence of a theory," when he recorded the result of his investigations. The correctness of Montagu's assertions has been verified by my own observations.

Mr. Ord has, moreover, doubted the accuracy of Montagu's observations on the changes of plumage in a Black Stork taken in England. The latter gentleman stated that this bird continued very gradually to moult during summer and winter; that in the month of March, the violet and purple feathers appeared on the back and that the whole upper parts had nearly assumed this beautiful plumage by the 1st of April. He supposes that it could not have moulted, although Montagu asserts that to be the fact, and considers this statement of Montagu as "affording one of the most apposite illustrations of the fact of a change of colour in mature plumage that could well be desired." I have not been able to regard it in this light. Admitting that all the species of Storks and Herons moult but once a year, of which we have no certainty, it does not appear difficult to account for the slow pace with which this individual assumed its bright colours. It will be recollected that it was a foreigner retained in captivity during winter in the moist, cold climate of England. Had it been left at liberty to pursue its migrations to Africa, where the species hyemates, it would probably have sooner attained full plumage. It is, moreover, highly probable that this stork was changing from the young to the adult plumage; in this case the process was similar to that of the *Plotus anhinga*, referred to in a former part of this paper, and of the Blue

Grosbeak now in my possession; the latter having been gradually moulting and receiving its bright colours during the winter and spring. Young birds always shed and renew their feathers more slowly and irregularly than the old; and in this instance, where we have a process so similar in other species, we cannot avoid giving credence to the declarations of a naturalist, who was an eyewitness of the facts he describes in preference to the mere conjectures of another.

Although it has been more my intention in this paper to record the facts which I have been able to collect on this subject than to frame a new theory or endeavour to restore an exploded one; yet it has often occurred to me that the advocates of the two opposite theories have run into extremes in consequence of the adherents of each with too much pertinacity to a preconceived opinion. Hence Montagu, having ascertained that some birds are subject to two moultings in a year, conceived that this was the only mode by which the colours of feathers were changed, asserting that "he had no conception of the feathers themselves changing colour." On the other hand, Dr. Flemming and those of his school, including the Rev. Mr. Whitear and Mr. Ord (the latter advanced a step beyond all the rest, in stating as his opinion that "all birds moult but once a year"), having observed that the colours in feathers change in the early stage of their growth, came to the conclusion that the same process must always be carried on in the old feathers and that a fresh moult being unnecessary would consequently not take place.

The observations which I have made thus far have led me to conceive that the truth lies between these two extremes. Although my investigations were not extended to as great a number of species as I could have wished or repeated through a succession of seasons, it has, notwithstanding, been very apparent to my mind that young feathers frequently change colour, particularly in autumn; but that in those cases where there is a semi-annual change of colours, there is in all or nearly all the species a semi-annual moult. That a much greater number of species change their feathers twice a year than is usually supposed and that our inquiries must be directed to individual species, rather than to genera, since in the same genus one species is subject to a double moult, whilst the others moult but once a year.

4. Having now seen in what manner nature performs her operations

in effecting the changes in the plumage of birds, I proceed to inquire whether the same laws may be applied to those variations in colour to which the hair of quadrupeds is subject. The vascular bulbs in which the roots of the hairs are inserted bear in analogy to the bulbs which contained the rudiments or sheaths of the feathers. The hair, like feathers, is nourished and receives its colour from the secretions of the animal and is exposed to the same sun, air, and moisture. It continues to grow during a certain period and drops off when it has arrived to full maturity, as do the feathers from the bird and the leaf from the tree. There is also a striking similarity between the covering of birds and quadrupeds in the changes to which they are subject from the young to the adult state and the mutations of others during the different periods of the year. As the majority of the species of young birds are of the colour of their parents, so are the young of quadrupeds. Some young birds remain for a length of time of a very different colour from that which they assume when they arrived at maturity, as is the case with some of the species of our *Fringillæ*, *Ardeæ*, and other genera. This is also apparent in some of our quadrupeds. The young of the Deer and Cougar (*Felis cougar*), for instance, are striped or spotted with white and red whilst their progenitors are of a uniform dun or fawn colour. Some birds are disguised into such opposite sides of colours in the course of the year that the gay visitor of spring can scarcely be recognised under his homely dress in autumn. So also our Ermine and two at least of our species of Hares are brown during six months of the year and of a snowy white pelage during the other six months. Bearing so strong a similarity in many other particulars, we are led to believe that in the process of moulting or shedding of the hair, this harmony of nature still continues to prevail.

Here, however, the same difference of opinion has obtained among naturalists. Probably much has been written on the subject in European journals which has not fallen under my observation. Dr. Flemming (see Philosophy of Zoology, vol. II., p. 24) advances an opinion, the result of personal observations, that "the change of colour in those animals which become white in winter is effected, not by a renewal of the hair, but by a change in the colour of the secretions of the *rete mucosum*, by which the hair is nourished or perhaps by that secretion of the colouring matter being diminished or totally suspended." This theory, as far as

I have been able to ascertain, has generally been adopted by physiologists. I have also noticed the remarks of Dr. Richardson on the changes in the colour of the American Hare (*Lepus Americanus*) in his Fauna Boreali-Americana, where he expresses a belief, founded on examination of many specimens, that "the change to the winter dress is produced, not by the shedding of its hair, but by a lengthening and blanching of the summer fur."

From these opinions I am obliged to dissent for the reasons already advanced, and the evidence I shall proceed to adduce.

There are but four quadrupeds yet found in our country, the Polar Hare, Northern and Prairie Hares (*Lepus pampestris*, Bach.), and the Ermine, in which these mutations are very striking. Let us now examine this peculiarity in some of these and one or two other of our quadrupeds, and we shall be able to judge how far this theory is calculated to maintain its ground among naturalists.

Lepus glacialis, Polar Hare. I have had no other opportunity of becoming acquainted with the changes of colour to which this fine Hare is subject than that afforded by a specimen which was kindly presented to me by Audubon; but this in itself affords sufficient evidence on which an opinion may be grounded with safety. The animal was purchased in the flesh by our distinguished American ornithologist from an Indian at Newfoundland on the 15th of August 1833. At that early season, then, in the cold regions of the north, this change from its summer to winter colours takes place. The specimen before me is in that interesting stage when the summer fur had commenced dropping off, and the white winter dress was fast advancing to resume its place. This, as far as has been ascertained, is the only specimen, in summer colour that exists in any collection. The specimens brought home by Dr. Richardson, Captain Parry, and the other adventurous explorers of our polar regions were all in the white pelage of winter. Its summer colour is grayish brown above with conspicuously black ears. In winter the hairs all become snowy white even to the roots. In the specimen now before me there is a large spot, nearly a hand's breadth, of pure white on the back extending nearly to the insertion of the tail; three or four white spots of about an inch in diameter also exist on the sides. The hairs forming these spots are shorter than the surrounding fur; a few longer hairs of the summer dress are still interspersed, which had not yet dropped off. The short white hairs are, in several places, seen pushing forward whilst the surrounding ones seem to have been thinning and falling. This, then, is undoubtedly the process of the change of colour in this large and interesting species.

Let us now examine how far other quadrupeds, subject to the same mutations, differ from the above in changing from the brown dress of summer to their white clothing of winter so much in unison with the snows around them, and which by concealing them from the view of a host of enemies is often the cause of their preservation.

Lepus Virginianus, Virginian or Northern Hare. This Hare, which is an exclusively northern species and not a resident of Virginia, seems to have been not only improperly named, but very imperfectly described by our naturalists. The habits, however, of quadrupeds and birds are, in general, only as alluded to in this article so far as it may enable us to throw some light on the subject now under discussion. I possessed favourable opportunities of witnessing the semi-annual changes of colour to which the *Lepus Virginianus* is subject; having in early life, whilst residing in the state of New York, had several of this species in a state of domestication, where they produced and reared their young. The notes made twenty-two years ago were mislaid, but I have a pretty distinct recollection that the following was the process. The animal shed its white fur in spring. The hair, although yellowish and soiled by age and exposure, indicated no appearance of a change of colour from white to brown before dropping off. The new hairs came out reddish brown, in which dress it continued till autumn, when the summer fur gradually dropped off, and the hairs composing the winter pelage became visible through the rest. In four weeks this summer dress had entirely disappeared. The new hairs did not, however, appear pure white, but of a light iron gray colour mixed with occasional white and black hairs. Gradually the hair grew longer and seemed to become whiter till in the course of a few more weeks, the change was complete. In this case, then, nature seems to pursue the same process as in effecting the changes of colour in some birds, by a gradual blanching after the moult. It will be observed that in this species that hairs are only white although broadly so at the points and not throughout their whole extent as in the *Lepus glacialis*.

Lepus Americanus, American Hare. The changes of this Hare I have also had an opportunity of witnessing in a warren. It cast its hair in the spring and became of a yellowish brown colour. In the autumn it again commenced shedding. Whether all of the hair dropped off or not I cannot say with positive certainty; new hairs, however, were continually adding; the points of these were white as they came forward. In the old hairs I could perceive no change. There were many black ones interspersed, but whether these were of a new or of a former growth, I had no means of ascertaining. I felt confident, however, that the light colour of winter was produced by the new hairs it had received in autumn. This Hare has by some of our authors been described as becoming white in winter. It should be observed, however, that the points of the hairs are so narrowly tipped with white, and the markings of brown and cinereous still so visible that it can in no part of the northern United States he described as a white.

Lepus palustris, Marsh Hare. For a description of this species, see Journal of the Academy of Natural Sciences, vol. VII., and an engraving in Audubon's Birds of America, vol. IV., pl. 366. This singular and almost aquatic species sheds its hair twice a year, as I have had an opportunity of ascertaining from having had one in confinement. Although much hair dropped off in autumn, I found it, however, difficult to satisfy myself that this change was as thorough as that in the spring. In the beginning of winter, the points of its hair instead of growing whiter, as in the American Hare, grow darker until they have become nearly black, thus proving that the effort of nature is to change the colours from brown to pure white in some species and to black in others.

Mustela erminea,* The Ermine. This animal, usually called Weasel in the northern states, is brown during six months of the year and in winter becomes white with the exception of the tip of its tail, which is black.

---

* Godman has stated, on the authority of Charles L. Bonaparte (Nat. Hist. vol. I., p. 193), that our common Weasel (*M. vulgaris*) has been proved to be the Ermine in the summer pelage. I had an opportunity of ascertaining from actual examination that there is some inaccuracy in this statement. I preserved several of both species in the same cage during a winter in the northern part of New York. The Ermines became white in autumn although some of them were still young and not more than two-thirds grown. The other species retained through the winter their brown colour. Richardson states that the latter

I was not aware until this paper had been nearly written that any one had published an account of a particular examination of the Ermine during those periods when it is subject to change its colour. I am indebted to my friend Professor Moultrie for a reference to a paragraph in vol. II., p. 24 of Flemming's Philosophy of Zoology, where the following remarks occur.

"The appearances exhibited by a specimen now before us are more satisfactory and convincing. It was shot on the 9th of May 1814, in a garb intermediate between its summer and winter dress. In the belly and all the other parts, the white colour had nearly disappeared in exchange for the primrose yellow, the ordinary tinge of these parts in summer. The upper parts had not fully acquired their ordinary summer colour, which is a deep yellowish brown. There were still several white spots and not a few with a tinge of yellow. Upon examining those white and yellow spots, not a trace of interspersed, new, short, brown hair could be discovered. This would certainly not have been the case if the change of colour is effective to buy a change of fur. Besides, whilst some parts of the fur on the back had acquired the proper colour, even in those parts, numerous hairs could be observed of a wax yellow, and in all the intermediate stages, from yellowish brown, through yellow, to white.

"These observations leave little room to doubt that the change of colour takes place in the old hair, and that the change from white to brown passes through yellow. If this conclusion is not admitted, then we must suppose that this animal casts its hair at least seven times in the year. In spring it must produce primrose-yellow hair; then hair of a wax yellow; and lastly of a yellowish brown. The same process must be gone through in autumn, only reversed, with the addition of a suit of white. The absurdity of this exposition is too apparent to be further exposed."

---

species also becomes white in high northern latitudes. This is certainly not the case in lat. 45°. There is another peculiarity which I had occasion to notice. Whilst the Ermine is much abroad during winter, its footprints appearing every where on the snow, the common Weasel is rarely, if at all, seen during that period. A large brood that had made the root of a tree their residence in winter were seen around this retreat almost every day during autumn till the ground became covered with snow. They could no longer be traced till the snows began to melt, when their holes were again opened, and they were seen as usual. Our *Mustela vulgaris* ought to be carefully compared with that of Europe, which goes under the same name.

This examination and the arguments which have been drawn from it by so able a naturalist as Dr. Flemming would appear at first sight to be conclusive. But on a closer investigation of the subject and a more careful inquiry into the operations of nature, doubts will arise both as to the accuracy of his investigations and the soundness of his theory. The Doctor's observations were made, it will be recollected, on the 9th of May, the very period when the Ermine is shedding its winter fur. A sensible writer in Rees's Cyclopœdia, on the article Hair, makes the following remarks. "As the pulp is intended for the nutrition of the hair, it is found to extend only to that portion of the hair which is in a state of growth; and in those which are deciduous, or are cast at particular seasons of the year, such as the hairs covering the bodies of quadrupeds, the pulp becomes entirely obliterated for the period of shedding the hair, and its root is converted into a solid pointed mass." If, then, the change as Dr. Flemming contends had taken place in the old hair, so much in opposition to the views of the writer above quoted, what must have been the process? In the summer, after it had cast its old hair, it must have acquired a coat of wax yellow, which in process of time changed to yellowish brown; then in autumn, this same hair would have turned white; now comes the spring, and even in this advanced age of the hair, it once more turns yellow and then yellowish brown. No harlequin assumes greater changes; and it may be inquired whether the Doctor's theory is not even more absurd than that which he assails with such apparently strong facts and powerful arguments.

It has, however, appeared to me that the changes in colour to which the Ermine is subject have not been closely observed, or accurately described. Admitting that this quadruped sheds its hair in spring and autumn, which I hope to prove is the fact, it will be discovered that in this respect the change is not more remarkable than that produced on several other species of quadrupeds and on many birds; among the latter, as instances, we may mention the Yellow Ground Warbler, Rice Bird and American Goldfinch.

At an early period of life I had an opportunity of witnessing the change in the pelage of the Ermine from white to brown. The following is the only memorandum of this occurrence that I can find in my diary. "April 17, 1814. My ermine has become a little tamer, but all its beautiful white hairs are dropping out, and it begins to look like a brown Weasel." I have endeavoured in vain to conjecture the possibility of my having in some way been deceived in my observations. The notes were made, I acknowledge, when I had no knowledge of natural history, and had read no work on the subject; but, on the other hand, I had no favourite theory to support. That the Ermine shed its hair at that early period, I feel confident, from another circumstance. I was in the habit of combing out its white hairs, which were continually falling off, and on one of these occasions it inflicted a wound with its teeth on my hand. Admitting, then, that my observations were correct, and they happened by a singular coincidence of circumstances to be made the very year and within a few weeks of the time when Dr. Flemming examined the same species, it will not be difficult to ascertain in what manner he had been deceived. The new hair, both in the Ermine and the Northern and Polar Hares, comes out in spots, sometimes only of a few inches in diameter. It grows so rapidly that it appears to attain its full length in less than a week. In the meantime the surrounding hair may not yet have commenced shedding and remains of its former colour. This may account for the white spots on the specimen examined by Dr. Flemming, which in all probability was a prepared skin. These parts of the animal had not yet moulted and therefore had undergone no change. With regard to the wax-coloured hair of which he speaks, they must have been the old, faded and soiled hairs of winter, as any one may easily ascertain by examining the stuffed skin of an Ermine. The white pelage soon assumes a yellowish cast.

In the autumn of 1823 I had an opportunity of witnessing the change of the Ermine from brown to its snowy mantle of winter. It had been brought to Charleston by an itinerant showman. The cage, I observed, was every where strewed over with brown hairs that had dropped off. The young hairs were white in appearance, but whether purely so at once, or became more blanched as the season advanced, I had no means of ascertaining, as the Ermine was wild and vicious and made formidable opposition to an examination. The man raised his price upon it as soon as he ascertained that it was becoming white.

If I were not deceived in my observations, it will, then, appear that the Ermine has only two instead of seven colours in a year. In the spring it casts off its white coat and becomes brown; the hairs may be a little lighter at first than they become a week after, as is the case with those of other quadrupeds. In autumn these yellowish brown hairs drop off and are replaced by the white fur of winter. In investigating the changes to which this animal, which so seldom comes under the inspection of the naturalist, is subject, we may derive some information from this process of nature in regard to species with which we are more familiar. If the old hair of the Ermine changes from white to brown, in spring, previous to its shedding, we may present that a similar process will be observed in the Virginia Deer. A pair of the latter animals are now shedding their hair under my daily inspection. The long, gray hairs of winter are continually dropping off; they have not undergone the slightest change. Reasoning from analogy, on the principle of Dr. Flemming, these ought to become red, the colour of the hair of the deer in summer. This, however, is not the case in this species, nor in any other with which I am acquainted.

Although I feel a tolerable degree of certainty that I was not deceived in my observations on the Ermine, yet as they were not made in reference to this mooted point, I have a strong desire to have the subject further investigated by some scientific naturalist.*

---

* The Ermine is easily captured in a box trap and can be fed on any kind of fresh meat, although it prefers birds and mice. The old, which I had at different times in confinement, although they did not seem to suffer in their health never became reconciled to captivity. One of this species, however, taken when about five months old, appeared in a few weeks to have lost all its wildness and ferocity, leaving the cage at my call, following me about my study, and taking food from my hand. It was occasionally let loose in the out houses and barns, where it made fearful havoc among the rats and mice. I observed it did not seize a rat, as this species is wont to do when it attacks poultry, by the neck, but pounced upon it suddenly, sinking its teeth into the skull, and then leaping off a few feet (as if to avoid being bitten), leaving it to struggle and die, without any further effort on the part of its enemy. When it had killed a considerable number, it was in the habit of dragging them on a heap, and covering them with straw. On these occasions it showed some reluctance to return to its place of confinement. After having tasted the sweets of liberty, it would often conceal itself for a day or two in the neighbourhood of its prey, and the calls of hunger alone would bring it back again to the house. On an occasion of this kind it disappeared, and it was supposed had been killed by a dog. It was not again observed by the family for six months. On my return from a college vacation, I once more occupied my former chamber. During the night I found some small animal creeping among the blankets, and on procuring a light ascertained to my great surprise and pleasure that my long-lost pet had come as if to greet my return. It had, in the mean-

How far other quadrupeds in a state of nature are subject to this biennial shedding of the hair is an inquiry which we must make not from books, but from a careful examination of the different species. The generally received opinion of quadrupeds shedding but once in a year, should be more carefully examined in regard to some of the species before it is fully adopted. The rule may admit of more exceptions than is generally supposed. The Deer (*Cervus Virginianus*) sheds its hair in Carolina in the month of April, and sometimes earlier. These, till autumn, are of a reddish colour, when the animal assumes a bluish dress, gradually fading to grey. In what manner is this change produced? The Harvest Mouse (*Mus leucophus*, Raff.) is a fawn colour above in summer and bluish ash in winter. The black variety of the common Gray Squirrel (incorrectly referred to *Sciurus Carolinensis*) is much more brilliantly black in winter than in summer. Can these old and faded hairs receive an additional blackness and brightness as the cold weather approaches? The fur of quadrupeds is thicker and warmer in winter than in summer, and if they do not shed their hair, they must at least receive an additional covering in autumn; and the same effort of nature which can produce a part, may restore the whole. Possibly it may on a more careful examination be found that the new hairs received by these and other species in autumn may so predominate as to conceal in part the rest. I can easily conceive that hairs may become redder under the influence of a burning sun; but that they should, after six months, again change and become plumbeous cannot be credited without a minute examination of facts. Our domesticated animals can afford no criterion by which we may judge with certainty of all other quadrupeds. The Horse, although it shares its coat freely in spring, is irregularly losing and restoring its hair throughout the year. A change in food or in health is continually accelerating or retarding this process.

---

time, changed its colour and was beautifully white. How long it had kept possession of the vacant chamber is unknown. I perceived it had found egress through a hole in the hearth. It was from this time no longer placed in confinement, but permitted to take its own course, as it had committed no depredations on the poultry. It attached itself to the premises for about two years afterwards, having formed a large nest of straw under the covering of a globe. In this situation it would sometimes live for two or three days as if dozing and disinclined to take exercise; at other times it was absent for several days in succession. It finally disappeared, from what cause I was never able to ascertain.

If some apology be requisite for these somewhat desultory remarks and for having entered so much into detail, it may be offered in defence that the subject on which it treats, although apparently trifling and unimportant, leads to many inquiries in philosophy and in natural science. In investigating the changes from brown to white, in some quadrupeds, we are led to inquire into the cause of this change. If it be answered that it is the effect of cold, the question will naturally arise, Why does not cold produce the same effect on other animals? Why do not the Fox, the Raccoon and the Bear residing in the same neighbourhood also become white in winter? Besides the change commences in the heat of summer. The Polar Hare, brought by Audubon, was undergoing the change on the 15th of August. The Ermine was becoming white in the middle of October in Carolina when the weather was still very warm. The Emperor of China is said to have preserved the *Lepus variabilis* in the warmer parts of his dominions, and even there it was subject to become white. This colour, then, rather anticipates than succeeds cold weather, and it would seem as if there was some constitutional predisposition of the animal to the change. A further examination into these mutations of colour to which birds are subject may extend our knowledge of physiology in regard to the development and growth of feathers, and the process by which their colours are imparted to. Although it is generally admitted that there is no circulation in hair or feathers, still this point does not appear to be fully determined. Bichat* supposed that there was a species of circulation in the interior substance of the hair by which he endeavoured to explain the changes of colour. This opinion, however, has been contradicted by modern physiologists who supposed that hairs to be constituted by a colourless, transparent epidermic or horny sheath, filled with a species of coloured pulp; and they contend that the hair is not, properly speaking, caniculated, the colouring matter being deposited by the papilla at the same time as the epidermic sheath.** The chemical composition of feathers is believed to be nearly the same as that of hair, nails, &c. consisting principally of inspissated albumen, united with small portions of gelatine and animal oil; and it is a subject of inquiry what chemical action takes place, which imparts colours

* Anatomie Générale, Tom. II., p. 788.
** Cruveilhier, Anatomie Descriptif, Tom. III., p. 420.

not only to the shafts of the quill, but to the barbs and even the minute barbules attached to these barbs. Are these changes in colour effected by the action of the external atmosphere, or by a vital operation? As the air contained in the feathers is exposed to the influence of the vascular pulp, may it not in this way produce a chemical action on the colour of the feathers? Some birds certainly moult twice a year with the exception of their wing and tail feathers, which in most species are only cast annually. As the latter are longer in coming to maturity, may not colours be imparted to them more slowly, and may not this account for the changes of colour which are progressively taking place in the bars on the tails of Hawks and other species?

These inquiries may be also attended with beneficial results to the science of ornithology and enabling us to discriminate the true from merely nominal species. By a little attention to this subject, Wilson was enabled to expunge from our nomenclature of birds many species which Catesby, Edwards, Pennant, Latham and others had multiplied to a great extent, and to extricate the science from difficulties which were continually leading the student of nature into infinite doubts and perplexities. Through this mean also, Bonaparte, Nuttall, and Audubon corrected some of the errors into which their predecessors from a want of opportunity for a fuller investigation of the subject had fallen. The work is not yet completed. By a further inquiry into the changes in plumage to which birds are subject, species that are now supposed distinct may be found to be identical. Many new species of Gulls that have been multiplying in Europe and America may prove to be the young or the winter plumage of species that have long since had a name.* The long disputed, often rejected, and as often readmitted Winter Hawk (*Falco hyemalis*)

---

* This genus requires a more careful revision. It will be found, on an attentive examination of the changes of plumage to which the species are subject, and on a comparison with those of the eastern continent that our species are unnecessarily multiplied and that some that are described as identical with those of Europe are distinct. *Larus minutes, Larus canus,* and *Larus fuscus,* as given in Bonaparte's Synopsis, will probably be found not to exist in the United States. Our species, which has gone under the name of *Larus minutes,* it is, I am inclined to believe, the *Larus Bonapartii* of Richardson; the *Larus capistratus,* the young of *Larus Bonapartii.* The *Larus argentatoides* of a young of *Larus marinus;* and the species which our authors have considered as *Larus Canus* is probably the immature bird of *Larus zonorhynchus* of Swainson.

might, on being preserved in confinement for a single year, be placed where he ought to be; probably by the side of the Red Shouldered Hawk (*Falco lineatus*). The Connecticut Warbler (*Sylvia agilis*) might claim identity with the once rejected, but now admitted Mourning Warbler (*Sylvia Philadelphia*). The Autumnal Warbler (*Sylvia autumnalis*) that is abundant in autumn, but has been sought for in vain in spring might then perhaps find a father in the Black Poll Warbler (*Sylvia striata*). Roscoe's Yellow Throat of Audubon might come to claim relationship with the *Sylvia trichas*; his *Sylvia Childrenii* might be found to be only another name for *Sylvia æstiva* and his *Sylvia Vigorsii* for *Sylvia pinus*. The difficulty in preserving several of the genera of birds in confinement for the purpose of studying their habits has been considerably overrated. Granivorous birds, as it is well known, may be fed on seeds, and in this manner thousands are annually brought from Africa and the East Indies. Warblers, Fly-Catchers, and even Swallows, the most difficult (except the Humming Birds) to preserve, are now kept in cages throughout the year in London, Paris, and especially at Rome. They are fed on vermicelli, and occasionally on chopped meat. A regular temperature in the room where they are confined is preserved throughout the winter. The various species of Ducks, which are easily caught in traps or nets, soon accustomed themselves to the food of the poultry yard and become domesticated. By this means several of our wild species may be made to minister to our pleasures and comforts. The Mergansers, Gulls, Lestris, Procellarias, Anhingas and Cormorants may be fed on fish, their natural food. The Sandpipers, and as in the instance of the Turnstone already noticed, may be preserved by being fed on various kinds of soft food easily to be procured. I have seen the Ruff of Europe (*Tringa pugnax*)*

---

* This species has recently been added to our ornithology by Mr. Nuttall in consequence of a fine specimen having been obtained in the neighbourhood of New York. I doubt whether we have a right to claim it as American. Birds well known on the eastern continent unless they exist in high northern latitudes ought to be admitted with great caution into our Fauna. The European Partridge has been killed in the middle states; no doubt it had escaped from the cage. I obtained some miles from Charleston a male Chaffinch (*Fringilla cœlebs*) in full song and afterwards saw its imported mate in a cage in the market. I had for some years in possession a European Turtle Dove (*Columba turtur*), which had escaped from confinement, and it was afterwards ascertained had flown on board of a vessel at sea 300 miles from the coast of France; and I received

for sale in a cage in the Charleston market. It was fed on soaked ship biscuit on its passage from Liverpool.

On revising this article, which has insensibly grown upon me as I proceeded, and which has extended to a length not originally anticipated, it has occurred to me that it might be construed by some as evidencing a disposition to cavil at the writings and undervalue the labours of the able and estimable naturalists of our country. I should reproach myself with ingratitude, could I for a moment conceive that I had been influenced by such unworthy motives. The number of American naturalists has been exceedingly limited; their labours have been great, and poorly requited in a pecuniary point of view or in what they regard of most value, fame. We are indebted to them for those exertions and sacrifices which have removed the obstructions in the path of science and have tended to interest, enlighten, and instruct us. Their errors were the result of circumstances beyond their control; and the best service we can render them or their memories is to supply the materials which will render their works immortal. In our researches after truth, in sci-

---

two years since, from my friend Mr. Nicholson, a European Kestrel (*Falco tinnunculus*), which had a lighted on the rigging of the ship several days' sail from Liverpool on his passage to America. Surely the species brought to our country by force or accident cannot be claimed as belonging to us. It will be a source of regret, if in this respect, we are led to imitate the example of European ornithologists, who in order to swell their list of birds publish every foreign species escaped from the cage or driven on their coast by a tempest.

The admission of species from specimens obtained from museums on doubtful authority is a still greater evil. In this way Mr. Nuttall, who exercised so much knowledge and caution in his botanical works, has in his ornithology admitted species to which I am inclined to think we have but a doubtful claim. His Reed Bunting (*Emberiza schœniclus*) of Europe, for instance, is given on the authority of specimens presented to Audubon by a keeper of a museum, who stated their having been obtained at Harrisburg, Pennsylvania. I had an opportunity of examining the specimens; the materials with which they were filled, and the English holly bush on which they were fastened, betrayed evidences of their having come from across the Atlantic, ready stuffed and perched. Mr. Audubon finally came to the same conclusion. It is difficult to expunge a species once admitted into books. The Willow Wren of Catesby and a little Spotted Grey Sparrow of Latham have caused many a poor ornithologist to wear out his shoes in a fruitless search. It might be advisable to act towards these perplexing species as is done in some colleges, where a name, if not answered to after having been called a certain number of times, be stricken from the rolls.

ence, where the wisest are liable to error, it is not only admissible, but a duty which we owe to mankind to point out, in the language of courtesy and respect, the mistakes into which we conceive our superiors to have fallen. Our distinguished naturalists we should claim as a portion of our country's choicest wealth, and in them, as in the railroads which are spreading comfort and sociality over every quarter of our land, we should all strive to have an interest and hold a share, in order that we in our turn may pass along with greater safety and comfort.

Specimens elucidating some of the Points treated of in this Communication.

No. 1. *Lepus Americanus*; obtained on the 20th of October. Examined the animal in the flesh. It was shedding its hair. On the back and a spot on each of the sides, the reddish summer fur still remains. On the sides and near the tail of the lighter winter fur is visible. The difference of colour is sufficiently apparent.

No. 2. *Fringilla Savanna*; 23d of February. Moulting around the throat, and extensively on the neck, breast and back. The feathers on the breast coming forward nearly white.

No. 3, 4. *F. Savanna*; May 4. The fresh moulting nearly completed and the summer colours restored. In the specimens the moult had extended to the larger feathers in the tail and wings.

No. 5. *Fringilla tristis*, male; 16th of March. The summer feathers appearing bright yellow on the breast. Other specimens at the feathers coming out of pure black on the crown.

No. 6, 7, 8, 9. *Fringilla Pensylvanica*, White Throated Sparrow; in various stages of moulting. Obtained in March and April.

No. 10. *Bombycilla Carolinensis*; 23d of February. Young feathers appearing on the breast.

No. 11. *Fringilla cyanea*, male; May 3d. Moulting extensively. The Young feathers coming forward nearly in bright colours.

No. 12, 13, 14, 15, 16, 17. *Sylvia coronata*; obtained from the 6th of March to the 10th of April. Every individual I obtained at the latter date, both male and female, was undergoing the change.

No. 18. *Sylvia coronata*; May 1st. Summer colours nearly restored. The moult has been general.

No. 19. *Fringilla palustris*; April. All the specimens procured at this season were in extensive moult.

No. 20. *Emberiza oryzivora*, Rice Bird; May 6th. Moulting on the breast. The Young feathers coming forward nearly black.

No. 21. *Ardea cœrulea*, Blue Heron; female. This bird was obtained from the nest in June 1835; it became very gentle. On the following March he commenced moulting irregularly; all the new feathers were tinged, more or less, with blue. In July 1836, when a little more than a year old, it was accidentally killed.

Portrait of the Rev. Dr. John Bachman
by John Woodhouse Audubon
(Courtesy of the Charleston Museum,
Charleston, South Carolina).

Bachman's Warbler from Audubon's *Birds of America*.
General Research Division, The New York Public Library,
Astor, Lenox and Tilden Foundations.

Turkey Buzzard from Audubon's *Birds of America*.
General Research Division, Rare Books Division,
The New York Public Library, Astor, Lenox and Tilden Foundations.

Eastern Gray Squirrel from Audubon and Bachman's *Viviparous Quadrupeds*.
Rare Books Division, The New York Public Library, Astor, Lenox and Tilden Foundations.

Virginia Opossum from Audubon and Bachman's *Viviparous Quadrupeds*.
Rare Books Division, The New York Public Library, Astor, Lenox and Tilden Foundations.

Beaver from Audubon and Bachman's *Viviparous Quadrupeds*.
Rare Books Division, The New York Public Library, Astor, Lenox and Tilden Foundations.

Bachman's Hare from Audubon and Bachman's *Viviparous Quadrupeds*.
Rare Books Division, The New York Public Library, Astor, Lenox and Tilden Foundations.

Benefits of an Agricultural Survey

In 1843 the Literary and Philosophical Society asked Bachman to consider the need for an agricultural survey of South Carolina, and as usual, he considered the problem comprehensively. This essay is another major example of his approach to the solution of complex problems. He considered what similar surveys had accomplished in other states and what they were likely to accomplish in South Carolina, but most of his essay went beyond what he was asked to consider and dealt with more practical ways that agriculture had been improved elsewhere.

Bachman's knowledge of northern agricultural practices had been gained while growing up in farm country in New York State and during subsequent visits. Since arriving in Charleston in 1815, he became well informed about southern agricultural methods. While in Europe in 1838, he had paid particular attention to its most successful agricultural practices. He was well qualified to assess needs, to make comparisons, and to recommend improvements.

As Bachman indicated in this article, the state of South Carolina had recently provided funding for an agricultural survey equivalent to surveys that had been completed or were in progress in at least twenty other states. Bachman considered the advantages to be gained from geological, agricultural, zoological, and botanical surveys and argued against a comprehensive scientific survey in favor of one intended primarily to produce results that could be readily applied by farmers. He preferred that a survey be made by "a highly intelligent practical man to a purely scientific one, who is unskilled in the practical application of the laws of agricultural chemistry." While supporting the survey that was planned, he considered it inadequate. He pointed out the greater needs for an agricultural college, for agricultural societies, for widespread experimentation, and for inexpensive publications explaining how results could be applied.

Bachman reviewed the great effects that European agricultural experiments had produced: superior seeds, balanced soils, fewer pests. He contrasted the increasing yields in northern states with the declining yields in South Carolina. He warned against the exclusive planting of cotton and rice and the extravagance of importing food and nearly all manufactured products. He emphasized the need for diversity, crop rotation, and soil replenishment. Most of this address was about applying knowledge that already existed and about how to persuade planters to make use of the knowledge that could be provided. There was no question in Bachman's mind about the need for a survey, but he doubted that its findings would be any better applied than existing knowledge unless planters and farmers could be persuaded to become involved in local agricultural organizations and unless the needed information was made available in publications that were both inexpensive and directly relevant to agricultural practices.

Bachman considered the creation of an agricultural school to be much more important than the need for a survey. He expected graduates of such a school to "lead the way to all these other aids to our knowledge and success." In order to produce graduates able to make the needed scientific experiments, Bachman argued for a curriculum in which the following subjects would be well represented: chemistry to study soils and ways to improve them; geology to discover sources of fertilizer; vegetable physiology to understand "the structure and vital character of plants"; simple mathematics for computation and surveying; physics ("mechanical philosophy") sufficient for sound construction and for selection of needed equipment; heredity to improve breeds; botany to improve crops through a better understanding of types, "properties, and uses"; ornithology to preserve birds that destroy reptiles, insects, and the seeds of weeds; and entomology to understand the causes of many crop failures and to find preventive measures.

Bachman concluded by stating that until South Carolina began to restore its exhausted soil and adopted the most successful practices in widespread use, it would continue to lose population and the fertility of its land. His recommendations were considered so important that they were printed at the Society's expense to send copies to the state's governor and to every member of its legislature. His essay was reprinted in the *Southern Agriculturist* and in the *Southern Quarterly Review*.

An Inquiry / into the / Nature and Benefits / of an / Agricultural Survey / of the / State of South-Carolina. / by / John Bachman. / Charleston: / Miller & Browne, Printers and Publishers, / Old Stand, No. 4 / Broad-street. / 1843.

Introduction.

The writer of this Essay submits a few words of explanation in regard to the circumstances that induced him to prepare and finally send it to the press. He has the honor of belonging to a Literary Club, composed of a limited number of gentlemen from the different learned professions, who meet weekly at each other's houses in rotation, for the purpose of interchanging sentiments and promoting sociality. A subject for discussion is selected at one meeting which forms the topic of conversation on the next. The question for the evening of the 28th December was "what benefits may be derived from an Agricultural Survey of the State." The leisure of a rainy day had enabled him to collect his thoughts on the subject, and in part commit them to paper. The Essay was therefore prepared and read without the remotest idea of publication. At a subsequent meeting the Club, under an impression that it might afford some information on a subject which had so recently been agitated at Columbia, requested its publication, and that a copy be sent to the Governor, and to each member of the two Houses of the Legislature. He has yielded his assent in deference to the wishes of his literary associates, and especially to the solicitations and liberality of his friends, the Hon. D. E. Huger, and the Hon. Mitchel King.

An Inquiry, &c.

The Legislature of our State has recently made an appropriation for an Agricultural Survey, and the question is naturally suggested what benefits are likely to result from this liberality of our State in fostering our Agricultural interests.

Within the last few years surveys have either been made or are in a state of preparation in no less than twenty States, and some of the Territories. Some are on a limited scale and are only confined to Agriculture, whilst others are more expensive. Some States include Geology and Mineralogy in their Agricultural Surveys. Some, in addition to the above, have appointed naturalists of known talents to give descriptions of every native production of the State in every branch of Zoology, whilst one State, that of New-York, has ordered not only detailed descriptions, but expensive engravings.

It would be well in the introduction of this subject to consider not only the relative terms, but the object of these surveys. Geology in a strict sense of the word is the science which illustrates the structure, relative position, and mode of formation, of the different organic, metallic, mineral, and other substances, that compose the crust of the earth. Without touching on that branch of the subject which relates to the various theories of the earth, which have in many instances given rise to a tissue of extravagant notions, the Legislatures of our different States seem to have wisely directed the researches of their scientific men to an examination of those products of nature which are within the reach of our observation and may be applied to practical purposes, being more intent on collecting valuable information than in an indulgence of speculations or the invention of theories. As yet we know but little in regard to the means which nature employs to form the very soil on which we tread by converting into mould the various animal and vegetable exuviæ. We are just beginning to learn how scanty are the genuine observations we possess on the process of alluvial deposits or on the depositions at the foot of mountains by means of the decomposition of the various rocks. We know but little of the process in producing petrefactions; and the world has only just commenced to apply to agricultural purposes the various mineral, as well as vegetable repositories.

An Agricultural Survey comprehends an examination of the various soils so as to enable the cultivator to ascertain what plants are best suited to each plantation or district, what ingredients are wanting in the soil to render it productive, and to offer suggestions for its improvement. This requires the skill and the practice of an able chemist, possessing also an acquaintance with the laws of vegetable physiology and a fund of practical agricultural knowledge.

It embraces an examination of the various localities where manures may be obtained together with directions for their judicious application. It points out the errors in the mode of cultivation and suggests such new improvements as have undergone the test of experiment. It is intended to direct the planter to such new objects of culture as may be safely introduced, when others have been found unprofitable. It extends to agricultural statistics, and to the management of animals and domestic use. In a word, it includes every department of agriculture.

A knowledge of several branches of Natural History is more or less intimately connected with agricultural improvements. The localities of plants indicate peculiar soils, the ranges of quadrupeds, and the migration of birds, afford us lessons in regard to temperature, nearly equal to those of the thermometer; and the study of the habits of insects, which are either a pest or a blessing to the farmer, is of very great importance.

In some of the districts in several of the States, great benefits have accrued from these Geological and Agricultural Surveys. In a few instances new localities of metallic deposits have been found, whilst in others the various mineral manures, limestone, gypsum, marl, &c., have been detected, which have converted whole districts into fertility. In addition to these, other discoveries have been made such as valuable clays, building stones, marble, materials for segments and localities where by boring springs of wholesome water have been conducted to the surface. On the whole, however, I am inclined to think that the mass of the community in the greater number of the States has been somewhat disappointed in the results of these Geological, Agricultural and Zoological Surveys. Too much no doubt was anticipated. Men hoped that the veins of gold would be found running under the surface of their farms, and that the quantity of silver which should be detected among the rocks would facilitate the great desideratum in our country, a specie currency. The farmer expected to be taught by the chemist how to double the product of his fields without any additional labor. These results did not generally follow, and men had no right to anticipate them. There can be no doubt, likewise, that in an undertaking so new to our country, some mistakes have been made in the selection of the individuals who carried on these surveys. Some of them having been incompetent to the task assigned them, and others having performed it carelessly, and more from a desire of obtaining pecuniary appropriations of the State than that of adding to its resources or of advancing their reputations among men of science. The reports on the surveys of the different States are now slowly and irregularly coming before the public. In general, they are characterised by those defects, which are incident to a new and difficult undertaking. Whilst some are very credible to their authors, others afford abundant proofs of carelessness, haste, and a want of knowledge.

In the State of Massachusetts, not only an Agricultural, but a Geological and Zoological Survey was ordered. The report on Agriculture by the Rev. Mr. Coleman is of very high merit. In the Zoological department, some information is given by Harris on Insects that may be beneficial to agriculture, and some additions made to Ichthyology by Dr. Storer that may aid the cause of natural science; on the whole, however, the papers on zoology betray evidences of imperfection and haste. Still, as some of these branches are but distantly connected with agriculture, and the works have been got up without much expense to the State. If they confer no extensive benefits on science, they can do it no harm. In the State of New-York, however, the Legislature proceeded on a more magnificent and expensive scale. In 1836, an act was passed, which was amended in 1840 and '42, ordering a survey of the State. Various distinguished individuals were appointed to give detailed descriptions of all the natural productions of the State in the departments of Zoology, Botany, Mineralogy, Geology and Palænontology. These were to be accompanied with expensive engravings. To what number of volumes the results of these labors will be swelled, we are not yet informed, or what will be the expense when completed, it is difficult to conjecture, only one volume having as yet appeared. In 1842, however, the State had already appropriated $130,000 to this object. There appears on the whole to have been a State pride in this lavish expenditure, not very credible to the wisdom of the Legislators of the Empire State, which may eventually produce a reaction, and finally occasioned more injury than benefit both to the cause of science and agriculture. It may reasonably be asked, what benefit can be conferred on a State by a publication of descriptions of well known Birds and Quadrupeds, not a single species being peculiar to the State, whilst the great majority have a range of several thousand miles, especially when they are well described and better figured by others, and when no new information can be imparted and no evidence can be exhibited of any improvements in art. I am not aware that in the most important branch, an Agricultural Survey, any thing was ordered to be done at the expense of the State, and no examination was instituted of the Insects that are either a blessing or an injury to the farmer. An examination of the minerals and organic remains was important, as the various localities of the State had not before been scientifically explored, and although no new discoveries of coal and other objects of anticipated wealth were made, it was well to ascertain that none existed. A simple list of the plants and their localities and of the Mammalia and Birds of the State indicating those which were resident or migratory, injurious or serviceable to the husbandman, with reference for descriptions to standard works seems to be all but the wants of the State and science required. I allude to these facts in order that our own Legislators in the important work we have undertaken may be guarded against lavish expenditures on secondary objects.

I come now to notice the recent appropriation by our Legislature for the survey of our State. In the present case, it cannot be said that the State has been hurried into the measure as it has been proposed, I think, by every Governor and agitated in both Houses of the Legislature for the last five years. The appropriation also involves but a moderate share of expense and is limited to a single object: an Agricultural Survey.

There can be no person of education and practical knowledge who has had an opportunity of witnessing the improvements in agriculture, in Europe and our Northern States who must not be decidedly favorable to the introduction of science into our system of agriculture, nor have we any room to doubt that when this is fully understood and carried into practical effect in our State, the product of our soil will be vastly increased; our country will be rendered more healthy, and our improvements in agriculture will advance manufactures and the mechanic arts; the number of inhabitants will be greatly multiplied, and a greater degree of intelligence as well as prosperity will be the inevitable consequence.

In England, Belgium, and some parts of France and Germany, agriculture is now pursued on scientific principles; and the preparatory study for the occupation of a successful farmer is the work of years. There are, however, advantages in all those countries possessed by the cultivators of the soil that enable them to introduce science into their modes of cultivation, which are not enjoyed by the farmers of our Northern States and only in a limited degree by our planters of the South. The European farmers are either wealthy landowners or rent large and extensive tracts of land amounting in most cases to many hundred and frequently to several thousand acres. The peasantry are in their employ—under their direction, and are obliged to adopt the modes of culture determined on by those who employ them. In the Northern United States, the farms are small compared to our Southern plantations or to the extensive domains of an English nobleman. Our American farmers not only superintend the concerns of their farms, but generally labor in the fields. Hence every small farm has its own system of agriculture according to the knowledge or caprice of its owner, and except in a few cases, science has lent but a feeble aid to agriculture. In the Southern States, although our plantations are much larger, and our operatives under the control of the master, yet we labor under many disadvantages owing to our climates and more especially to our great deficiency in agricultural knowledge. Whereas, in the European kingdoms, I have mentioned, the soils of each district and frequently of each farm have been thoroughly analyzed, and the intelligent farmer is fully acquainted with the kind of cultivation best adapted to his lands. He has been taught by a system of underdraining how to diminish a redundancy of moisture; and by irrigation how to render an arid soil fertile. Science teaches him how to apply manures to correct a superabundance of clay, how to use the various formations of lime, and when and where to withhold them; and he is guided by the lights of science and experience, and the selection of those manures best adapted to the roots and plants he is desirous of cultivating. For the last half century at least, this system of agriculture on scientific principles has been maturing in the minds of the Europeans. Manures have been dug from the bowels of the earth, gathered from the sea, and imported by ship-loads from the battle-fields, and other depositories contained in foreign lands. The Physiology of plants has been carefully studied, and every year is adding to their knowledge in this important branch. An acquaintance with the laws of Chemistry has become more general. Botany is no longer regarded as a merely amusing, but a practical and beneficial science. Their knowledge of Ornithology teaches them to know what birds should be preserved to aid them in diminishing the number of depredating insects; and Entomology, one of the most important, but most neglected branches of science, has been so far studied as to enable them to guard in a great measure against the depredations of insects which infest their grains, fruits, and trees. In these various departments, the conquests of science have been such, but the cultivation of the soil on scientific principles, and the study of natural science, as a part of the system is no longer viewed as a doubtful experiment; on the contrary he who rejects the lights of science is regarded by the most intelligent and most successful cultivators of the soil, a half a century behind the knowledge of his fellow man in this age of improvement.

In our own country, few farmers have adopted those modes of culture, which the experience and science of Europeans have discovered to be most productive, and the planters of the South are in this particular behind the farmers of the North. Cotton and rice, the rich staples of our State, has so far banished other cultures, that we have now to import the corn we use in Charleston from North-Carolina, Maryland, and Virginia; our flour from the Middle States; our hay from New-York and New-England; our butter, cheese, and Irish potatoes, from the same prolific source; and our horses, beeves, and hogs, from Kentucky and Tennessee.

Hundreds of thousands of acres of our former inland rice-fields are now wholly abandoned and have become the habitation of the frog and the alligator. The soil in many of our districts has been exhausted by bad cultivation as is the case in some parts of St. Paul's, St. Andrews, Christ Church, and other parishes. Many of their former inhabitants have gone to Alabama and the West, where by a similar system, they have in many cases been equally unsuccessful, and some of them or their sons have, after years of absence and deprivation, returned to become overseers over the lands they once owned; like Ruth, to take the gleanings of fields once their own; may they prove as fortunate and as deserving. Our mountains abound with metallic wealth, but until recently the iron of the plow-share that turned up the soil, resting on beds of the finest iron ore in the world, was imported from the North, and the iron bars of our Rail-Road came from Liverpool. Marl exists in hundreds of localities in our lower country; and lime-stone in our mountains, and even in our middle districts, sufficient to enrich the soil to the end of time. Our rivers and our sea-shore abound with ingredients of inestimable value to the planter, but we have not availed ourselves of these rich manures which nature has so bountifully provided. The Hessian fly and the chinch-bug destroyed our wheat; the weavil our corn and rice; the army worm; the rock and the rust, our cotton; the sawyer of our pines; and the curculio, the coccus and aphis, our fruits so that we lose one half of the products of our fields, gardens and forests; and yet there is scarcely a man in our Southern States that is acquainted with the habits and character of a single one of the species of these depredators; and of course till its habits and modes of propagation are known, it will be impossible to suggest an antidote.

From this admitted defect in our knowledge of agriculture, the important question arises, how can the evil in question be best remedied, and in what way may an Agricultural Survey be rendered beneficial under present circumstances? No one acquainted with the subject can deny the benefits which would result from a survey, conducted on scientific principles, provided it can be rendered available to practical utility.

There are, however, immense difficulties in the way of success; these should be candidly stated in order that they may be met and overcome. If the survey is to be conducted on purely scientific principles, founded on a careful analysis of soils, and a thorough knowledge of vegetable physiology, it is to be feared that the individual suitably qualified to perform this complicated and arduous task cannot be found in the country; and even should we be successful in obtaining such a person, the agricultural knowledge in the community is not sufficiently advanced to enable our cultivators to be acquainted with the mode of applying the results to practical purposes. Besides, if the whole work is left to a single individual, unaided by Agricultural Societies and men of science, he would not be able to survey the whole State during the term of a long life. The survey of a single county of England required in some instances four years aided by Agricultural Societies as well as the intelligence, advice, and personal aid of nearly every landholder in the county. Different individuals employed in these surveys arrived at different chemical results. Errors were corrected but slowly. New tests were resorted to, new surveys made, and the subjects were discussed from week to week for a succession of years. There is another subject which we ought not to overlook. The chemical analysis of the soil is one thing; the application of the knowledge thus derived to the plain purposes of agriculture is another. The chemist may be correct in his statements of the various ingredients in the soil submitted to his examination; but he must be either acquainted with the practical operations of these results to the purpose of agriculture or the agriculturalists to which he submits them must have sufficient knowledge of chemistry and vegetable physiology to carry them into successful operation. Let us take, for instance, the able scientific analysis of the soils made by Professor Shepard from eight localities of a plantation on the Edisto Island, and let us inquire how many planters can be found in South-Carolina to have a sufficient knowledge of agricultural chemistry to be guided by that analysis in ascertaining what ingredients are wanting to render the soils more fertile or what causes have been operating in producing sterility? Even the admirable report which accompanies it (see Southern Cabinet, 1840, p. 449), drawn up with great care and research, by a committee of intelligent practical planters, although it contains much valuable information in regard to various manures, does not afford us those plain and practical instructions which the unskilled planter is so desirous of possessing.

The difficulties moreover, which attended the process in making such an analysis of the soil, as will be available to practical purposes are greater than the practised chemist is willing to admit. Sir Humphrey Davy believed that "neither much time, nor a minute knowledge of general chemistry were necessary for pursuing experiments on the nature of soils and properties of manures." To him, who was thoroughly acquainted with the subject, the work was simple enough; but to men who know nothing, even of the first principles of the science, there are difficulties which are for a long time insuperable. Even Davy, Lavosier, Chaptal, Decandole, Liebig, Dana, Coleman and Jackson, the lights of the world in the science of agricultural chemistry, have often differed, not only in regard to their experiments, but in their practical application. It is to be greatly feared that our planters have not received that preparatory education, which would enable them to derive immediate benefit from a purely scientific survey. In England, Agricultural Societies and agricultural education preceded these surveys by half a century. This was in a great measure the case in the Northern States of our own country. It has been observed that when this deficiency existed among the people, they derived no immediate benefits from these lights of science; but in those counties where men had been long trained in these preparatory schools of agriculture, they immediately profited by those aids which science presented. It is to be feared, moreover, that sectional jealousies and dissatisfaction may arise from the fact, that the agricultural Surveyor does not possess the power of ubiquity and is obliged to confine himself for a considerable time to one portion of the State in order to render his labors of any value. In this stage of our progress, should the overwrought anticipations in regard to the great advantages of such a survey result in disappointment, a reaction might be produced and cause a delay beyond the proper time; for although such a scientific survey of the State, may perhaps at present be rather premature, yet it would in a few years hence, when the public mind has become sufficiently enlightened be productive of immense advantage.

Fortunately these difficulties, which are here presented in order that they may be guarded against, are not insuperable.

Much will depend on the individual this important work shall be entrusted. In his selection, all party feeling and personal attachments should be disregarded. He should not only be a man of science, but of practical experience in agriculture. He should be satisfied with our peculiar institutions and have some knowledge of the culture of the staple articles of our State as well as of those productions which are essential to our food and furnish pastures for our cattle. He must be a man of an enlarged mind and if possible free from those strong prejudices which so often prove a barrier to the reception of truth. I have often met with managers of large estates of cotton or rice who had been eminently successful in a mode of culture adapted to a particular region and a particular plant so wedded to the mode of cultivation they had adopted that no arguments could convince them that a different plant, another locality and soil, required a very different treatment. An agricultural surveyor should know enough of chemistry to enable him to analyze the soils and be able to detect deposits of marl, limestone, and those other ingredients, which should be used as manures. In the present limited state of our agricultural knowledge, I would prefer a highly intelligent practical man to a purely scientific one, who is unskilled in the practical application of the laws of agricultural chemistry. He must, moreover, be a man of labor and patience for he will have to experience some deprivations and encounter a host of difficulties. Such an individual may gradually prepare the way for a more thorough and scientific survey of our State. He might encourage and give a proper direction to the labours of our agricultural societies and call forth latent talents in every part of the State. Some such unpretending practical examinations should be made of the agriculture of our various districts, as we have seen from time to time in the labors of Ruffin and Legare. The time may not yet have arrived when we can be much benefited by such surveys as were made by Coleman of Massachusetts and Jackson of New-Hampshire and Rhode-Island and unless they are rendered far more plain and practical than those contained in their scientific reports; but he may prepare the way and give a new stimulus to agriculture. The results of his labors should be regularly published in so cheap a form that they may find their way to every family in the State.

Much reliance must also be had on the public in aid of this important undertaking. Men must not expect too much or become impatient. A work has been commenced which, to prove beneficial, must be continued for years. Sectional jealousies must be avoided, and we must regard ourselves as belonging to Carolina rather than to one of its parishes. The minds of our planters must be more directed to those agricultural studies on which their prosperity so much depends, and being now about to engage a teacher, they must become industrious scholars.

I will now proceed to offer a few suggestions in regard to some of the means of instruction of which we might avail ourselves in order that an improvement in our agriculture may be effected. These indeed should have preceded this survey by many years, if it is to be conducted on really scientific principles, or may now be rendered important auxiliaries, if it is intended to be merely an examination of the products of and modes of culture in the different districts of the State.

I. I would suggest the establishment of Agricultural Societies in every district of our State, the fee of admission to membership should be so low that not only planters, but overseers and men in every walk and occupation of life may be encouraged and induced to become members. These Societies should be active and hold their meetings not once a year at a club-house, to eat a dinner and talk politics, but monthly or weekly, and interchange sentiments on the results of their several modes of agriculture. There will always be in every association of this kind a few men of education who read the agricultural publications of the day and who are possessed of sufficient zeal and industry to submit to the test of practical experiment the information imparted by agricultural journals. If I am asked, whether in order to carry on the process of cultivating the earth on the principles of science, I regard it as a necessary that every planter should be a chemist and physiologist and be at the same time acquainted with those branches of natural history, Botany, Entomology, &c. which are so closely connected with it, in a word, whether every culturalist must be a man of learning, and of science, I answer unhesitatingly, No. As in government, a few leading men give a tone to the politics of a State—so in agriculture, science and practical success of a few prominent planters in the State will be a perpetual practical lesson to the districts around them, and men will adopt their practice without knowing much of the principles of science by which they have been governed. Man is an imitative animal and is not slow in adopting the improvements of his neighbors where he sees how much his own interest is concerned. When the celebrated Arthur Young in 1767 commenced his valuable and well directed labors and pointed out to his countrymen an improved mode of husbandry, they adopted his mode of culture although they only looked at the effects and were unacquainted with the scientific views which had governed him in carrying on his successful experiments. It has been ascertained that in those counties of England where Agricultural Societies were first established, the products of the earth have been trebled within the last thirty years. The Highland Society, which has existed for sixty-four years—the most prominent, active, and most efficient in the world, whose meetings are held at Edinburgh—has by the stimulus it gave to industry on the principles of science rendered a once barren soil in an inhospitable climate equal in many of its counties to the best portions of England itself. The Lothians are covered with the most luxuriant crops of wheat, barley, beans and other products. On the Meadows, the most valuable grasses are cultivated; the mountains, even to their very summits, are covered with rich pastures, and I observed herds of cattle and sheep grazing on the very top of Ben-Lomond, and other high peaks of that romantic lands. All this I contend has been effected by a practical application of scientific knowledge, diffused by means of an Agricultural Society.

A fact or two in elucidation will be mentioned. Surveys were made in each county of Scotland as well as of England, the soils were analyzed, the materials in each vicinity for manuring were examined, and a printed and detailed account was placed into the hands of every landholder, which would serve as a guide in the management of his farm. It is, moreover, not generally known that Scotland furnishes more than a fourth of Europe and a portion of America, with genuine undegenerated seeds of many of the grains, melons, garden and flowering plants that are usually cultivated. How is this effected? Botanists have discovered that a superior variety of seed immediately degenerates on being planted near those of other species or varieties of its own or a kindred genus and that on the second or at farthest the third year, the original and valuable character of the plant has in a great measure disappeared. Hence it is that in Carolina, when we plant our imported cantalope melon seeds in the vicinity of our common melons, squashes, &c., all their original, valuable properties disappear on the second year; so also, our cauliflower becomes a mongrel cabbage, as I have ascertained; and I am inclined also to think, that the generally received opinion in Carolina that all Indian corn, when planted near our sea-board, whatever may have been the original variety is converted into what is called flint corn by the peculiar character of our soil and climate may be erroneous and that this peculiarity may be traced to the near approximation of our abundantly prevailing fields of flint corn, communicating their farina to the small patches of new varieties of corn on which these experiments are making. But how is this evil remedied in Scotland and why are the seeds of their grains and vegetables preserved without the slightest degeneracy from age to age? In raising seeds for planting or exportation, no two varieties of the same species or even genus are suffered to grow within miles of each other lest the winds might waft the fructifying farina of another plant and produce degeneracy in any approved variety.

Some of the benefits, then, which we would have every reason to anticipate from well conducted Agricultural Societies, would be the following:—

1. Such Societies would bring to a closer intercourse of a few educated and scientific man and a vast number of industrious practical agriculturalists, who by an interchange of their different modes of culture would be equally benefited by the details of failures as well as of successful experiments.

2. They would not long exist before the members would be made sensible of the importance of analyzing the soils of their several districts, and thus ascertaining whether there is a deficiency of those ingredients, which are necessary to the nourishment of the plants cultivated, what manures should be applied, and what modes of culture should be pursued.

3. They would be able to ascertain the causes which have converted the once fertile plantations of Carolina into old fields, grown up with broom-grass, and no longer yielding sustenance to man or beast. They would learn the importance of a rotation of crops as it is now well ascertained that different plants not only feed on different substances contained in the soil, but that there are peculiar exuviæ from each, which would be injurious where the same plant reared on the soil for a secession of years, but would be a source of nourishment to plants of a different genus.

4. They would be able by this increased intercourse and knowledge of culture, not only to augment the quantity of the staples now in cultivation, but introduce other valuable products to which our soil and climate are well adapted.

5. They could scarcely fail to direct their attention to the introduction of some of those grasses which would answer as substitutes for the herd's-grass and clover of the North, which do not succeed in our Southern climate except in particular soils and situations. The introduction into Carolina of a perennial grass, suited to pasturage and hay, would confer a greater benefit on the State than the discovery of the richest gold mine.

6. By this additional stimulus to industry and by the better draining in cultivation of our land, not only the wealth, but the health of the country would be improved. I could not fail to be forcibly struck with a remark made by Liebig in his Agricultural Chemistry, although I am aware that physicians have adopted contrary opinions in regard to this

theory: "Plants (says he) improve the air by the removal of carbonic acid and by the renewal of oxygen, which is immediately applied to the use of man and animals. Vegetable culture heightens the healthy state of a country, and a previously healthy country would be rendered quite uninhabitable by the cessation of all cultivation." The truth of Liebig's remark is verified by the increased unhealthiness of our Southern country since our own water-courses have been obstructed by decayed vegetable matter and our fields suffered to remain uncultivated. Sixty years ago, the planters did not find it necessary to remove from their plantations on account of any apprehensions from fever, and many of our oldest inhabitants still living were born and reared in situations where there would now be imminent danger in remaining only a single night during summer; and the question is of momentous importance, what process would render our climate of Carolina more healthy than it is at present. We may learn something on this head by looking at the effects of cultivation in other countries. The boggy fens of England were once the fruitful sources of fever. They have been drained; the Peat Moss has been converted into fuel; the lands are cultivated in grain; the peasant's cottage now stands on its borders, and he enjoys uninterrupted health. The time was when the pontine marshes were traversed even in the day time at the risk of life; we are informed that those portions which are drained, embanked, and cultivated, are now comparatively healthy. The low grounds of Holland and Belgium were once as sickly as Carolina is at present; in the autumn 1838, I slept several nights in their vicinity, and I was informed that since they cultivated their grounds more carefully, their former fevers had disappeared. The sluggish waters were still in their dykes, but decayed vegetation was no longer steeped in them. Every foot of land was cultivated; the borders of their ditches were planted with nursling trees, which were to become the future pride of their forests; and the cabbage and cauliflower plants along the public highway nearly touched the wheels of our carriage. Thus, the plants inhaled the unhealthy carbonic acid gas, renewed the oxygen, and the improvements in agriculture rendered countries healthy that had formerly been very sickly.

II. The auxiliary to our improvements in agriculture, I would suggest, is *cheap* and widely circulated *Agricultural Papers*. This is a subject so self-evident that it is unnecessary to offer any remarks on its importance. Agricultural Societies without a publication of their transactions would be as inefficient as a rail-road without a locomotive.

III. I would above all recommend a *School* where those branches are especially taught, which appertain to Agricultural and Horticultural pursuits. Schools of this class first had their origin, I think, in Germany; they were next introduced into France and Switzerland and are now springing up in every part of Europe. The Renssellaer School near Albany in New-York is also an agricultural one. The most complete Institution of this kind, I had an opportunity of examining, is called the Institute of Agriculture and Forestery at Hohenheim near Stutgard [Stuttgart, Germany]. I observe that it is characterised in the British Farmer's Magazine as "the most complete Agricultural School in Europe." Here, in addition to all the studies usually pursued in academies, all the operations of agriculture and horticulture are performed by the Students in the open air under the supervision of Teachers qualified to undertake, note down, and record every observable fact, and traceable cause. Here are delivered regular courses of lectures on Geology, Mineralogy and Chemistry, on soils, water, moisture, vapour, fermentation, gases, their extraction, mutual attraction, condensation and results. Instructions are given and elucidated by experiments on light, heat, electricity, galvanism, magnetism, &c. These are all employed by Nature and are in incessant operation. They constitute the class of great natural agents. Botany in the most comprehensive sense of the term forms a very important feature, which extends to the physiology of plants, their uses, medical and other virtues. Entomology is also taught as a science connected with agriculture; and the habits of Insects as well as Birds and Quadrupeds are studied in order to guard against their depredations or be benefited by their labors.

The establishment of an *Agricultural School* on a model of which the above is a faint outline, which may be modified in some particulars to adapt itself to the wants of our country, I most certainly believe to be of greater importance to our agricultural interests than even an Agricultural and Geological Survey, than Agricultural Societies, or Agricultural Papers, inasmuch as such a school would inevitably lead the way to all these other aids to our knowledge and success.

I will not venture on the details necessary to the establishment, support, and successful operation of such a school. I will leave to politicians the settlement of the disputed point whether the State has or has not the constitutional right to expend some of its funds in promoting our agricultural interests as well as the aids it now affords to our College and our Military School. The suitable Professors may be obtained, although perhaps at present with some difficulty. The expenses would be less than those of a Military School. If the State cannot be induced to lend its aid in such an undertaking, it may be worthy of inquiry whether united individual effort might not be made available. In a short time the School under judicious management would support itself. The term for those who had previously received a good English education should be about two, at farthest, three years. Our planters, I should suppose, would prefer having their sons educated in such a seminary, after suitable instructions in some of our grammar Schools, to that of sending them to our Northern Colleges or even to West Point. However highly I estimate the value of the higher branches of mathematics and the modern languages taught in the latter, I cannot conceive that even such a School will confer half the benefits on our country as would inevitably be derived from a well regulated Agricultural School on the principles of science. Fifty young men thus educated would disseminate a knowledge of the science of agriculture, which would give a stimulus and serve as guides to the whole State.

I am fully aware of the objections which many successful planters urge against the scientific cultivator. He is regarded as a theorist and a speculator, and it is predicted that he will eventually be unsuccessful. It is admitted that a man may have very correct ideas of agriculture, and yet, if he does not carry his knowledge into practice by constant attention to his planting interests, all his scientific knowledge will be unavailing. On the other hand, he who has become successful as a self-taught planter might have reached this eminence many years earlier and promoted his pecuniary interest to a much greater degree had he possessed the benefit of previous knowledge. All self-taught men who have risen to any high degree of eminence have subsequently lamented the disadvantages under which they had labored owing to the want of previous education. What would we think of a lawyer, a physician, a merchant, or a mechanic who would attempt to exercise his profession without having made himself acquainted with any of those previous studies, which the world regards as essential to his success in the profession he has chosen? At present, our young planters are engaged for years in their professions before they have learned even the first principles of agriculture, and they acquire a knowledge of planting more frequently from their past failures than by accidental instances of success.

I contend that nearly every improvement in agriculture as well as nearly every discovery of importance to mankind has been the result, not of the accidental discoveries of the ignorant, but of a previous knowledge of some of the sciences, and guiding these gifted and studious men onwards, in their researches after truth. If Newton derived his first idea of gravitation from the fall of an apple, it required such a mind as that of Newton to make the practical application. The cook has seen the steam issuing from the spout of the tea-kettle from early times, but such minds as those of Watt and Fulton were requisite to apply this knowledge to any available purpose. Every school-boy can fly a kite, but it required the scientific knowledge of a Franklin to render it the medium of conducting to earth the disarmed lightnings of heaven. The labors of such men as Sir Humphrey Davy, Arthur Young, Lavoisier and Liebig have done a thousand fold more for the comfort and happiness of Europe than all the Legislators that thundered in their Senates or all the Heroes, whose names are enrolled on the pages of history and whose monuments fill the niches of Westminster Abbey or adorn the romantic grounds of Père le Chaise. And when the political excitements in our country shall have happily subsided, such names as those of Judge Buel, Skinner, Ruffin and Seabrook, will be held in grateful remembrance whilst those of our noisy political patriots will only be handed down to posterity through the musty streams of a forgotten newspaper.

I will here enter a little into detail, on the nature of those studies which should be pursued in an Agricultural School, and on their unofficial results.

1. The first and most important is *Chemistry*, a branch of physical science which analyzes and investigates the composition of inanimate bodies. This claims our special attention, not only on account of the manner, but the variety of ways by which it may be applied. Soils, we know, must

differ widely in their various component parts since even in one part of the same field the product is double to that of another part. A field may be admirably adapted to one kind of culture, which would produce but a scanty crop of another current. Now, this deficiency in the latter case arises from the fact that the soil is wanting in some elements necessary to the growth of plants or possesses some ingredient which is positively injurious. In order to correct this defect in the soil, the culturalist must first be convinced that the evil in question is occasioned by some deleterious substance or by the absence of some necessary one. How can he ascertain the fact? Soils are so blended that we cannot be aided in the investigation by the examination of our senses without chemical tests. Sir Humphrey Davy in his Agricultural Chemistry, mentions the following fact which is in point: "A soil of good apparent texture, from Lincolnshire, was put into my hands by Sir Joseph Banks, as remarkable for sterility. On examining it, I found it contained sulphate of the oxide of iron, and I offered the obvious remedy of a top-dressing of lime, which converts the sulfate into a manure." Here was a soil, the causes of whose sterility could not be conjectured, even by so close an observer as the evident Naturalist Banks; yet, by the application of chemical tests, the whole mystery was solved probably in a few moments. It is a well known fact that whilst the farms on Charleston Neck are admirably adapted to the culture of the Irish potato, turnips, carrots, and the whole cabbage tribe, they will scarcely produce the sweet potatoe of large size. There must then be some deficiency in the soil necessary to the production of this vegetable, which a chemical analysis both of the soil and the potatoe itself would no doubt point out. A planter of this facility, desirous of improving his lands on which he was planting a crop of corn two years ago, placed in each hill a quantity of fresh bog earth from an adjoining old and abandoned rice-field. This was immediately covered in with the grains of corn. It produced scarcely 7 bushels to the acre, and he came to the conclusion that swampy mud was rather an injury than a benefit to the corn. He was unacquainted with chemistry and had no great regard for the opinions of scientific men as guides to agriculture. It was suggested to him that he had applied to his plants that which in its then state was poisonous and was advised to open the hills to expose the still undecayed mass of swamp mud to the operation of air, light, and atmospheric electricity, and replant on the following spring in the same hills and with the same manure, which would then have undergone chemical action and be in a fit state to afford sustenance to the plants. This, after necessary ploughing, was somewhat reluctantly done. In this second experiment he was more successful, having made 30 bushels to the acre instead of seven, the product of the former year. Manures which are beneficial to some kinds of land will be positively injurious to others. Putrescent vegetable matter, salt, and various alkalies are used as manures; and yet, some lands will be benefited by the one and would be rendered less productive where the other applied. The same may be said in regard to plants. A familiar instance may be mentioned. The rich soil which would cause the geranium to flourish in our flower-pots would, were it applied to our japonicas, azelias, and rhododendrons, cause them to deteriorate and finally to perish. Hence, the importance of chemical, as well as physiological knowledge, in enabling us to analyze the character of our different soils and ascertaining their adaptation to the various plants and grains we are desirous of cultivating.

2. The next important branches that should be taught in an Agricultural School, are *Geology* and *Mineralogy*. These are so closely connected in many particulars, that they can scarcely be separated, still some are alone applied to the arts, whilst others are more directly beneficial to agriculture. A beneficent Providence has scattered mineral and organic wealth through every portion of the earth. Our mountains abound with it, and on our sea-board there is no deficiency, although it appears in another form. The different minerals can be distinguished from each other by forms obvious at once to the senses. The mineralogist can at a glance detect the characteristics of each species and is thus enabled to ascertain what may be profitable to man and reject what is of no value.

How much labor is annually bestowed, how much money expended, and how much deception practised on the simple and credulous which might all be prevented by a knowledge of mineralogy. How many of speculation in a gold mine, which has brought ruin on the family of the purchaser, might have been avoided had he possessed this knowledge. I have seen a poor family thrown into extacies and dreaming for months over their anticipated wealth because an ignorant pretender had discovered on their lands an ore which he called gold and pronounced the mine of immense value, but which a mineralogist detected as nothing more than sulphuret of iron. I well recollect the sensation produced among the farmers of New-York, when in consequence of the last war, they were unable to import from the British province of Nova Scotia, their Gypsum or Plaster of Paris, which was in universal use. The plaster had risen in price to thirty dollars per ton and at last could not be procured on any terms. Man is full of expedients and without the restraining influences of religion is not over honest. A man in the north-western part of the State (whether the public finally decided him to be a nave, or an ignoramus, I do not recollect) pretended to have discovered a bed of plaster, which he offered for sale at ten dollars per ton. The farmers from Renssellaer, Albany, and adjoining counties in a circuit of an hundred miles went in their sleighs to lay in their store of plaster. The article was as hard as granite, and many a mill-stone was broken in the operation of grinding. The following summer it was discovered that they had spent their money and labor for nought as the material was of no more benefit to their land than so much sand. But the same science that detected the deception discovered also a remedy for these temporary losses and vexatious impositions. A careful examination was made of the surrounding country, and not 30 miles from the above locality, an inexhaustible bed of Gypsum was in reality discovered, which has proved a greater blessing to that part of the State than a mine of gold. Marl is now universally acknowledged to be of immense value as manure. The maritime districts of our State abound with it, and it has been discovered in many localities widely remote from each other. In fact, scarcely a planter in the State knows what it is, is sufficiently evident from the specimens they are constantly sending to Charleston from all quarters, which they either believe or hope to be marl, but which most generally proves a very different material and of no value as a manure. The knowledge, industry, and zeal of Ruffin, and having directed the planters of lower Virginia, to the use of marl cannot be too highly appreciated. In his scientific researches and judicious instructions, he is rescuing from sterility a soil which Washington had defended with his sword. His admirable Essay on Calcareous Manures, should be in the hands of every planter on our Southern sea-board.

3. *Vegetable Physiology* is another subject which holds an important

rank among the studies that should be pursued in an agricultural School. This science investigates and explains the structure and vital character of plants. Vegetation is either favorably or injuriously affected by the seasons, the changes of weather, the soil, and temperature. The organization differs so widely in the several species that some will flourish only in an atmosphere impregnated with salt particles whilst others inevitably perish if planted in the same locality. In my garden, a part of which is occasionally overflowed with salt water, the cabbage and cauliflower sustained no injury, but are rather improved by a moderate watering of this nature, whilst on the contrary, the bean, corn, spinage, and several other species, remain diminutive in size, become sickly in growth and generally perish. In particular regions of our country, certain plants are only found where that Clay soils prevail, but can never be detected beyond the line which separates this region from the adjoining sandy and loam formations. Some plants only flourish in limestone regions. The winds, the birds, and the hand of man, scatter the seeds far and wide that they refuse to vegetate in soils that are not adapted to their growth. The Baccharis, Salicornia, Salsola, and several other plants, as well as the interesting foreign shrub, the Tamarix gallica, thrive only in the neighborhood of salt-water and are not found beyond these prescribed limits; yet let but an atmosphere and a soil be prepared for them, either by nature or art, and they will easily grow and flourish there. Thus, the salt springs of Onondago, are, I think, several hundred miles from the sea, yet I have observed several species of these maritime plants flourishing there although not a single specimen can be found growing in all the intermediate regions. Plants too produce varieties often infinitely more valuable than the originals from which they sprung and are so improved by culture that nearly their whole characters are changed. Still all these changes take place in accordance with the fixed and invariable laws of nature. The object of the vegetable physiologist is to investigate these laws in order that by acting in accordance with them, he may bring the vegetable kingdom under his control and render it subservient to his views. The experiments of Van Mons on fruit trees—the infinite varieties of Japonica, Dahlia and Rose that have been produced of late years by peculiar modes of culture—are sufficient evidences of the triumphs of science in this department.

4. The next subject of importance to be taught in an Agricultural School, although not in the regular order of succession, is that of *simple Mathematics*, including Arithmetic, Geometry, Levelling, Surveying, &c. Without a general knowledge of these, the planter will often be at a loss in pursuing chemistry and other studies with profit and pleasure. The agriculturalist must be a careful calculator. He must be able to ascertain what kinds of produce will be most valuable for him to cultivate, taking into consideration his pecuniary means, his locality, and the peculiar character of his soil. He must be able to keep his accounts correctly. A knowledge of Surveying will often save him from litigation, and the art of Levelling is all-important, in enabling him to act with judgment and to save expense in digging ditches, or canals. A striking instance of the importance to the planter of the simple art of Levelling came under my notice during the last year. A gentleman engaged several Irish laborers to dig a boat canal of 2 miles in length at a cost of about $2000. The Tidewater was to convey the boats to both ends of the canal. The workman ridiculed the idea of having a scientific level taken inasmuch as by carrying the tide water with them, they would be certain of finding a true water level. The canal is dug; the water overflows its banks at the mouth, but scarcely extends half the distance up the canal, and the evil must now be remedied either by a lock or an immense labor of digging down from the summit, which appears to be six or seven feet above the level of the lower part of the canal.

5. Another department in an Agricultural School is *Mechanical Philosophy*. The planter is a man of all kinds of work and should understand of the general principles of every addition or improvement that is going on under his eye. A knowledge of mechanics will often save him from being imposed on by the workman he employs and will enable him to have his house, out-buildings, and fences, constructed for comfort with a due regard to durability, economy, and taste. He is constantly using machinery and various implements of husbandry; these should be so constructed as to produce the greatest effect with the smallest expense of power. New inventions are constantly pouring in upon him. And he should possess sufficient mechanical knowledge to be able to decide what he ought to appropriate to his use as beneficial and what he should reject as worthless. Such are the complicated duties of the planter that

some knowledge of mechanics seems almost indispensably necessary to enable him to carry on successfully the various operations that pertain to his profession.

6. The *rearing of Animals in domestic* use should be regarded as a subject of sufficient importance to hold some rank in an institution of this kind. In Europe much attention is paid to the subject. In this department Great Britain is far in advance of the world. Their fine breeds of horses,* horned cattle, sheep and swine, have been produced not accidentally, but by a thorough knowledge of the peculiarities of those varieties from which new and improved breeds were to be produced. In England, they have separate breeds of horses adopted to the various services required of them; and the strong unwieldy dray horse, the carriage horse, the hunter, and racer may be distinguished at a glance. Those varieties of horned cattle are selected which are best adapted to the pastures on which they are to feed; thus, in the rich level counties of England and the Lothians of Scotland, the heavier breeds of cattle and sheep are preferred; but, in the mountainous regions, breeds not inferior in size, but equally profitable, are reared, and the black cattle and the black faced sheep from the highlands of Scotland supply the markets of Edinburgh, where they are regarded as fully equal in flavor with the larger breeds. I need not say how negligent our planters have been in regard to their live-stock of every kind; cows turned into the woods to feed on Broom-grass are not likely to contribute much to the dairy, and we need not be surprised if our hospitable planters, who own an hundred head of cattle, sometimes find some difficulty in procuring milk for their coffee. Long legged hogs, which appear to have been selected

---

* The finest collection of horses I ever beheld was at Regensburg (Ratisbon) in Bavaria in 1838. They were owned by the wealthy and luxurious Prince of Taxus, whose expensive stables were more magnificent than many of the palaces of Europe. They were fitted up with marble troughs, fountains for bathing, with the different names of the animals, their countries and pedigrees, placed in gilt letters on the wall. Each horse had his groom, and they were daily exercised in a magnificent circus. Among these were not only horses of approved varieties from Mecklenburg, Saxony, Austria, Turkey, and France; but several of the famed barbs from Arabia, and a number from England. To my eye, which I confess is unskilled in these matters, the English courser appeared not only the most elegant in form, but was admitted by better judges than myself to be more active and fleet, than those of Arabia itself.

rather for speed than weight with long noses to root in the pine lands, require as much corn in the fattening season as would purchase a much greater quantity of bacon ready cured.

I was highly gratified at Edinburgh on observing a crowd of intelligent young farmers listening with intense interest to and taking notes on the Lectures of Professor Lowe on the rearing and management of horses, horned cattle, sheep and swine—on the varieties adapted to different localities—their diseases and modes of cure. His Lecture-room was embellished with fine paintings of various animals that were esteemed for their valuable properties.

Nor should the rearing of Poultry be regarded with indifference by the planter. We frequently hear the remark, "I do not raise poultry because I do not know the art." This only proves that he has still something to learn before he has made himself fully acquainted with the duties of his profession.

7. The next study which is very important in an Agricultural School is *Botany* in the broad sense of the term, which includes not only the names and systematic arrangement of plants, but their properties and uses. The world is infinitely more indebted to science for the introduction of those productions which minister to the comfort, the wealth, and power of man, than men are generally disposed to allow. For the introduction of nearly every grain, vegetable, fruit, and flower that minister to the support, comfort, or pleasure of man, we are indebted at first to the Botanist for its introduction and afterwards to the scientific culturalist for its improvement. The seeds of our valuable plants were collected by the Botanist often at the risk of life in wild distant regions; wheat, barley, rye, rice, the potatoe, cotton, the ground-nut, &c., are all the productions of distant lands, where the world is in most cases lying in barbarism. But we are as much indebted for the improvement of plants to the scientific cultivator as we are to the Botanist that first introduced to them. Such have been the improvements in the various varieties of wheat, rye, rice, and barley, that Botanists themselves are now at a loss to designate the original plants from which they sprung. The potatoe, which is now the food of millions, had its origin from a bitter root which grows wild in the mountains of South America and is not larger than a bean, as I have satisfied myself by examining specimens from

Montevideo. It was never used as food in that country till by scientific culture in Europe, where varieties were produced from the seed, it had become so improved that it has now been sent back to its native soil to prove a blessing to the barbarians that hitherto were unacquainted with its value. The carrot originated from a wild and poisonous plant growing on the rocky cliffs of England. The acrid and disagreeable Apium graveolens has been transformed into delicious celery. The wild cabbage (Brassica oleracea), a plant not weighing half an ounce, has been improved into cabbages, whose leaves alone weigh many pounds and into the cauliflower, the most delicious of vegetables. Our fine Swedish turnips have been produced by improved culture from the (Brassica napus), the common and worthless rape; as has our fine turnip descended from the turnip rape (Brassica rapa), which in its wild state is small, fibrous, bitter, and wholly unfit for use. The same may be said of our fruits. The bitter wild crab of Europe is the origin of our apples, and the wild pear, which I saw in the forests of Hungary not larger than a musket ball and as acrid as the unripe persimon, has by long and scientific culture produced our delicious pear.

There is another advantage which the planter would derive from a moderate share of Botanical knowledge. He would not be so frequently imposed on by having seeds of plants palmed on him as new species of inestimable value, which would subsequently prove some common worthless weed. The cunning Italian has chuckled over the oft repeated hoax he played on the Americans, who seized with avidity on and paid an exorbitant price for the seed of the Italian mulberry labeled Morus multicaulis. The famous Florida coffee was sent to the seed stores in Charleston as a newly discovered species that would soon shut up the coffee trade of Havana and Mocha. High prices were paid for the seed; acres were planted by the farmer and magnificent results were anticipated. A single glance of the Botanist detected in it a troublesome and worthless weed (Cassia occidentalis) growing in most of our plantations in Carolina, which the hoe of the servants had long been engaged in destroying, but which the simplicity of his master was now requiring him to replant as infinitely more valuable than cotton or corn. For the want of this Botanical knowledge, ludicrous and sometimes mortifying mistakes have occurred, where no blame could be attached to the vendor of the seeds or plants, but were simply occasioned by an ignorance of Botany in the purchaser. I have seen many American plants that had been ordered from Paris and London by our florists with no higher recommendation than the exorbitant prices marked on the Catalogue; and when these plants were received, their new owners were not a little surprised to find that they had imported Azelias, Rhododendrons, Lobelias, Kalmias, Bignonias, and Magnolias, which they could have obtained by cart loads in their own woods. Some fifteen or twenty years ago, my esteemed neighbor, James Nicholson, obtained from Missouri the seeds of a grass on which the wild cattle and buffalo were said to feed and arrive. He disseminated the seed among his friends, who cultivated it under the name of Lewis and Clarke's grass. Among the rest, I sowed a bit of it in my garden where I had for many years been making experiments on some European grasses. The plants came to maturity, and I was not a little surprised to find the far famed Lewis and Clarke's grass to be an old and well-known European species, the meadow soft grass (Holcus lanatus), which I already possessed. The soil in which it was sowed was not probably suited to its growth, and I finally regarded it as inferior to the bluegrass (Dactylus glomerata), to which I have given the preference as a winter grass. I made allusion to these experiments in the October number of the Southern Agriculturalist of 1834, page 520, of which I was the temporary Editor during the absence of the friend who so ably conducted it. Recently, a distinguished Agriculturalist in Europe sent to me for some of the far-famed Muskeet grass from Texas, of which I had heard a favorable notice from some of my friends, but had overlooked the printed accounts. The seeds were accordingly obtained from Columbia. I hesitated in sending them immediately, inasmuch as I suspected in them an old and familiar acquaintance, which on being forwarded to Europe would be like sending "coals to Newcastle." An imperfect specimen of the mature plant kindly sent by my friend Joseph O'Hear rendered it quite certain that I had once more, under a new name, obtained the Holcus lanatus, the famous Lewis and Clarke's grass of Mr. James Nicholson, described by Linnæus, Curtis, Schreber, Knapp, and a whole host of European Botanists—a native of the woods and fields of Europe, experimented on by agriculturalists, and supposed less valuable than many other grasses and cultivation; introduced into

New England by our Pilgrim forefathers and carried by migratory birds to the far West and all over the land. Here now are seeds of a grass that have passed through the hands of the most intelligent, patriotic, and purest and of the State, liberally and generously disseminated over the country, and yet sad mistakes have originated from a want of Botanical knowledge. The grass, I trust, may yet prove as beneficial to the State as the most sanguine could desire. That which is unsuited to one soil and climate may be admirably adapted to another. But had those who imported and disseminated the seed possessed a small share of scientific knowledge, they might have been instructed by the experiments made on it for a century past in Europe and America and obtained it far short of Texas, even in many of their own fields and gardens.

The advantages of the study of systematic Botany may be further seen in removing the evils which result from the use of the common names given to plants, which are not uniform even in different neighborhoods of the same state. How many species of plants are there that are called Snake-root by the people of the country, which are indiscriminately applied to the cure of diseases although their characters and medicinal properties differ very widely, some being cathartics, some emetics, and others tonics. Various species of the three very distinct genera of Gentiana, Aristolochia, and Asarum are called by this unmeaning name. But more especially do we find this confusion existing in the common names given to Grasses; I will illustrate this by a familiar instance. The farmer of New-York possesses an invaluable grass called the Timothy grass, named it is supposed after the man who is said to have first introduced it into cultivation. But he has heard of other fine grasses on which he is also desirous of making an experiment. He sends to England for the Meadow Cats-tail, the finest grass in Europe, and to Maryland or New-England for the far-famed Herd's-grass. When these seeds have been received, after much expense and trouble, he will discover that these are but different names for the Timothy grass he already possesses and that he could have acquired the specific name of Phleum pratense, by which it is known among Botanists all over the world, much easier than he has learnt the various vulgar names by which such confusion has been created.

8. Another department in Natural History, which ought not to be

altogether overlooked in an Agricultural School, notwithstanding it is regarded as a study wholly unconnected with Agriculture, is *Ornithology*. Nature has wisely provided that one race of animals should serve as a check upon the too rapid increase of others. This uniformity is seen in every department of her works. The bird is a blessing to the husbandman by destroying the reptiles and insects, which would otherwise be an annoyance to him, and by ridding the earth of a superabundance of the seeds of weeds and grasses. The Stork in Holland, the Turkey Vulture in Carolina, and the Rooks of England are familiar instances, where man has acknowledged the benefits derived from some of their species; and indeed the boxes provided for the Purple Martin, and the calabashes for the house Wren all go to testify that the farmer is not wholly unmindful of the benefits he derives from some of the feathered race. I would we could say this of many other species that have been either neglected or misrepresented. So ignorant are our current lists of Ornithology that they know not what birds should be destroyed as nuisances or preserved as benefactors. Old Kalm tells us a story that will bear frequent repetition; that in Virginia, in his day, a bounty was given for the destruction of the little Crow, meaning no doubt some species of what are usually called Black birds, of which there is a considerable number of different Genera and habits, a few doing some injury to the corn, but the great majority, the Cow Bird (Leterus pectoris) especially, being decidedly beneficial to the farmer; but they were destroyed, (of how many species there was no Ornithologists to tell), and the consequence was that such was the increase of destructive insects that they, after a great expenditure of money, would have brought back the murdered birds at any price. In the days of our forefathers, a Governor of New-England offered three-pence a head for the Purple grackle; but it is said the insects multiplied so rapidly, that the herbage was destroyed, and the inhabitants were obliged to import hay from Pennsylvania and England. In fact, a single bird of this species by destroying the grubs that feed on the young corn save some more corn than would feed an hundred Grackles for a whole year. Even our Hawks, against which the world has declared an exterminating war did not deserve to be indiscriminately denounced inasmuch as a considerable number of the species are not only harmless, but beneficial to us. Four species that visit Carolina feed on insects, one on fish, one on serpents, and one on frogs and lizards. "I have made a good day's work" said a planter to me, "for after watching and crawling nearly all day, I have shot these two Hawks, which is as good as a dozen fowls saved." I thought he looked a little disconcerted when I informed him that he had killed a pair of his best friends, the Mississippi Kite, a species which feeds alone on insects, and is so little carniverous, that he would not even pounce on a sparrow. Some of our Owls feed exclusively on mice, others on small birds; and, of all our Southern species, the great horned or Virginia owl is the only one which is injurious to our poultry, and this is exceedingly rare. We have two species of Crow on our seaboard, one of which, and the rarest is principally injurious to corn; the other, feeding on worms, berries, and fish, is comparatively harmless, yet they are seldom known as distinct, by the farmer, who denounces vengeance indiscriminately against friend and foe.*

We have in Carolina about 250 of the 500 species of American birds found North of Mexico, the majority of which are migratory; but a large minority either remains with us sometime during spring and summer or are permanent residents; of all this number, we have only about a dozen species, that can by any possibility be regarded as decidedly injurious to the planter. All our Thrushes and Orioles, our Fly Catchers and Warblers, are useful aids in destroying worms and insects. The tyrant

---

* In the State of New-York, the following plan has been successfully adopted for thirty years in preventing the crows from pulling up the newly planted Indian corn. Boiling water is first poured on the seed corn in sufficient quantities to fill the vessel in which it is placed. When the water has stood on a few hours to become perfectly cool, it is poured off and half a pint of boiling coal tar to every bushel of corn is poured on the seat, which is carefully stirred in till every grain is covered with a thin coating of tar. It is then rolled in Gypsum, which is used as a valuable manure, but is of no particular use in keeping off the crows. The hot water does not affect the vitality of the grain, on the contrary, it hastens its germinative powers. When this plan was first adopted in the Northern States on a few of the farms, it did not immediately answer the expectations of the farmer inasmuch as the fields were visited by a succession of crows, and the new comers were all obliged to submit to a tarred mouth, before they could be induced to desist, and their thievish propensities were encouraged by other fields of untarred corn. But since the practice has become universal, the crow-minder has been found to be superfluous. A few of our Carolina planters have within the last few years adopted this plan, adding saltpetre to the hot water, omitting, I regret to say, the Gypsum, and they speak favorably of it. I have no doubts, were it to become general, it would be equally efficacious.

Fly catcher or Bee bird is destroyed by our American youths (who appear to have the organ of destructiveness largely developed) because he is said to kill bees. It is true he may kill the bee that falls in his way, but he makes amends by destroying thousands of noxious insects and is, moreover, a great protection to the poultry yard, for he never suffers a Hawk to come within a quarter of a mile of his nest—makes war against the Crow and Vulture and even pounces on the back of the Eagle, the emblem of our pride and glory. The whole tribe of Warblers feed on caterpillars and worms, each individual consuming several hundreds in a day. During the last spring, I had a large bed of cabbages and cauliflowers, which were so infested by the small green cabbage worm that their leaves were perforated like a honey-comb, and I was obliged to have the worms picked every morning, amounting often to a tea-cup-full. About this time, I observed a nest of the orchard Oriole in a garden adjacent to mine; the old birds found their way into my cabbage yard and so thoroughly kept down the worms for three weeks, that they proved better scavengers than my servants and saved us further labor during that time. But the nest was soon discovered by the little lads of the neighborhood, who seized on the young and caught the old in a trap cage, and now the cabbage worm re-appeared and remained a pest during the whole summer. The countless millions of Sparrows that visit us in winter merit our protection and gratitude on account of their devouring the seeds of weeds and grasses that would, otherwise, overrun our fields; and I even doubt, whether our Rice-bird, whose delicious flesh should disarm our hatred against it on account of the depredations it commits on our late crops of rice, does not, in another particular, make amends for its hasty autumnal meal by its return to the same fields on the following spring in its harlequin dress to pick up the scattered grains of rice left on the ground and thus saves in part at least the labor of picking out the stems of volunteer or red rice.

9. *Entomology.* This important study must yet be pressed into the service of the culturalist and can only be pursued with advantage by men instructed in the schools of science. Until we know the characters of insects, their modes of propagation, and peculiar habits, we can never find a remedy against their depredations. A proper knowledge of Entomology will save us much useless labor and be productive of incal-

culable advantage. Although this science was not regarded in Europe as of great importance to agriculture until within the last twenty years; yet such successful results have been produced that the farmer now regards the Entomologist with respect and welcomes him as a benefactor. As far as I have been able to ascertain, there is not a single insect in that country whose depredations were formerly dreaded that has not either been compelled by the power of science to pause in its career of mischief or been exterminated from the farms. It must, however, be recollected that we are living in another hemisphere, which, as far as Natural History is concerned, is emphatically a new world. There is not a solitary species of native quadruped, land bird, plant or insect in Carolina that is identical with any species in the old world. Their science can, therefore, only aid us in the manner in which our experiments must be conducted. It is left to us to study the species and discover the remedies. The pecuniary losses which are sustained by insects in a single year in two of our staples alone, wheat and cotton, would be sufficient permanently to endow an Agricultural School in every State and Territory of the Union. There is no greater mystery in the insects that lie imbeded at the joints of the wheat stalk or that perforate the bowls of cotton than there is in many better known species, whose habits have been determined and whose depredations have been effectually checked. We should also recollect that situated as we are so near the tropics, where insects so abundantly abound, we are constantly liable to fresh importations of foreign species that may become permanent residents if we are not timely guarded against them. Within the last five or six years, a minute insect has appeared on the stems of our Fig-trees, which if left unmolested, destroys the largest tree in a single season and at one time threatened the destruction of these, the most valuable fruit trees on our Southern sea-board. Its character, however, has been investigated, and we are induced to hope a remedy, which is now in the course of the experiment, has been discovered. On the Orange trees of Carolina, an insect, which when removed from its envelop is no larger than a pins point, made its appearance some twelve or fifteen years ago. How it was brought here or from whence it came, no one can tell as it appears not to have been described. The evil might have been checked in the bud, but the insidious foe was too insignificant in size to create apprehension, and in a few years our beautiful orange groves presented the appearance of a forest through which the fires had passed. A severe frost succeeded and relieved us of the unseemly sight by destroying effectually the few trees in which there still existed the lingering remains of life. Unfortunately, the same pest (not the less formidable on account of its minute size) has very recently been introduced into the Orange groves around St. Augustine. If the fires of Heaven had blasted every tree and herbs and consumed every dwelling and out-house in those regions of East Florida, where the orange is an article of commerce, the injury the inhabitants would have sustained would have been an hundred fold less than they are now doomed to experience from this insect. I perceive by the fruits obtained this season from Cuba that this enemy has also found its way into that beautiful Island, and unless the neglected and despised science of Entomology comes forth to the rescue, we may a few years hence look in vain for an Orange, Citron, Lemon, or Lime from Florida or Cuba. If the facts I have stated are not yet sufficient to convince the agriculturalist of the evils which arise from the neglect of this study, I will add a few others that have fallen under my personal observation.

About thirty-five years ago, in my native State of New-York, a large spiny caterpillar appeared on the Lombardy poplar trees, the only shade trees then existing in our Northern villages. Men became alarmed for it was reported that a dog had been stung and died, rumor soon magnified it into a child that had been killed in another part of the State. The press issued many strange conjectures and crude speculations. Consultations were held by the fathers of the land on this new danger from a recently imported and dreadfully poisonous scorpion, as it was supposed to be. It was urged in vain that the so much dreaded insect was only the larva of a gaudy and well known species of Butterfly (Papilio antiope) and that it would be an act of vandalism to destroy the beautiful shade trees for harboring a harmless caterpillar. But the science of Entomology, which had scarcely been heard of, could not stem the torrent of prejudice and ignorance, and it was resolved that the trees should be cut down forthwith; the axe was soon at work, and in the course of three weeks, the stately poplars in many of the villages and along the public highways of New-York as well as in many towns in New-England were felled and burnt. I am glad that their posterity have grown wiser and re-planted them.

A farmer on Charleston Neck two years ago solicited my advice in regard to the Tomato worm, which in spite of the most careful attention on his part had so increased during the season that his large crop of Tomatoes, on which he principally depended for the markets, was in danger of being wholly destroyed. "I pick them carefully (said he) with my own hands every morning, and bury them. I am sure that at least none of the large ones escape, yet the more I destroy, the more they multiply." I carried him to his little grave-yard, where he had buried his worms, showed him a quart of the remains of his enemies, which had already been transformed into a chrysalis state, assured him that those which had attained nearly their full size would have buried themselves in the ground if he had not saved them the trouble; that to use a vulgar phrase, he had carried the "Rabbit to the briar-bush," that this chrysalis would soon become a moth (Sphinx carolinus) and in its crepuscular flight would, by depositing its numerous eggs on the leaf of the Tomatoe, lay the foundation of ten thousand more of his formidable opponents. There was a faint smile of incredulity on his countenance and thus we parted.

A few weeks ago, I observed in the district of Lexington, near the borders of Edgefield, an extent of country fifteen miles in diameter where nearly all the Long-leaved pines had been killed by a worm usually called Sawyer or borer, producing as they supposed (no doubt correctly) disease and mortality among the inhabitants from the decay of vegetable matter and the exposure of the unshaded earth to the rays of the sun. The ravages of this destructive insect have been occasionally observed for many years past in particular localities between Florida and North-Carolina. Many of the inhabitants were at a loss to account for the cause of this destruction of their finest timber. Some supposing it to have been occasioned by the heavy rains of the last season, which they thought might have caused the roots of the trees to perish. The pileated Woodpeckers were by hundreds busy in extracting the marauders from the bark in the various stages of the larva, chrysalis, and the beautiful perfect, coleopterous Insect, and the lads were shooting them in all directions as many of them believed from the tens of thousands of perforations in the bark that these useful birds were destroying their trees.

How frequently have we seen in our maritime districts vast labor expended by the lovers of fine fruit in placing layers of oyster shells or a pavement of bricks around the roots of our Peach trees in order to guard the fruit from the depredations of that pest the Peach curculio. A little knowledge of Entomology would have convinced them that the Insect as it drops from or with fruit buries itself in the earth often at a considerable distance from the roots, where it undergoes its transformations and in the spring crawls up the tree to renew its depredations and that all their labor and expense is perfectly useless as a preventative and positively injurious to the growth of the tree. How few farmers are there who can be convinced that the worm (Ægeria exitiosa) which causes the gummy exudation from the stem is in no wise connected with the destroyer of the fruit and belongs to an entirely different family of Insects. The cause of the slavering of horses, that feed on the young growth of clover in the Middle States during summer, rendering the animals poor and the pastures almost useless, has occasioned infinite contradictory and some very absurd conjectures; and even now, it has not been satisfactorily determined whether the evil was occasioned by a poisonous weed or an Insects or whether it was to be traced to the clover itself. The same difficulties exist in regard to what is called the milk sickness in Alabama and Mississippi, where thousands of cattle annually fall victims to this disease and from which we are told, even man himself is not exempt.

But it may be asked, can science find a remedy against the depredations of Insects? I answer, this has already been done in a large number of cases in our own country even with our limited knowledge of the subject, possessing as we do scarcely a dozen Entomologists to seventeen millions of inhabitants; and man will prove himself unequal to his high destination, if he does not triumph over every opposing difficulty. The God of Nature has appointed him Lord over his lower creation. In obedience to his laws, he has conquered the Lion, the Tiger, and the Bear, subdued the native ferocity of the Elephant and rendered him obedient to his will and subservient to his use. He has measured the very heavens and is now beginning to traverse the earth with the speed of the bird, and surely he will not suffer himself to be discomfited by an Insect.

In conclusion, whilst I sincerely hope the contemplated Survey may by the judicious selection of the individual to whom the work is entrusted and by the forbearance, intelligence, and cheerful aid of our enlightened planters be productive of much benefit to our agricultural interests, I cannot refrain from expressing the desire and belief that in due time we may hear of efficient Agricultural Societies, Agricultural Publications, and an Agricultural School.

The melancholy facts are undoubted, that whilst all our Northern States have advanced in agricultural improvements, South-Carolina has deteriorated. Notwithstanding our mild climate, capable of producing two crops a year, the number of products to which our soil is adapted, the fine streams for mill-seats and manufactures with which our State abounds, and the facilities to a market afforded by our navigable rivers and rail-roads, the increase in our exports is confined to cotton and rice alone, and of the latter, we do not export a much larger quantity than we did half a century ago. Formerly, we exported grains; we are now importing from other States the materials that serve as food for ourselves and our cattle. In other years, the boats from Camden and Columbia brought corn and oats to Charleston; now these boats are employed in conveying these important and necessary articles of food as well as hay from our city (where they have been received from the States North of us) to Columbia and Camden. I have somewhere read a statement made by Mr. Seabrook, the accuracy of which cannot be questioned, that during the last twelve years, South-Carolina had on an average imported 350,000 bushels of corn annually. How greatly would this list of our imports be swelled were we to add to the above the oats, hay, flour, butter, beef, pork, lard, and other articles of food which might be produced in our own State. Under such wasteful drains on the pecuniary resources of the State—with so many enemies preying on the vitals of her prosperity—under a system of husbandry that is yearly rendering her soil more sterile—confining herself to the culture of cotton, which has greatly fallen in price, and of which more is grown that the world can consume—with rivals in Egypt, India, and Brazil, whilst the South-

Western States and Texas, can produce the article at a cheaper rate—how long will it be before South-Carolina will become wholly impoverished?

I have alluded to these humiliating facts not from any desire of indulging a spirit of disaffection or censoriousness, but with deep and unmingled regret that the State of my early choice—whose institutions I love, with whose prosperity in my best interests are associated, and for which my most fervent aspirations ascend—should by a neglect of her agricultural interests have permitted her neighboring States, posessing fewer natural resources, to outstrip her in the race of improvement and in that agricultural knowledge on which our prosperity and power so manifestly depend.

There is a consolation, however, in the belief, that our people are learning a salutary lesson from adversity. The resuscitation of a soil rendered sterile by improper modes of cultivation, it has now been ascertained, can be easily effected by patience, judgment, and a small share of industry. Great-Britain had once an unproductive soil, which by scientific culture has become the Garden of the World. I well recollect when the farms of the State of New-York were cultivated in the mode which has so long and so ruinously been adopted in Carolina. Successive crops of wheat were sown on the same field. In four or five years, the lands were exhausted and scarcely produced 6 bushels to the acre. They were then thrown out as older fields; the trees in the neighboring woods were deadened, the under-brush burnt, the new lands cultivated in the same slovenly way till they, in their turn, became unproductive, and the farmers went to seek for richer lands in the western counties. Now, these old and once abandoned fields have by a judicious mode of culture, by a rotation of crops, by the use of clover, plaster, lime, and other manures, been rendered more productive than when the lands were originally cleared and yield on average from twenty-five to thirty bushels of wheat or eighty bushels of corn to the acre. In Carolina too, it has been ascertained that although we cannot raise such large crops of corn as are produced in the Northern States, we can at least treble the product of former years. The results of the several experiments recorded in the Southern Agriculturalist, are sufficient evidences of this fact. Mr. O'Hear during the last season raised forty-six bushels, Mr. Coward fifty-two and three quarters, a writer under the signature of Z. fifty-seven and a half bushels, and Mr. B. R. Smith sixty-seven bushels and eighteen quarts to the acre.

Within the last few years, I have observed a spirit of inquiry and increased knowledge among our young planters. I can enumerate at least twenty in the limited list of my acquaintance who by judicious and scientific modes of culture have doubled the products of their fields. If the schoolmaster is at work, let us hail him as a benefactor and send him scholars. If there is a spirit of inquiry let us give it a right direction. If South-Carolina ever recovers her proud pre-eminence among her sister States, it will be through the means of Agricultural knowledge. When this is effected, she may hope to win back her sons who have not already permanently established their families or found graves in the far West. Our impoverished soils will be renovated, our people will be contented with their native homes, and the future destiny of our State will fulfil the glowing anticipations of those who found it a wilderness, rescued it from the savage, defended it against a foreign foe, and left it as a rich legacy to their descendents.

American Beaver: A Chapter from Audubon and Bachman's *Quadrupeds*

The *Viviparous Quadrupeds of North America* was published in six volumes: a total of 150 illustrations were published in three folio volumes in 1845, 1846, and 1848, and the text was published in three octavo volumes in 1846, 1851, and 1854. For the illustrations by the Audubons, Bachman helped primarily by identifying the species to be included, by supplying scientific names, and by supplying some of the animals to be painted. He had the primary responsibility for preparing the three volumes of text.

The introduction to the first volume of text states, "the habits of our quadrupeds was obtained by both authors, either from personal observation or through the kindness of friends of science on whose statements full reliance could be placed. For the designation of species and the letter-press of the present volume, the junior author is principally responsible" (p. 12). In 1852 Bachman wrote, "the figures were made by the Audubons, and the descriptions and letter-press were prepared by myself" (Bachman 1888: 279). The text was largely written and was entirely edited by him. He prepared the scientific portions for each entry, and he furnished much of the information for the life histories. Entries usually included a summary of the most relevant published accounts by numerous naturalists. Entries sometimes included long quotations from the journals of John James Audubon, information from the journals of John Woodhouse Audubon, and unpublished notes by other writers. Bachman evaluated and selected the best available information, identified the sources, organized the material coherently, and did most of the actual writing as well as furnishing much of the information himself from his own previous publications and unpublished observations.

The history of this project has been covered in the introduction. The difficulties involved in the long collaboration on such a major project are documented in Bachman's letters to the Audubons at the end of this volume. Some aspects of this project have also been considered in more detail by other authors (including Tyler 2000).

Bachman's previously published articles provided much of the scientific information that was needed for the text of the *Quadrupeds*. In 1837 his articles on hares and on shrews were published in the *Journal of the Academy of Natural Sciences of Philadelphia*. He described eight hares including four new ones and thirteen shrews including eight new ones. In 1838 his monograph on squirrels in the *Proceedings of the Zoological Society of London* (included in this volume) described fourteen squirrels including seven new ones. In 1839 he described six new species of quadrupeds in the *Journal of the Academy of Natural Sciences of Philadelphia*: one shrew mole, two mice, one marmot, and two ground squirrels. Thus, in 1837–39, Bachman had created a framework for the study of three genera of North American mammals (*Lepus, Sorex,* and *Sciurus*), and he had described species of five other genera. The American naturalist Charles Pickering called this work "the most valuable contribution that has ever been made to our mammalia" (Schuler 1995: 148). Although subsequent research has required revision, Bachman's previous articles and his own additions and corrections for the *Quadrupeds* provided a largely sound basis for later researchers to build upon (Cahalane 1967; for an updated list of the species in each of Bachman's articles, see Waddell 2005).

The genera Bachman knew best included species that were among the most difficult to distinguish from one another. Of the 147 species represented in the *Quadrupeds*, 127 were small mammals: squirrels, ground squirrels, shrews, shrew-moles, hares, rats, mice, and a mole. The twenty larger animals were ones that Audubon knew best and had recorded in most detail in his journals. For example, returning from his 1843 expedition on the Missouri River, Audubon wrote, "I have the best accounts of the habits of the Buffalo, Beaver, Antelope, Big Horns &c. that were ever written," and while this was undoubtedly true, Bachman wrote back, "I am afraid the broad shadows of the Elk, Buffalo, and big-horn [sheep] hid all the little marmots, squirrels, jumping mice, rats and shrews" (Peck 2000: 84–89).

In 1842, the year in which the first plates were published for the *Quadrupeds*, Bachman and Audubon coauthored an article that described twenty-three new species of quadrupeds in the *Journal of the Natural Sciences of Philadelphia*. These species represented eight genera (bats, weasels, moles, mice of two genera, marmot, squirrels, and marmot-squirrels). Together with the forty-seven species Bachman had previously described, these twenty-three additional ones raised the number of descriptions already prepared to seventy, nearly half of the total number needed for the *Quadrupeds*. Most of the remaining species had been previously described by other authors, but had been variously named and had often been incompletely described. All previously described species needed to be correctly identified and distinguished from one another, to be described as uniformly as possible, to have their life histories given as completely as possible, and to have their distributions summarized.

In 1842–43, Bachman summarized the information available on five species of the genus *Scalops* (shrew-moles), and he added two additional species to the one he had previously described and to another Audubon and he had described together. In 1852 they published what they considered to be a new species of genus *Vulpes* (fox), but their taxon was invalid. Two more new species were described in their *Quadrupeds*.

For his *Birds*, Audubon had been able to rely on Alexander Wilson's *Ornithology* as a good basis for determining the species of approximately half of the total number, but no equivalent work had been published on North American mammals. Although the *Quadrupeds* included only about one third as many species as the *Birds*, few mammals had been correctly named or accurately described, and most had been misnamed. Some published accounts were so vague that it was impossible to identify the species being described. Bachman had to find copies of every potentially relevant description, to ascertain which scientific name deserved priority, to determine which descriptions actually referred to the species he was working on, and to evaluate the reliability of all facts previously stated about each species. In the case of the beaver, Bachman was convinced that it was one of the few species of North American mammals that was represented in Europe and Asia, and he referred to it as *Castor fiber*, the name Linnaeus had assigned to all beavers. It is now considered a distinct species and is called *Castor canadensis*, and the name *Castor fiber* is reserved for the Eurasian species. In-vitro fertilization could settle the question whether these beavers were one species or two, and it could settle many similar questions definitively.

Most birds could be seen during the day, and since many migrated, they could be seen in various places at different times of the year. Since many mammals are nocturnal and live underground, mammals were found with greater difficulty and presented even more difficulties than birds.

Initially, Bachman argued for the creation of a comprehensive study of all mammals of North America including bats, seals, whales, and porpoises, and Audubon had reluctantly agreed. Audubon made a number of superb paintings of bats, but eventually the study had to be limited to quadrupeds.

The entries Bachman prepared for each genus and species were more comprehensive and systematic than the entries for Audubon's *Birds*. Each entry for a genus ordinarily includes a dental formula and discussion, and each entry for a species includes scientific and most common names, an illustration, the principal characteristics given in Latin and English, names used by other writers (synonymies), a detailed description, color, dimensions, habits, geographical description, and general remarks. Bachman had provided similar information as early as 1837 in his article on the genus *Sorex*.

In 1844, Bachman wrote Victor about what he planned to do and the help he would need:

> I enclose my plan. I wish always, a month before the time that you would give me notice of the [five] species you intend to put into the hands of the engraver and send me, at the same time the specimens. I cannot describe without it; I will guess at nothing. . . . Writing descriptions is slow and fatiguing work. I cannot in the careful manner that I am doing them write more than three in a week. . . . The following is my daily practice: I am up at 4 A. M. and work till breakfast, and recently, when parochial duties would permit, have kept on until 3 P. M. (Bachman, 1888: 202–3)

On his 1843 trip up the Missouri and Yellowstone rivers, Audubon found few new species of mammals, and he never reached the Rocky Mountains or the Pacific. In late 1845 and early 1846, John Woodhouse Audubon traveled in Texas to procure quadrupeds, but ran out of money and found that hostile

Indians blocked the way to the Pacific. Later in 1846, he went to London for a year to paint in its zoo and museums to add many mammals that were not represented in collections in the United States. In 1849 he joined the gold rush to California and finally reached the Pacific, but the folio edition of the plates had been completed in 1848. Five plates were subsequently added to the octavo edition.

Usually, it is possible to tell which portions of each entry were prepared by Bachman or Audubon. This is evident from their respective interests and from references to specific locations, and it is usually possible to distinguish their styles of writing. Audubon's descriptions of the prairies were among the most vivid and memorable he ever wrote. He contributed many details about the beaver.

Bachman was more systematic, thorough, and reflective, and although Audubon could be all these things, he was better at painting and at describing dramatic incidents than at scientific description. He had needed assistance in preparing the scientific part of the text of his *Ornithological Biographies*. Audubon rarely made systematic observations over long periods (as he did with the Pewee) or experimented (as to determine that American vultures detected prey by sight rather than smell). Bachman frequently did both, making, for example, over a three-year period the first accurate observations on the gestation of the Opossum.

When each genus is introduced, a headnote gives its dental formula, a brief description, discussion of how the genus differs from closely related genera, the number of species known worldwide and in North America, and the meaning of the name. The introduction acknowledged that the genera were "easily ascertained, by the forms and dental arrangements peculiar to each." The principal source for identifying genera was "F. Cuvier, on whose judgment, in regard to characters founded on dentition, we would sooner rely than on that of any other naturalist... (*Des dents des Mammifères*, 1825)." The dental formula is the number of different types of teeth on one side of the mouth (incisor, canine, premolar, and molar), and the numbers for the upper jaw are written before the numbers for the lower jaw. Bachman sometimes disagreed with Cuvier about the definitions of genera, but more often about the genera to which various species belonged. A total of forty-eight genera were considered valid at the time, and the number of species included for each genus has been listed elsewhere (Audubon 1989: index).

Reviews of the *Quadrupeds* were highly favorable (Stephens 2000: 176, n. 16). In 1846 an anonymous but well informed reviewer in the *American Review* summarized the efforts of Audubon and Bachman's predecessors in detail, and he concluded that the quadrupeds had reached "the head of Illustrative Mammalogy in the world." Similar praise by Agassiz and later by Cahalane have been noted in the introduction for this volume.

The *Quadrupeds* went far beyond all previous publications on American mammals in its depth, quality, comprehensiveness, readability, and illustrations, but its coverage was necessarily somewhat uneven and limited. As its authors acknowledged, the eastern half of the United States was far better represented than the western half, and some species had been described on the basis of fewer specimens than others. Most species were represented by a single set of measurements. However, the basic work of defining and interrelating the principal species had been accomplished, and the finest of the illustrations have rarely been equaled. Of the 147 species included in the *Quadrupeds*, naturalists of the American Museum of Natural History found 119 to be valid in 1967. Although four out of five continued to be accepted, twenty-eight of the mammals described as species are now considered varieties (Cahalane 1967: xvi).

Considering that nearly 450 species of mammals are now recognized in North America north of Mexico, Audubon and Bachman's contribution might seem to have been less significant than it was (Wilson and Ruff 1999). They did well to reduce the number of species that had been proposed by their predecessors, and the species they retained and added were a far greater proportion of the present total than had been described previously and were far better described and illustrated. Moreover, even though they planned to, they did not include bats, whales, porpoises, and seals (which are included in the total that is now recognized); they were able to obtain relatively little information available about the western half of the United States; and their work was prepared by only three persons.

Following publication, the number of known species was almost immediately enlarged very substantially. In 1859, Baird's *Mammals of North America* included 256 species, but he benefited from the work of about one hundred naturalists who were involved in collecting and analyzing specimens found during surveys for the transcontinental railroad. That averages out to just over two species per naturalist compared to the twenty new species that Bachman alone had identified and described. Baird's illustrations consisted primarily of line drawings of skulls, and his views of mammals were without settings. A recent summary of nearly 450 mammals in North America was prepared with the collaboration of two hundred specialists, and photographs were used for its illustrations (Wilson and Ruff, 1999: ix, xi, xiii).

The text of the *Quadrupeds* offers the general reader as much as it offers the scientist, and it has often been reprinted. Its best entries deserve to be made widely available in a separate volume. The best in terms of insight, comprehensiveness, and general interest are about the mole, northern hare, musk rat, lynx, gray rabbit, wolverine, northern gray squirrel, common American skunk, beaver, badger, otter, brown rat, buffalo, white weasel, raccoon, elk, harvest mouse, opossum, antelope, deer, cougar, caribou, grizzly bear, and black bear. The beaver is the only one of the 147 descriptions that could be included in a volume that is intended to provide a balanced view of Bachman's accomplishments and that also includes his articles on the opossum and on most species of squirrel.

The / Viviparous Quadrupeds / of / North America. / by /
John James Audubon, F.R.S., &c., &c. / and / the Rev. John Bachman, D.D., &c., &c. / vol. I. / New-York: / Published by J. J. Audubon. / M DCCC XLVI.

Genus Castor.—Linn.

Dental Formula.
Incisive 2/2; Canine 0-0/0-0; Molar 4-4/4-4 = 20.

Incisors very strong. In the upper jaw their anterior surface is flat and their posterior surface angular. The molars differ slightly from each other in size and have one internal and three external grooves. In the lower jaw of the incisors present the same appearance as those of the upper, but are smaller. In the molars there are three grooves on the inner side with one on the external.

Eyes small; ears short and round; five toes on each foot. On most fore-feet the toes are short and close; on the hind-feet long and palmated. Tail, large, flat, and scaly. Mammæ, four, pectoral: a pouch near the root of the tail in which an unctuous matter is secreted.

There is but one well-established species known to belong to this genus.

The generic name is derived from the Latin word *Castor*, a beaver.

Castor Fiber.—Linn.
(Var. Americanus.)
American Beaver.
Plate XLVI. Two-thirds natural size.
C. Arct. Monace major, supra badius, infra dilutior; cauda plana, ovate, squamosa

Characters.

Larger than the ground-hog (Arctomys monax); of a reddish-brown colour, with a short downy grayish fur beneath; tail, flat, scaly, and oval.

Synonymies.

CASTOR FIBER, Linn., 12th ed., p. 78.
CASTOR, Sagard Theodat, Canada, p. 767.
BEAVER, CASTOR, Pennant, Arc. Zool., vol. i., p. 98.
CASTOR ORDINAIRE, Desm., Mamm.
CASTOR AMERICANUS, F. Cuvier.
CASTOR FIBER. Lewis and Clarke's Expedition, vol. i.
THE BEAVER, Hearne's Journal, vol. viii., p. 245.
BEAVER, Cartwright's Journal, vol. i., p. 62.
    "   Catesby, App., p. 29.
CASTOR FIBER, Harlan, Fauna, p. 122.
    "   "   Godman, vol. ii., p. 21.
    "   "   AMERICANUS, Richardson, F.B.A., p. 105.
    "   "   Emmons, Mass. Reports, p. 51.
    "   "   Dekay, pl. 1, p. 72.

Description.

The shape of the body bears a considerable resemblance to that of the musk-rat; it is, however, much larger, and the head is proportionally thicker and broader. It is thick and clumsy, gradually enlarging from the head to the hips, and then is somewhat abruptly rounded off to the root of the tail.

Nose, obtuse and divided; eyes, small; ears, short, rounded, well clothed with fur, and partially concealed by the longer surrounding hairs; mustaches, not numerous, but very rigid like hogs' bristles, reaching to the ears; neck, rather short. The fur is of two kinds. The upper and longer hair is coarse, smooth, and glossy; the under coat is dense, soft, and silky. Fore-feet, short and rather slender; toes, well separated and very flexible. The fore-feet are used like hands to convey food to the mouth. The fore-claws are strong, compressed, and channelled beneath. The middle toe is the longest, those on each side a little shorter, and the outer and inner ones shortest.

The hind-feet bear some resemblance to those of the goose. They are webbed beyond the roots of the nails and have hard and callous soles. In most of the specimens we have seen, there is a double nail on the second inner toe. The palms and soles are naked. When walking, the whole heel touches the ground. The Beaver is accustomed to rest itself on its hind-feet and tail; and when in the sitting position contracts its fore-claws in the manner of the left-hand figure represented in the plate. The upper surface of all the feet with the exception of the nails, which are naked, is thickly covered with short adpressed hairs.

The tail is very broad and flat, tongue-shaped, and covered with angular scales. The root of the tail is for an inch covered with fine fur. The glandular sacs containing the castoreum, a musky unctuous substance, are situated near the anus.

Colour.

Incisors, on their outer surface, orange; moustaches, black; eyes, light-brown. The soft under down is light grayish-brown. The upper fur on the back is of a shining chestnut colour; on the under surface, and around the mouth and throat, a shade lighter. Nails, brown; webs between the toes, and tail, grayish-brown. We have seen an occasional variety. Some are black; and we examined several skins that were nearly white.

Dimensions.

Male, represented in the plate.—Rather a small specimen.

From nose to root of tail,	23 inches.
Tail,	10 do.
From heel to end of middle claw,	5½ do.
Greatest breadth of tale,	3¼ do.
Thickness of tail,	⅞ do.

Weight, 11¼ lbs.

Habits.

The sagacity and instinct of the Beaver have from time immemorial been the subject of admiration and wonder. The early writers on both continents have represented it as a rational, intelligent, and moral being, requiring both the faculty of speech to raise it almost to an equality, in some respects, with our own species. There is in the composition of every man, whatever may be his pride in his philosophy, a proneness in a greater or less degree to superstition or at least credulity. The world is at best but slow to be enlightened, and the trammels thrown around us by the details of the nursery are not easily shaken off. Such travellers into the northern parts of Sweden, Russia, Norway, and Lapland, as Olaus Magnus, Jean Marius, Rzacynsky, Leems, &c., whose extravagant and imaginary notions were recorded by the credulous Gesner, who wrote marvellous accounts of the habits of the Beavers in Northern Europe, seem to have worked on the imaginations and confused the intellects of the early explorers of our Northern regions—La Hontan, Charlevoix, Theodat, Ellis, Beltrami, and Cartwright. These last, excited the enthusiasm of Buffon, whose romantic stories have so fastened themselves on the mind of childhood, and have been so generally made a part of our education, that we now are almost led to regret that three-fourths of the old accounts of this extraordinary animal are fabulous; and that with the exception of a very peculiar mode of constructing its domicile, the Beaver is in point of intelligence and cunning greatly exceeded by the fox and is but a few grades higher in the scale of sagacity than the common musk-rat.

The following account was noted down by us as related by a trapper named Prevost, who had been in the service of the American Fur Company for upwards of twenty years in the region adjoining the spurs of the Rocky Mountains and who was the "Patroon" that conveyed us down the Missouri river in the summer and autumn of 1843. As it confirms the statements of Hearne, Richardson, and other close observers of the habits of the Beaver, we trust that although it may present little that is novel, it will from its truth be acceptable and interesting to our readers. Mr. Prevost states in substance as follows.

Beavers prefer small clear-water rivers and creeks and likewise resort to large springs. They, however, at times, frequent great rivers and lakes. The trappers believe that they can have notice of the approach of winter weather and of its probable severity by observing the preparations made by the Beavers to meet its rigours; as these animals always cut their wood in good season, and if this be done early, winter is at hand.

The Beaver dams, where the animal is at all abundant, are built across the streams to their very head waters. Usually these dams are formed of mud, mosses, small stones, and branches of trees cut about three feet in length and from seven to twelve inches round. The bark of the trees in all cases being taken off for winter provender before the sticks are carried away to make up the dam. The largest tree cut by the Beaver seen by Prevost measured eighteen inches in diameter; but so large a trunk is very rarely cut down by this animal. In the instance just mentioned, the branches only were used, the trunk not having been appropriated to the repairs of the dam or aught else by the Beavers.

In constructing the dams, the sticks, mud, and moss are matted and interlaced together in the firmest and most compact manner; so much so that even men cannot destroy them without a great deal of labour. The mud and moss at the bottom are rooted up with the animal's snout, somewhat in the manner hogs work in the earth, and clay and grasses are stuffed and plastered in between the sticks, roots, and branches, in so workmanlike a way as to render the structure quite water-tight. The dams are sometimes seven or eight feet high and are from ten to twelve feet wide at the bottom, but are built up with the sides inclining towards each other so as to form a narrow surface on the top. They are occasionally as much as three hundred yards in length, and often extend beyond the bed of the stream in a circular form so as to overflow all the timber near the margin, which the Beavers cut down for food during winter, heaped together in large quantities, and so fastened to the shore under the surface of the water, that even a strong current cannot tear it away; although they generally place it in such a position that the current does not pass over it. These piles or heaps of wood are placed in front of the lodges, and when the animal wishes to feed he proceeds to them, takes a piece of wood, and drags it to one of the small holes near the principal entrance running above the water, although beneath the surface of the ground. Here the bark is devoured at leisure, and the wood is afterwards thrust out or used in repairing the dam. These small galleries are more or less abundant according to the number of animals in the lodges. The larger lodges are in the interior about seven feet in diameter and between two and three feet high, resembling a great oven. They are placed near the edge of the water, although actually built on or in the ground. In front, the Beavers scratch away the mud to secure a depth of water that will enable them to sink their wood enough to prevent its being impacted in the ice when the dam is frozen over, and also to allow them always free egress from their lodges, so that they may go to the dam and repair it if necessary. The top of the lodge is formed by placing branches of trees matted with mud, grasses, moss, &c., together, until the whole fabric measures on the outside from twelve to twenty feet in diameter and is six or eight feet high, the size depending on the number of inhabitants. The outward coating is entirely of mud or earth, and smoothed off as if plastered with a trowel. As Beavers, however, never work in the day-time, no person we believe has yet seen how they perform their task or give this hard-finish to their houses. This species does not use its fore-feet in swimming, but for carrying burthens: this can be observed by watching the young ones, which suffered their fore-feet to drag by the side of the body, using only the hind-feet to propel themselves through the water. Before diving, the Beaver gives a smart slap with its tail on the water, making a noise that may be heard a considerable distance, but in swimming, the tail is not seen to work, the animal being entirely submerged except the nose and part of the head; it swims fast and well, but with nothing like the speed of the otter (*Lutra Canadensis*).

The Beavers cut a broad ditch all around the lodge so deep that it cannot freeze to the bottom, and into this ditch they make the holes already spoken of, through which they go in and out and bring their food. The beds of these singular animals are separated slightly from each other, and are placed around the wall, or circumference of the interior of the lodge; they are formed merely of a few grasses or the tender bark of trees: the space in the center of the lodge being left unoccupied. The Beavers usually go to the dam every evening to see if repairs are needed and to deposit their ordure in the water near the dam, or at least at some distance from their lodge.

They rarely traveled by land unless their dams have been carried away by the ice, and even then they take the beds of rivers or streams for their roadway. In cutting down trees they are not always so fortunate as to have them fall into the water or even towards it, as the trunks of trees cut down by these animals are observed to lie in various positions; although as most trees on the margin of a stream or river leaned somewhat towards the water or have their largest branches extended over it many of those cut down by the Beavers naturally fall in that direction.

It is a curious fact, says our trapper, that among the Beavers there are some that are lazy and will not work at all, either to assist in building lodges or dams or to cut down wood for their winter stock. The industrious ones beat these idle fellows and drive them away; sometimes cutting off a part of their tail and otherwise injuring them. These "Paresseux" are more easily caught in traps than the others, and the trapper rarely misses one of them. They only dig a hole from the water running obliquely towards the surface of the ground twenty-five or thirty feet, from which they emerge when hungry to obtain food, returning to the same hole with the wood they procure to eat the bark.

They never formed dams and are sometimes to the number of five or seven together; all are males. It is not at all improbable that these unfortunate fellows have, as is the case with the males of many species of animals, then engaged in fighting with others of their sex, and after having been conquered and driven away from the lodge, have become idlers from a kind of necessity. The working Beavers, on the contrary, associate, males, females, and young together.

Beavers are caught and found in good order at all seasons of the year in the Rocky Mountains; for in those regions of the atmosphere is never warm enough to injure the fur; in the low-lands, however, the trappers rarely begin to capture them before the first of September, and that they relinquished the pursuit about the last of May. This is understood to be along the Missouri and the (so called) Spanish country.

Cartwright, (vol. i., p. 62) found a Beaver that weighed forty-five pounds; and we were assured that they have been caught weighing sixty-one pounds before being cleaned. The only portions of their flesh that are considered fine eating are the sides of the belly, the rump, the tail, and the liver. The tail, so much spoken of by travellers and by various authors as being very delicious eating, we did not think equalled their descriptions. It has nearly the taste of beef marrow, but is rather oily and cannot be partaken unless in a very moderate quantity except by one whose stomach is strong enough to digest the most greasy substances.

Beavers become very fat at the approach of autumn; but during winter they fall off in flesh so that they are generally quite poor by spring, when they feed upon the bark of roots and the roots of various aquatic plants, some of which are at that season white, tender, and juicy. During winter, when the ice is thick and strong, the trappers hunt the Beaver in the following manner. A hole is cut in the ice as near as possible to the aperture leading to the dwelling of the animal, the situation of which is first ascertained; a green stick is placed firmly in front of it, and a smaller stick on each side about a foot from the stick of green wood; the bottom is then padded or beaten smooth and even, and a strong stake is set into the ground to hold the chain of the trap, which is placed within a few inches of the stick of green wood, well baited, and the Beaver, attracted either by the fresh bark or the bait, is almost always caught. Although when captured in this manner, the animal struggles, diving and swimming about in its efforts to escape, it never cuts off a foot in order to obtain its liberty; probably because it is drowned before it has had time to think of this method of saving itself from the hunter. When trapping under other circumstances, the trap is placed within five or six inches of the shore and about the same distance below the surface of the water, secured and baited as usual. If caught, the Beaver now and then cut off the foot by which they are held in order to make their escape.

A singular habit of the Beaver was mentioned to us by the trapper, Prevost, of which we do not recollect having before heard. He said that when two Beaver lodges are in the vicinity of each other, the animals proceed from one of them at night to a certain spot, deposit their castoreum, and then return to their lodge. The Beavers in the other lodge, dissenting this, repaired to the same spot, cover it over with earth, and then make a similar deposit on the top. This operation is repeated by each party alternately until quite a mound is raised, sometimes to the height of four or five feet.

The strong musky substance contained in the glands of the Beaver is called castoreum; by trappers, bark-stone; with this the traps are baited. A small stick, four or five inches long, is chewed at one end, and that part dipped in the castoreum, which is generally kept in a small horn. The stick is then placed with the anointed end above water, and the other end downwards. The Beaver can smell the castoreum at least one hundred yards, makes towards it at once, and is generally caught.

Where Beavers have not been disturbed or hunted and are abundant, they rise nearly half out of water at the first smell of the castoreum and become so excited that they are heard to cry aloud and breathe hard to catch the odour as it floats on the air. A good trapper used to catch about eighty Beavers in the autumn, sixty or seventy in the spring, and upwards of three hundred in the summer, in the mountains; taking occasionally as many as five hundred in one year. Sixty or seventy Beaver skins are required to make a pack weighing one hundred pounds; which, when sent to a good market, is worth, even now, from three to four hundred dollars.

The Indians occasionally destroy Beaver-dams in order to capture these animals, and have good dogs to aid them in this purpose. The Mountain Indians, however, are not trappers.

Sometimes the Indians of the Prairies break open Beaver lodges in the summer-time as during winter they are usually frozen hard. The Beaver is becoming very scarce in the Rocky Mountains, so much so that if a trapper now secures one hundred in the winter and spring hunt, he is considered fortunate.

Formerly, when the fur was high in price and the animals were abundant, the trading companies were wont to send as many as thirty or forty men, each with from six to twelve traps and two good horses: when arrived at a favourable spot to begin their work, these men erected a camp, and each one sought alone for his game, the skins of which he brought to camp, where a certain number of men always remained to stretch and dry them.

The trappers subsist principally upon the animals they kill, having a rifle and a pair of pistols with them. After a successful hunt, on meeting each other at the camp, they have a "frolic" as they term it.

Some old and wary Beavers are so cunning, that on finding the bait they cover it over, as if they were on the ground with sticks &c., deposit their own castoreum on the top and managed to remove the trap. This is often the case when the Beaver has been hunted previously. In places where they have remained undisturbed, but few escape the experienced trapper. The trappers are not very infrequently killed by the Indians, and their occupation is one involving toil and hazard. They rarely gain a competence for their old age, to say nothing of a fortune, and in fact all the articles they are of necessity obliged to purchase in the "Indian country" cost them large sums as their price is greatly increased by the necessary charges for transportation to the remote regions of the West.

When at Fort Union, we saw a trapper who had just returned from an unfortunate expedition to the mountains; his two horses had been stolen, and he lost his gun and rifle when coming down the river in a slender canoe and was obliged to make for the shore, dig a hole where he deposit a few furs he had left, and travel several hundred miles on foot with only berries and roots for his food. He was quite naked when he reached the Fort.

The Beaver which we brought from Boston to New York was fed principally on potatoes and apples, which he contrived to peel as if assisted with a knife, although his lower incisors were his only substitute for that useful implement. While at this occupation the animal was seated on his rump in the manner of a ground-hog, marmot, or squirrel and looked like a very large wood-chuck, using his fore-feet as squirrels and marmots are wont to do.

This Beaver was supplied every day with a large basin filled with water, and every morning his ordure was found to have been deposited therein. He generally slept on a good bed of straw in his cage, but one night having been taken out and placed at the back of the yard in a place where we thought he would be secure, we found next morning to our surprise that he had gnawed a large hole through a stout pine door which separated him from that part of the yard nearest the house and had wandered about until he fell into the space excavated and walled up outside the kitchen window. Here he was quite entrapped and having no other chance of escape from this pit into which he had unluckily fallen, he gnawed away at the window-still and the sash, on which his teeth took such effect that on an examination of the premises we found that a carpenter and several dollars' worth of work were needed to repair damages. When turned loose in the yard in the day-time he would at times slap his tail twice or thrice on the brick pavement, after which he elevated this member from the ground, and walked about in an extremely awkward manner. He fell ill soon after we had received him, and when killed, was examined by Dr. James Trudeau, who founded that he would shortly have died of an organic disease.

It is stated by some authors that the Beaver feeds on fish. We doubt whether he possesses this habit as we on several occasions placed fish before those we saw in captivity, and although they were not very choice in their food and devoured any kind of vegetable and even bread, they in every case suffered fish to remain untouched in their cages.

The food of this species in a state of nature consists of the bark of several kinds of trees and shrubs and of bulbous and other roots. It is particularly fond of the bark of the birch (*Betula*), the cotton-wood (*Populus*), and of several species of willow (*Salix*); it feeds also with avidity on the roots of some aquatic plants, especially on those of the *Nuphair luteum*. In summer, when it sometimes wanders to a distance from the water, it eats berries, leaves, and various kinds of herbage.

The young are born in the months of April and May; those produced in the latter month are the most valuable, as they grow rapidly and become strong and large, not being checked in their growth, which is often the case with those that are born earlier in the season. Some females have been taken in July with young, but such an event is a rare occurrence. The eyes of the young Beaver are open at birth. The dam at times brings forth as many as seven at a litter, but from two to five is the more usual number. The young remain with the mother for at least a year, and not infrequently two years, and when they are in a place of security, where an abundance of food is to be procured, ten or twelve Beavers dwell together.

About a month after their birth, the young first follow the mother and accompany her in the water; they continue to suckle some time longer, although if caught at that tender age, they can be raised without any difficulty, feeding them with tender branches of willows and other trees. Many Beavers from one to two months old are caught in traps set for old ones. The gravid female keeps aloof from the male until after the young have begun to follow her about. She resides in a separate Lodge till the month of August, when the whole family once more dwell together.

Geographical Distribution.

According to Richardson the Beaver exists on the banks of the Mackensie, which is the largest river that discharges itself into the Polar Sea: he speaks of its occurring as high as 67 ½ or 68° north latitude and states that its range from east to west extends from one side of the continent to the other. It is found in Labrador, Newfoundland, and Canada, and also in some parts of Maine and Massachusetts. There can be no doubt that the Beaver formerly existed in every portion of the United States. Catesby noticed it as found in Carolina, and the local names of Beaver Creek, Beaver Dam, &c., now existing, are evidences that the animal was once known to occupy the places designated by these compounds of its name. We have, indeed, examined several localities, some of which are not seventy miles from Charleston, where we were assured the remains of old Beaver dams existed thirty-five years ago. Bartram in his visit to Florida in 1778 (Travels, p. 281) speaks of it as at that time existing in Georgia and East Florida. It has, however, become a scarce species in all of the Atlantic States, and in some of them has been entirely extirpated. It, however, may still be found in several of the less cultivated portions of many of our States. Dr. Dekay was informed that in 1815 a party of St. Regis Indians obtained three hundred Beavers in a few weeks in the St. Lawrence county, N.Y. In 1827 we were shown several Beaver-houses in the North-western part of New York, where, although we did not see the animals, we observed signs of their recent labours. Dekay supposes (N.Y. Fauna, p. 78) that the Beaver does not at present exist south of certain localities in the State of New York. This is an error. Only two years ago we received a foot of one, the animal having been caught not twenty miles from Asheville in North Carolina. We saw in 1839 several Beaver-lodges a few miles west of Peter's Mountain in Virginia on the head waters of the Tennessee River and observed a Beaver swimming across the stream. There is a locality within twenty miles of Milledgeville, Georgia where Beavers are still found. Our friend Major Logan, residing in Dallas county, Alabama, informed us that they exist on his plantation and that within the last few years a storekeeper in the immediate vicinity purchased twenty or thirty skins annually from persons residing in his neighbourhood.

We were invited to visit this portion of Alabama to study the habits of the Beaver and to obtain specimens. Some years ago we shot one near Henderson, Kentucky in Canoe Creek; it was regarded as a curiosity and probably none have been seen in that section of the country since. We have heard that the Beaver was formerly found near New Orleans, but we never saw one in Louisiana. This species exists on the Arkansas River in the streams running from the Rocky Mountains and along their whole range on both sides; we have traced it as far as the northern boundaries of Mexico, and it is no doubt found much farther south along the mountain range. Thus it appears that the Beaver once existed on the whole continent of North America, north of the Tropic of Cancer and may still occur, although in greatly diminished numbers, in many localities in the wild and uncultivated portions of our country; we are nevertheless under the impression that in the Southern States the Beaver was seldom found in those ranges of country where the musk-rat does not exist, hence we think it could never have been abundant in the alluvial lands of Carolina and Georgia as the localities where its dams formerly existed are on pure running streams and not on the sluggish rivers near the sea-coast.

General Remarks.

It is doubted by some authors whether the American Beaver is identical with the Beaver which exists in the north of Europe; F. Cuvier, Kuhl, and others described it under the names of *C. Americanus, C. Canadensis*, &c. From the amphibious habits of this animal and its northern range on both continents, strong arguments in favour of the identity of the American and European species might be maintained, even without adopting the theory of the former connexion of the two adjacent continents. We carefully compared many species (American and European) in the museums of Europe and did not perceive any difference between them except that the American specimens were a very little larger than the European. We saw a living Beaver in Denmark that had been obtained in the north of Sweden; in its general appearance and actions it did not differ from those we have seen in confinement in America. It has been argued, however, that the European animal differs in its habits from the American; that along the banks of the Weser, the Rhone, and the Danube, the Beavers are not gregarious; and that they burrow in the banks like the musk-rat. The change of habit may be the result of altered circumstances and is not in itself sufficient to constitute a species. Our wild pigeon (*Columba migratoria*) formerly bred in communities in the Northern States; we once saw one of their breeding places near Lake Champlain, where there were more than a hundred nests on a single tree. They still breed in that portion of the country, but the persecutions of man have compelled them to adopt a different habit, and two nests are now seldom found on a tree.

The banks of the European rivers (on which the Beaver still remains although scarcely more than a straggler can be found along them now) have been cultivated to the water's edge, and necessity, not choice, has driven the remnant of the Beaver tribe to the change of habit we have referred to. But if the accounts of travellers in the north of Europe are to be relied on, the habits of the Beaver are in the uncultivated portions of that country precisely similar to those exhibited by the animal in Canada. We consider the account of these animals given us by Hearne, (p. 234) as very accurate. He speaks of their peculiarly constructed huts, their living in communities and their general habits. In the account of Swedish Lapland by Professor Leems, published in Danish and Latin, Copenhagen, 1767, we have the following notice of the European species (we quote from the English translation in Pinkerton's Voyages, vol. i., p. 419.): "The Beaver is instinctively led to build his house near the banks of lakes and rivers. He saws with his teeth birch trees, with which the building is constructed; with his teeth he drags the wood along to the place destined for building his habitation; in this manner one piece of timber is carried after another where they choose. At the lake or river where their house is to be built, they lay birch stocks or trunks covered with their bark in the bottom itself, and forming a foundation, they complete the rest of the building, with so much art and ingenuity as to excite the admiration of the beholders. The house itself is of a round and arched figure, equalling in its circumference the ordinary hut of a Laplander. In this house the floor is for a bed, covered with branches of trees, not in the very bottom, but a little above, near the edge of a river or lake; so that between the foundation and flooring on which the dwelling is supported, there is formed as it were a cell filled with water in which the stalks of the birch tree are put up; on the bark of this, the Beaver family who inhabit this mansion feed. If there are more families under one roof, besides the laid flooring, another resembling the former is built a little above, which you may not improperly name a second story in the building. The roof of the dwelling consists of branches very closely compacted, and projects out far over the water. You have now, reader, a house consisting and laid out in a cellar, a flooring, a hypocaust, a ceiling, and a roof, raised by a brute animal, altogether destitute of reason, and also of the builders art, with no less ingenuity than commodiousness."

It should be observed that Leems, who was a missionary in that country, gave this statement as related to him by the Laplanders who reside in the vicinity of the Beavers and not from his own personal observations. This account, though mixed up with some extravagancies and the usual vulgar errors (which we have omitted) certainly proves that the habits of the Beaver in the northern part of Europe are precisely similar to those of that animal on the northern continent of America.

Generation of the Opossum

The Virginia Opossum is the only marsupial in North America, and at the time Bachman did research on it, most aspects of its reproduction were uncertain. As for his studies of migration and of changing colors, he approached this problem comprehensively and provided new insights of lasting value. He made extensive observations during three breeding seasons to be sure his conclusions were correct.

This article consists of three parts, two of which are by Bachman; the third part is a supplement by another author. Bachman evidently submitted his manuscript in either two or three separate parts, but they were received in time to be published together. The first part consists of observations he made from March 1, 1846, through February 13, 1848. In its second part, "Further Observations," he presented information discovered on February 15, 1848, that enabled him to draw more definite conclusions. In an appended letter, Middleton Michel added more related information.

Bachman noted that during the spring of three successive years, he had as many opossums as possible brought to him around the time he suspected they were being impregnated. He was trying to find females during the earliest period of gestation. He found that the uterus is divided into two parts and the penis is bifurcated, that opossums copulate on their sides, and that the period of gestation is from about fifteen to seventeen days.

On February 14, 1848, the day after he had submitted his first report, Bachman received a number of additional opossums. One had an approximately equal number of fetuses in each part of its uterus, and others had newborn in their pouches. He established conclusively that the fetus was not born attached to a teat. Contrary to other accounts, the fetus at birth was already largely formed and was breathing. The newborn remained in the separate pouch for two or three weeks before emerging.

Middleton Michel's letter notes further that the number of young was from six to thirteen and that the male and female face one another during copulation. Neither Michel nor Bachman could detect a placenta or umbilical cord.

Bachman's article on the opossum was originally published in 1848 in the *Proceedings of the Academy of Natural Sciences of Philadelphia*; in 1851 it was published in the *Archiv für Naturgeschichte*. The manuscript is in the Academy of Natural Sciences of Philadelphia (Stephens 2000: 277, n. 23). Bachman's findings shed rare light on this animal: "As an authority on *Didelphis virginiana* noted a century later, Bachman 'made astute observations on the opossum and . . . brought up to date the existing knowledge about that animal. . . .' Seventy years passed before the problem of marsupial birth was again taken up" (Stephens 2000: 54 [citing Hartman, *Possums*, 88, 94]).

Notes on the Generation of the Virginia Opossum (*Didelphis Viginiana.*)

By John Bachman, D.D.

Under an impression that the following extracts from notes made at intervals during the last few years may throw some additional light on the natural history of one of the most interesting of American quadrupeds, I communicate them for the information of the Society.

March 1st, 1846.—Received to day five female opossums, captured last night. One of these had ten young in the pouch; another nine; the third had eleven & the fourth fourteen. They were all very diminutive and appear to be nearly the same age—about two or three days. The fifth was a small animal of the preceding autumn, and I was doubtful whether she had been impregnated.

March 3d.—On the evening of this day, I examined my small female opossum. The mammary organs were considerably distended, and I began to suspect that I had erred in my previous conjectures, and concluded to dissect her on the following day.

March 4th.—At 7 o'clock this morning, when prepared to commence my dissection of the opossum, I discovered three young in the pouch, and supposing that so small a female would produce no additional number, I concluded that I would spare her life. She was confined in a box in a room where I was writing. When I occasionally looked at her I found her lying on her side, her body drawn up in the shape of a ball; the vulva appeared to reach the pouch, which was occasionally distended with her paws. At 6 o'clock in the afternoon, as she had appeared very restless for several hours, I was induced to examine her again, when I discovered that she had added four more to her previous number, making her young family now to consist of seven. With no inconsiderable labour, and the exercise of much patience, I removed three of the young from the teat, one of which perished under the process. The three weighed twelve grains, averaging four grains each. I replaced the two living ones in the pouch; at 9 o'clock examined her and found the young again attached to the teats.

The young were naked, blind, ears protuberances covered by an integument; mouth closed, with the exception of a very small orifice sufficiently large to receive the small attenuated teat. Tail ¼ inch in length.

March 11.—Weighed the largest of the young and found that it had increased to 30 grains. Length of body 1¼ inch, tail ½ inch. The nostrils were now open. The young were very tenacious of life, as on removing two they remained alive through a cool night in a room containing no fire, and still evidenced a slight commotion at 12 o'clock on the following day. The teats of the mother, after the young had been gently drawn off, measured an inch in length, having been much distended, and appeared to have been drawn into the stomach of the young.

March 16th.—The dark colour of the eye can be seen through the transparent skin, but it is still perfectly closed. A few hairs have made their appearance on the moustache. The orifice of the ears beginning to be developed. Nails visible and sharp. The pouch of the young females is quite apparent, and the sexes may be determined as soon as born. They voided urine and excrement—used their prehensile tails, which were seen in twine around the necks of others even at a week old.

February and March, 1847. Made a number of observations on a large number of females. As they, however, all had young in their pouches before I procured them, I will only notice one experiment made in order to ascertain the manner in which the young became attached to the teats.

March 11th.—Conjecturing that the young were aided by the mother in finding the teat and believing that she would not readily adopt the young of another or afford them any assistance, I removed six of the ten which composed her brood—returned two of her own to the pouch, together with three others, fully doubled the size that had been obtained from another female. She was soon observed double up with her nose in the pouch and continued so for an hour, when she was examined, and one of her own small young was found attached to the teat. Seven hours afterwards she was again examined, and both the small ones were attached, but the three larger ones still remained crawling about the pouch.

March 12th.—The mother seemed now to have adopted the strangers, and the whole family of different sizes were deriving sustenance from her.

February 11th, 1848.—Having received from the country a large female that appeared to be impregnated, I this day dissected her. As soon as the uterus was removed from the body of the animal, which had just been killed and was yet a warm, I observed the whole mass in irregular motion. There were nine young that would evidently have been produced in one or two days. Three were contained in one department and six in the other. They lay embedded in a thick dark-brown mucus substance, which filled and greatly distended the sacs. They possessed more life and motion than I had previously been led to suppose. One of them moved several inches on the table and survived two hours. I attempted to weigh this uterine fœtus, and as far as I can ascertain with an imperfect pair of scales, it weighed 3 grains.

Although naturalists at the present day could scarcely entertain a doubt that the process of generation in this species did not differ materially from those of the Kangaroo and other Marsupialia, yet I am not aware that the young of the Virginian Opossum had been previously detected in the uterus.

The short period of gestation, the reluctance of many of them in copulating in a state of confinement, unless perfectly domesticated, rendered the discovery one of considerable difficulty. I have moreover found that during the period of gestation the females, like those of some other species, particularly the Bear, can seldom be found.

In February, 1847, by offering premiums to servants, I procured from various localities in three nights 35 opossums, and there was not a single female in the whole number. As soon, however, as the young were contained in the pouch, I received more females than males.

February 14th, 1848.—Dissected a small female that had been captured six days before. She proved impregnated, but in a much earlier stage of development than the one I examined three days ago. On opening the uterus, I found five on one side and seven on the other. These were nearly the size of a garden peak, and resembled pellucid vesicles. Under a microscope the germinal membrane represented a cellular structure as in other animals. The corpora-lutea corresponded with the number of ova.

The manner in which the act of copulation is effected is no longer a subject of conjecture, although I have not personally observed it. An intelligent coloured man in whose veracity I place great confidence, was requested five years ago to watch this process. He assured me that he had observed the female receiving the embraces of the male while lying on her side. Within the last few weeks, Dr. Middleton Michel of this city, an intelligent and close observer who has devoted much time to the investigation of the subject, has observed that this process with two female opossums which he has preserved in a domesticated state. He informs me that they received the male whilst lying on their right side.

From various observations I have made for the last three years, I have set down the period of gestation in the opossum at 17 days. I received a female, said by the servant to have been captured in the act of copulation. She produced her young on the seventeenth day. I had, however, placed her with a male that I kept in confinement at the time; but she exhibited such a savage temper towards him that for the sake of peace, I was compelled to separate them after three days. Dr. Michel, however, informs me that a female in his possession produced young on the fourteenth day. Although I was at first confident that the time period was 17 days, I think it's probable that from the superior advantages Dr. Michel has possessed with his animals in a state of domestication, he may have approximated nearer to the true time than myself.

In the second volume on American Quadrupeds, now in the course of publication, the history of this animal will be treated more in detail.

Further Observations on the Generation of the Opossum.
By the Rev. Dr. Bachman.

February 15th, 1848.—On the morning of this day I received five female opossums from the country, three of whom I was informed by Col. Hall (who zealously and successfully interested himself in procuring specimens for my examination) had produced young in the box in which they were confined a day or two previous to their having been sent. There several pouches contained eight, nine, and eleven young. There were two, as he informed me, in the state in which I was anxious to obtain them; as they had not yet produced their young. On examination, however, I discovered that one of the two had evidently brought forth amid the joltings consequent on her conveyance from the country. Five young were in the pouch. I am served, on examination, that a sixth was lying at the bottom of the box and was still living. Supposing it possible that all the young had not been excluded, I concluded to sacrifice the mother; and was repaid for an apparent cruelty, exercised very reluctantly, by discovering that the female was still in the act of parturition; a remaining young one was found in the vagina within half an inch of the external surface. It was moving, head downwards, among a reddish-brown mucus mass such as had been previously observed in the uterus of a female already referred to. There was not even the rudiment of a placenta. If it had previously existed, it must have been ruptured in the passage of the fœtus, and escaped my most careful search. I was however under an impression that I discovered the slight rudiment of an umbilical cord. The nostrils were open; the lungs were filled with air; and on a subsequent experiment, they were observed to float on the surface of water. On dissecting the uterus it was found flaccid and nearly empty, a slight brown mucus on the sides only be visible.

On the afternoon of the same day I had the remaining female destroyed. On dissecting down to the uterus, I found it greatly distended—full of young, and, as I then supposed, near the period of production. There was a constant but irregular motion in the various parts; and I felt confident that I would now be furnished with the long sought for opportunity of making a thorough investigation of the various particulars that required farther elucidation. I concluded, however, previously, to have a drawing made of the uterus as it presented itself in this state; this consumed the remainder of the evening. As the weather was warm I made a slight incision in the parts, and placed the whole in alcohol. On the following morning, when, with a scientific friend, we entered on the examination, I was greatly disappointed and mortified, to find that the hole had been so much dissolved by the alcohol that we could make no satisfactory examination. The young were lying in broken fragments in the midst of the unctuous and now considerably diminished mass. I now can scarcely suppose that the motion I had observed for an hour while the drawing was in progress could have been any other than a muscular contraction and dilation of the different parts of the uterus itself, and not of the young, which were evidently not sufficiently advanced to have occasioned it.

I would here observe that where the outward integuments of animals are so very tender as those of young opossums a few days previous to their birth, it is advisable to dilute the alcohol to more than half its original strength, as I find the young one that was fully formed, taken from the mother a few moments before birth by the Cæsarean operation referred to, has been preserved in good order in alcohol thus diluted.

In conclusion I will yet add a brief summary of the present state of our knowledge of the natural history of an animal, whose anatomical structure and peculiar habits have led to the adoption of many vulgar errors, and produced several contradictory theories among physiologists. We will thus be enabled to see what important points still remain for farther investigation, and will at the same time be gratified to observe that, although our progress in the investigation of a singularly perplexed subject has been very slow, yet there has been a gradual advance in our knowledge leading us to the conviction that in a very few years the history of the opossum will be as correctly and familiarly known to the community at large as that of the hare or squirrel.

1. The interesting group of the Marsupialia has recently been arranged by Owen into five tribes and families and sixteen genera: these include about seventy known species, to which additions are continually making; the Virginian opossum being, however, the only species known in the United States. The onteological characters of the latter species

have been so accurately described and delineated that little remains to be added in this department.

2. The organs of generation being found perfect and adapted to their peculiar uses—the double uterus to the bifurcated organ of the male—should have and themselves been sufficient to have thrown doubts on the assertions of our early authors—Marcgrave, Pison, Valentyn, Beverly, the Marquis of Chastellux, Pennant, and others—that "the pouch was the matrix of the young opossum, and that the mammæ are, with regard to the young, what stalks are to their fruits."

3. The mode of copulation, although differing from that of the majority of quadrupeds, is far from being the only exception to a general law; our porcupine (Hystrix dorsata) may be cited as another instance. In this respect the actions of animals correspond with their peculiar organization, and the structure of the genital organs, as well as the whole anatomy of the opossum, are in accordance with this habit.

4. The question propounded in 1819, to naturalists, by Geoffroy, "Are the pouch to animals born attached to the teats of the mother?" is satisfactorily answered.

5. The period of just Station [gestation] being between fifteen and seventeen days, is in this respect shorter than that of any other known species (that of the Kangaroo being thirty-eight days) suggests the idea of the probability of some modification of uterine structure, approaching in some respects that of the birds and ovo-viviparous reptiles.

6. Although the period of gestation is so short, the young are far more perfectly developed at birth then has been usually supposed. The views of Blumenbach, who likens them to abortions, as well as those of Dr. Barton (I quote from Griffith as I have not recently seen the original) appear in this particular surprisingly inaccurate. "The Didelphes," he says, "put forth, not fœtuses, but gelatinous bodies; they weigh at their first appearance generally about a grain, some a little more, and seven of them together weighed ten grains." My observations have convinced me that they are far from being merely "gelatinous bodies," but that they are pretty well developed, indeed nearly as much so as the young of the white-footed mouse and several other species of Rodentia. They are covered by one integument—nor rushed by the mammæ—breathe

through nostrils—are remarkably tenacious of life, and are capable of a progressive movement at the moment of their birth. Hence I am not fully satisfied with the accuracy of the terms used by De Blainville and Dr. Barton—when they speak of two sorts of gestation—one uterine and the other mammary. It is admitted that for so large an animal as the adult opossum, the young are not only very small, but feeble and are for several weeks sustained in a kind of secondary domicil, termed the pouch, where they receive warmth, and that they continued during this period firmly attached to the teats, which they do not relinquish till they are pretty well grown. It will be recollected, however, that there is in several of our animals and approach to this latter peculiarity. The white-footed mouse (Mus leucopus); the Florida rat (Neotoma Floridana), and several species of Bats are known, the two former to travel, and the latter to fly about for one or two weeks, with their young attached to their teats, and that these young are not only blind and naked, but nearly as helpless as those of the opossum. It will be farther recollected that there are several species in the extensive group of Mammals to which the opossum belongs, that are destitute of the pouch, the young in these cases adhering to the teats like those of the Florida rat, &c., exhibiting an approach to species of a different conformation.

7. The manner in which the young are placed into the pouch and attached to the teats, I have referred to in my observations on the female that brought forth her young in the room where I was sitting on the 4th March 1846 (although I was not at the time aware that she was in the act of parturition). She was reclining in the corner of the cage, a little on one side, with her shoulders somewhat elevated; her body was much doubled, the vulva nearly reaching the pouch, the latter being occasionally opened by her paws. She was busily employed with her nose and mouth licking, as I thought, her pouch, which I afterwards ascertained was her young. I came to the conclusion that she showed them into the pouch, and with her nose or tongue moved them to the vicinity of the teats, where by an instinct of nature, the teat was drawn into the small orifice of the mouth by suction. I am served subsequently that the well-formed young I extracted from the vagina, which I rolled in warm cotton, was instinctively engaged in sucking at the fibres of the cotton, and had succeeded in drawing into its mouth a considerable length of

the thread. I may here remark that on the 21st of February of the present year a female opossum was sent to me late in the evening. She had been much wounded on being captured, and died in consequence a few days afterwards. On the morning after I received her I perceived in her pouch seven young; they had not been attached and were dead; abortion had taken place, and they had evidently been placed in the pouch by the mother's uncontrollable attachment to her offspring even after they were dead.

8. The opossum is one of the most prolific of our quadrupeds. I consider the early parts of the three months of March, May and July as the periods when they successfully bring forth; it is even probable that they breed still more frequently [inasmuch] as I have observed the young during all the spring and summer months. I find in my notes the following memorandum: "May, 1830. In searching for insects I was removing with my foot some sticks composing the nest of the Florida rat. I was startled on finding my boot unceremoniously and rudely seized by an animal which I soon ascertained was a female opossum. She had in her pouch five a very small young, whilst seven others, about the size of full grown rats, were detected peeping from under the rubbish, and were captured."

9. An interesting inquiry remains to be answered. Is the opossum a placental or a non-placental animal? If I am to understand by this term, whether the opossum has or has not a placenta, I can readily answer in the negative. In these intricate matters the naturalists should, if possible, see with his own eyes and speak at all times as feeling himself firm on his own feet. I have had all the opportunities I could have desired of perfectly satisfying my own mind on this subject, but can only state that in all the examinations I have made I could never find the slightest appearance of a placenta, and I do not believe that one exists.

I am, however, far from being equally satisfied on another point to which I confess my observations were not directed until it was almost too late to make the necessary investigations. Although I do not believe that a placenta exists or that there is any attachment of the fœtus to the parietes of the uterus, it does not from hence follow as a necessary consequence that there is no allantois. If an animal has a placenta there is a sure evidence of the pre-existence of an allantois; but there is in many animals and especially among the smaller species of Marsupialia a modified structure in these parts; and the allantois, umbilical cord, as well as the omphalo-mesenteric arteries and veins may exist in the absence of a placenta. In the very unsatisfactory examinations I have been enabled to make on the subject, I came to the conclusion that there was some reason to believe that an allantois existed, and that there were some traces of the omphalo-mesenteric vessels running through the mucous substance in which the young lay imbedded. It is proper, however, to observe, that my friends Prof. Hume and Drs. Harlbeck and Michel, who subsequently examined the well preserved specimen of the 15th, and the imperfect remains of the contents of the other uteri, came to the very opposite conclusion. I nevertheless hazard the conjecture that these appendages may yet be found in the uterus at an advanced stage of pregnancy. This suspicion, however, remains either to be confirmed or refuted by a more favourable opportunity for examination. Owen, in describing, in 1834, the fœtus and membranes of the Kangaroo at apparently the middle period of uterine gestation, found its conditions such as obtains in the viper and other ovoviviparous reptiles, there being no trace of the existence of an allantois. In 1837, however (see Magazine of Nat. Hist., p. 481), having received another specimen in a more advanced stage, he founded numerous ramifications of the umbilical vessels constituting a true allantois. The umbilical cord extended three lines from the abdominal surface of the fœtus. Having seen and examined that specimen, I may have unconsciously formed a theory which has misled me in conjecturing that I had observed a similar organization in the opossum.

Letter from Middleton Michel, M.D., of Charleston, S.C.

To the Rev. John Bachman, D.D.

Dear Dr.—You will oblige me by adding the few facts which I am able to state concerning the habits and generation of our Opossums to your valuable communication, addressed to the Philadelphia Academy of Natural Sciences.

1st. I have first noticed their mode of copulation, which though singular in itself, finds its explanation in the position and structure of the penis. The female, after repeated solicitations on the part of the male, which are conducted as among other animals, finally reclines upon her left side, being drawn into this position by the male; his front legs are employed in securing her, while the hinder ones are made to pass on each side of the loins of the female over and between her hind legs. The penis, measuring 2 inches and more, is thus brought into more immediate relation with the sexual organs of the female. Copulation lasted five minutes. The sperm passes along the lateral canals, its only possible course, as the bifurcated organ of intromission is received to some distance into them.

2d. I have further determined, that the period of gestation is not twenty and twenty-two days as has been believed. I placed a female with the male on the 27th of January, and on the 28th at 7½ o'clock, A.M., I witnessed them engaged in the act. She was left three days with the male, then isolated, and on the 12th of February, fifteen days after the first coitus, had her young, six in number in the pouch. Admitting that the period may vary from fifteen to seventeen days, the having settled this point I regard of paramount importance in answer to another question to be presently examined.

3d. The rut begins in January and continues till June, as I have seen young just received into the pouch during these months.

4th. The number of young is from *six* to *thirteen*. I have had a female with *thirteen* in the pouch; never less than *five*.

5th. The size of the young at birth is four lines in length, two in breadth; weight four grains.

6th. The structure of the male and female organs has been well described by *Cowper, Tyron, De Blainville, Home*, and others. But I would remark that there is no communication between the uterine extremity of the lateral tubes (or the sinus, as I would term it) and the vagina as figured by Home and others.

7th. This leads me to mention that parturition takes place as follows: the young pass down through the lateral tubes, there being no other exit for them, and immediately after parturition these canals are very much enlarged.

The mode of transmission to the pouch is a part of the process hitherto unknown, which I have recently witnessed as well as the nature of the circumstances would permit. The female stood on her hind legs, and the body being much bent, the young appeared at the vulva; they were linked into the pouch. They were born without any trace of an umbilical cord. The pouch was not interfered with for some time, when her mouth was introduced into it while her front paws held it open; after this manœuver was completed, the little ones were all found attached to the teats. I would further remark that this attachment is an instinctive act on their part, as it is impossible to conceive of any interference of the mother effecting it. The mouths of the embryons present but an infinitely small opening compared with the size of the teat, and with the hand is an almost impossible attempt to attach them.

8th. The ova in the vesicles are larger in proportion than in other mammalia; the vitellus is enveloped by a thin vitelline membrane. The germinal vesicle is, however, in the same position as in other mammals; the transformation in the tubes, where I have met with one, after fecundation, appears the same as in the rabbit. In the uterus, the germinal membrane has the same structure and appearance as in the rabbit. This stage I witnessed, through Dr. Bachman's kindness, as he gave me the uterus to examine.

9th. Whether these animals be placental or non-placental is a question which I cannot positively decide until I have finished the series of observations proposed, but the inference that they are not placental is rendered legitimate, first, by the peculiarities in the structure of the brain and other organs, which show their close proximity to the bird; second, by no allantois attached and conveying blood vessels to the chorian; third, by the short period of gestation; for the ova were discovered in the uterus on the ninth day, and the period of gestation being fifteen or sixteen days would render such a structure needless.

Unity of the Human Race

In *The Doctrine of the Unity of the Human Race Examined on the Principles of Science*, Bachman demonstrated that all races are a single species by showing how much the varieties of human beings have in common with one another and with other varieties of domesticated animals. He concluded that human varieties constitute a continuous spectrum. Since the offspring of every combination of races is fertile, all races must be considered a single species by the only definition that can legitimately be used to distinguish a species from a variety or a hybrid.

Bachman's inductive reasoning was exemplary for dealing with the solution of a problem that required consideration of a vast amount of biological and historical evidence. His *Unity* systematically summarized all types of evidence that related to the question, and he showed in every case that the most credible interpretation supported the unity of the human race. By doing so, he accomplished more to confirm this unity than any individual had previously.

Bachman also showed flaws in the evidence and arguments of the polygenists, who had argued for the multiple creation of different species of human beings. Indeed, "it was in large measure because John Bachman had been engaged in species identification and studies of hybridization that he was so well placed to attack the polygenists" (Bernasconi 2002; cf. Stanton 1960).

In the *Quadrupeds*, Bachman had been explicit about the definition of a species: "When a male and a female, however different in size and colour, unite in a wild state and their progeny is prolific, we are warranted in pronouncing them of the same species" (Audubon 1989: 114). In the eighteenth century Buffon had proposed this rule, and in the early nineteenth Kant had endorsed it, but "by the middle of the nineteenth century, in large measure as a result of

[Samuel George] Morton's studies [in physical anthropology], Buffon's rule had lost much of its force" (Bernasconi 2002).

To explain why various types of human beings seemed to differ so much, Bachman showed how they were subject to the same influences as every kind of domesticated animal. He discussed domesticated plants and animals one by one to show that they were varieties of wild species rather than separate species, and he traced every major farm animal and crop to a specific origin. He noted that in isolation, varieties could maintain their distinctive characteristics indefinitely, but that when reunited with their parent stock, these characteristics disappeared (being reabsorbed as recessive traits). Aristotle had argued that all human beings were a single species, but that some were wild and others domesticated (*Historia Animalium* 488a30; see also *De Partibus Animalium* 643b5 and the Aristotelian *Problemata* 895b24). Bachman took for granted that "in his physical nature man is an animal" and is consequently subject to the same processes and natural laws as all other animals (117). Bachman demonstrated the correctness and importance of Aristotle's observation. Darwin later relied heavily on domestication to reveal the ways that varieties are created.

Bachman pointed out the unfairness of polygenists such as Agassiz and Morton of always comparing the extreme ranges of races rather than comparing the most characteristic type of any race and of ignoring continuous links that connect every variety. As he demonstrated, a full range of intermediate varieties indicates that gaps exist only in isolated groups. Moreover, in groups that are not isolated, there is often a great range of variation: "there are many wide departures from all of these types, not only in neighbouring races, but in the races themselves, and even in the children of the same parents" (30). Thus, the significance of apparently extreme differences had been overemphasized, and the ranges of normal variation had been ignored.

For comparison, Bachman considered in detail the range of differences in domesticated animals including the horse (ranging from Shetland ponies to "the great dray-horse"), cows (some with and some without horns), swine (with adults ranging from fifty to twelve hundred pounds), sheep (with straight hair, fine hair, or wool), the dog (from the lap dog to the St. Bernard), and the pigeon (with enough varieties to fill a printed page) (124–36). He showed that "from the history of every domesticated species of quadruped or bird that they all, without a solitary exception, have branched out into striking and multiplied varieties and that these have become more numerous and striking in proportion to their wide dispersion and change of climate and habit" (136). He also pointed out that "the dog has produced infinitely more striking varieties than have been exhibited in the varieties of the human race" (174). He insisted that the characteristics these species had in common had to be taken as the basis for determining species more than how they differed: "if we are only guided by external forms, we will find it very difficult to separate the true species from the varieties" (137).

Likewise, "the cereal grains, such as wheat, oats, rice, maize, millet, etc., have produced innumerable varieties. We observed at Edinburgh one hundred and six varieties of wheat, for sale in one seed store" (139). It was well established that these and other varieties of plants were bred out of a parent stock to which they would return if not kept separate.

Bachman's evidence from natural history and agriculture was firsthand; he was "expressing no views on any subject which we had not personally investigated" (216). However, he did not consider himself or any American scientist as well informed about the internal structure of human beings as many Europeans, and in the case of anatomy, he relied on the opinions of European specialists. He summarized scientific evidence indicating that neurologically there was no significant difference between races (219–31).

He cited the accounts of numerous anthropologists to demonstrate how imperceptibly varieties of human beings blend into one another worldwide:

> We cannot . . . fail to perceive that in passing onward step by step, till we meet with the Mongolian, on the North-east, the African on the South, and the white man on the West, we have been carried from the centre to the extremes, from one tribe to another, so insensibly, that we have scarcely marked the difference of the links in the great chain by which these extremes of the races were bound to each other. (162)

He noted how varied Blacks were in different parts of Africa and how much more closely some types resembled Whites than did the American slaves who had largely come from one part of Africa (its southwest coast; 166, 225). He emphasized that "there are very dark if not black men among all the races that are arranged under the Caucasian family" (232). He quoted Humboldt's observation that with increasing knowledge, "'the greater part of the supposed contrasts to which so much weight was formerly assigned, have disappeared'" (167).

Bachman's *Unity* could only have been prepared by someone who had spent most of his life reading widely and solving complex problems. His approach to the problem of whether all types of human beings were one or more species required broad interests, specialized knowledge, and great reasoning power. He pulled together all the facts he knew relating to a single problem of great complexity and produced a coherent argument. The *Quadrupeds* best reflects his scientific expertise, but the *Unity* required an even wider range of knowledge and reflects even more insight.

Much of the first third of the *Unity* is about what all varieties of human beings have in common, and his summary is included here as the first of two selections. The second selection is about the extent to which acquired characteristics can become instinctive, and he presented experimental evidence that some could be. His experiments deserve to be repeated and the results need to be explained. In 1850 he had no way of knowing that all new characteristics are accidentally introduced and that the only ones to persist have survival value for a species. These two selections provide a good indication of how he approached the problem. Since his inductive approach required bringing together many other types of evidence and innumerable examples, it was the weight of the evidence as a whole that was persuasive rather than one or a few parts of the book. The entire book deserves to be read as a classic work of scientific methodology.

The Doctrine / of the / Unity of the Human Race / Examined / on the Principles of Science. / by / John Bachman, D.D., / Prof. Nat. Hist. College of Charleston; Corresponding Member of the Zool. Soc.; / Hon'y Member of the Entomol. Soc., London; Cor. Memb. Royal Botanical Soc., Saxony; / Royal Soc., St. Petersburgh; R.S.A. Copenhagen; Acad. Nat. Sciences, Philad.; N.Y. Lyceum; N.H. Soc., Boston, N. Haven, and Toronto; / National Institute; American Assoc., etc., etc. / Charleston, S.C. / C. Canning, 29 Pinckney-Street. / 1850.

[Preliminary conclusions on pages 147–51]

We will now, in the conclusion of this chapter, sum up the evidence which we have produced in various parts of this Essay or which are self-evident and require no further proof in favour of the unity of the species.

1. There is but one true species in the genus homo.

In this he does not form an exception to the general law of nature. There are many of our genera which contained but a single species in the genus. Among American quadrupeds the musk ox (*Ovibos moschutos*), the beaver (*Castor fiber*), and the glutton or wolverine (*Gulo luscus*) and among birds the wild turkey (*Meleagris gallipavo*) are familiar examples. The oscillated turkey, which was formerly regarded as a second species, has recently been discovered not to be a true turkey; in addition to its different conformation, it makes its nest on trees and lays only two eggs, possessing in this and other particulars the habits of the pigeon.

2. We have shown that all the varieties evidence a complete and minute correspondence in the number of teeth and in the 208 additional bones contained in the body.

3. That in the peculiarity in the shedding of the teeth, so different from all other animals, they all correspond.

That they are perfectly alike in the following particulars:

4. In all possessing the same indirect stature.

5. In the articulation of the head with the spinal column.

6. In the possession of two hands.

7. In the absence of the intermaxiliary bone.

8. In the teeth of equal length.

9. In a smooth skin of the body, and the head covered with hair.

10. In the number and arrangement of the muscles in every part of the body, the digestive and all the other organs.

11. In the organs of speech and the power of singing.

12. They all possess mental faculties, conscience, and entertained the hope of immortality. It is scarcely necessary to add that in these two last characteristics of man is placed at such an immeasurable distance above the brute creation as to destroy every vestige of affinity to the monkey or any other genus or species.

13. They are all omnivorous and are capable of living on all kinds of food.

14. They are capable of inhabiting all climates.

15. They all possess a slower growth than any other animal and are later in arriving at puberty.

16. A peculiarity in the physical constitution of the female, differing from all the other mammalians.

17. All the races have the same period of gestation, on an average produce the same number of young, and are subject to similar diseases.

If an objection is advanced against the rules by which we have been governed, and we are told that we have been blending specific and generic characters, we answer that in all the genera a species is selected and described as a type of the genus: hence there being but one species in the genus we have, in accordance with the rules by which naturalists are governed, selected the species as a type.

18. We have shown that man, as a domestic animal, is subject to the same changes which are effected in all domesticated animals; hence as species are taken in a different acceptation in wild and domestic animals, our examinations of the varieties in men must be subjected to the same rules of examination. That these changes in men are constantly taking place is evident from the fact that great variations have occurred in several of the branches which we admit to be Caucasians, whilst wild animals with few exceptions have not undergone the slightest change. We have shown that from the many intermediate grades of form and color and a being more subject to varieties than in any known species of animals, we can find no specific character so permanent as to warrant us in separating the varieties into distinct species. We insist on the right of applying the rules of classification to man as a domestic species. If our opponents urge the right of comparing him with wild animals, then they must first prove that men, like wild species are not subject to produce varieties. This is an experiment on which we think they will not venture. The human species cannot, therefore, be compared with wild animals that with few exceptions present a perfect uniformity. Place before you a hundred specimens of any wild species of quadruped or bird with a few exceptions above as alluded to, and there is scarcely a variation among any of the specimens. The descriptions of Aristotle are as applicable now as they were in his day. On the other hand, look at the countenances even of our neighbours and the members of our

own families, gathered together around the social circle, and you see the most striking differences in the color of the eyes, the hair, and the complexion, in size, in form, in the length of nose, shape of the head, the volume of the brain, etc. These peculiarities are so striking that we can every where recognize those whom we have previously seen. On the other hand, the countenances of the individuals even in domestic animals can seldom be distinguished from each other. The eccentric poet, Hogg, or as he was proud to call himself the Ettric Shepherd, was able, as he stated and no doubt correctly, to distinguish the individuals of the flock which he daily carried to the hills; but this talent even in distinguishing the countenances of domesticated animals is possessed by few others; on the contrary the very child learns to distinguish individuals of the human race by their countenances; no two individuals even in the same family can be found possessing the same set of features. Man must, therefore, be compared and examined by the same rules that govern us in an examination of domesticated animals. Let us compare him with any of these species. Take those about whose origin a difficulty exists; the horse for instance, the only true species in the genus, for naturalists have now classed all the others under the asses and zebras; or take the hog, whose origin is admitted by all naturalists; first examine the characteristics of Sus scrofus, then all the races which have sprung from it; apply the same rule first to the species and then to the varieties of man, and by these fair and legitimate rules of science, we are willing to enter into a comparison, and I bide by the decision. The most eminent naturalists of all past ages have with a unanimity almost unsurpassed already decided the question, and those who are now entering into the field, about whose qualifications, as judges, the world as yet knows nothing and is therefore unprepared to pronounce an opinion are bound to give some satisfactory reasons for their dissent.

19. The varieties in men are not greater than are known to exist among domestic animals.

20. That all the varieties of men produce with each other a fertile offspring which is perpetuated by which new races have been formed; and that this is not the case with any two species of animals.

21. That the insects which are found on the surface and the vermes

within the body as far as they have been examined are the same in all the varieties of man and that where peculiar parasites infested men in particular countries they are equally found in all the races.

Until our opponents have proved that these propositions are not in accordance with the laws of science, or in violation of truth, we must regard their new theory as founded in error.

Chapter VIII.
Intellectual and Moral Predispositions in Men and Habits and Instincts in Animals, the Result of Domestication and Acclimatisation, are Transmissible.

Thus far we have attempted to show that animals and men are not only subject to great variations in form and colour, but that these become constitutional are transmitted to their progeny and finally produce permanent races. We will now present such facts as are calculated to prove that these variations and transmissions are not confined to the physiological changes, but that a state of domesticity evolves another phenomenon in what is usually termed the psychological character by which are designated habits, instincts, &c., in the lower animals and intellectual and moral attributes in man.

A volume might be written on the subject affording interesting subjects of philosophical inquiry; we will, however, limit ourselves to the statement of only a few facts that have an important bearing on the mysterious and inexplicable property in the transmission of these qualities from parents to their descendents.

We obtained eggs from the nest of our common quail or partridge (*Perdrix virginianus*), which we placed under a bantam hen; the young when first hatched were so wild that for several weeks we were obliged to confine them in a box with the foster mother to whom they did not for two weeks become reconciled. In time, however, they became so gentle that they followed her until they were nearly full-grown and then attach themselves to us, came into our study, and often amused themselves by taking hold of the pen with which we were writing. On the following spring these now domesticated birds laid eggs. These were placed under the same bantam hen together with similar eggs from wild birds. The young from the domesticated partridge were nearly as gentle as chickens and at once followed the foster parent whilst those from the wild birds immediately darted off into the shrubbery and became dispersed and lost. The young of our domesticated turkey are so gentle that they feed from the hand immediately after they are hatched whereas the young from the eggs of the wild turkey, although hatched by a tame bird, are as we have often observed so wild at birth that they instinctively run off and conceal themselves. It requires weeks of confinement in a pen to reconcile them to the tame turkey and even with the most careful treatment nine out of ten are known to perish. The young turkeys produced from parents of the first domesticated pair are less wild, and we have observed that it is not until about the fourth generation that the young at birth are as gentle as those from the tame turkey. This peculiarity was exhibited in many species of birds which we have from time to time subjected to the process of domestication. Each succeeding generation became more gentle than the previous one.

We observed moreover that a taste for particular kinds of food is hereditary. The young of the tame turkey immediately pick at cornflower and other kinds of broken grains whilst the young of the wild turkey for weeks refuse any other food except insects. In quadrupeds there is a considerable difference in the dispositions of the young of different species. The young of the common deer, the moose, the prongs-horned antelope, and the buffaloe become gentle in one or two days; on the other hand, the common wild cat (*Lynx rufus*), the cougar (*Felis concolor*), and several other species are always restive under restraint and are tamed with great difficulty. We have, however, observed that the descendents of the latter in the third generation become almost as gentle as a cat. This is also the case with the lion and tiger.

In these cases and in a vast number that occur to us at this moment, we perceive in what manner of domestication during successive generations produces an influence on the dispositions of the progeny. We will mention a few facts which have come under our immediate notice although we are aware that we will scarcely escape the charge of credulity. A male pet lamb that had been brought up with the cows was taught by the boys at the farm-house to butt with his horns, and for this acquired propensity he had subsequently to be confined. It was observed that his offspring became so pertinacious that the owner was finally obliged to change his stock of male sheep. Dogs that have acquired the vicious habit of killing sheep seem to transmit this propensity to their progeny, and we have known at least one instance where a whole race of these dogs was exterminated from a neighborhood on account of this habit. The pointer and setter of good breeds stand at the game without being taught. The dog called the retriever has but recently been introduced and is a cross between two other varieties. It was originally taught to fetch the game, which is its only occupation. We were in England and Germany shown several of these dogs that had on the first time after being carried into the fields brought to the game without having been taught. The descendents of the shepherd's dogs are born with such an innate propensity to guard the sheep that they require scarcely any training; their predecessors, however, were subjected to a tedious process of training. The dogs in Belgium and France are accustomed to be harnessed and draw small wagons, etc. We observed that young dogs of this breed possessed so strong a propensity to this kind of employment that they often ran up to these small carriages, seized the traces, and endeavored to lend a helping mouth to an operation that was inherently natural and pleasant to them. The cow in her natural state yields milk no longer than a few months until the calf is able to provide for itself. On the other hand, breeds of cows have been produced by weaning the calf and constant milking that yield milk for several years without an intermission. If peculiarities and disposition, temper, and instincts in the lower animals are capable of being transmitted to their posterity, may we not from hence be led to form some conception of the manner in which peculiarities in the dispositions and characters of races of men may become perpetuated, and may we not also learn in what manner in a succession of generations an instinctive hereditary propensity may be directed in a different channel and nations of hunters or shepherds may in time become nations of agriculturalists? Our Indian tribes have been proverbial for a natural attachment to the chase, to a love of plunder and of war. It has been frequently remarked that their young men educated in civil life have had such a strong natural propensity to the life of the forest that in many instances they returned to this, their first love. We perceive Dr. Morton offers this as one of the evidences of essential distinctiveness of the race. We are not however to forget that at certain periods in the histories of nations, every race has had its ages of barbarism. The tales of war, rapine, and love of the chase in relation to our own Caucasian family should be an instructive lesson to us. The traveller along the beautiful banks of the Rhine from Schaffhausen at the falls down to Cologne and even onwards towards Holland witnesses the most lovely scenes of industry, rural simplicity, and a love of agricultural pursuits. Such are the characteristics of the present race of inhabitants, but he has only to turn his eyes upwards and he will have the monuments of the psychology of their forefathers staring him in the face. On every high and almost inaccessible peak of the surrounding mountains he sees in the ruined castles and dilapidated towers, the characteristics of their forefathers. These eagles' nests seated on the summits of the beetling rocks give evidence of the manner in which the ancient Barons with their plundering serfs watched the surrounding vallies and the river for opportunities to pounce down and levy black mail on all who ventured to traverse the plains or navigate the Rhine. It was a nation of rude, cruel, and lawless warriors, of hunters and robbers. Such also were France, England and Scotland, as history abundantly attests that our American aborigines have rather degenerated than improved is admitted, yet the ruins of ancient temples &c. give evidence that some of the tribes regarded by Dr. Morton himself as having belonged to the same race were not only agricultural in their occupations, but had made considerable advances in civilization and architectural knowledge. The early Cherokees were a tribe of barbarians who lived by hunting and plunder. Generation after generation succeeded, their habits have been undergoing a change, and now many of them have devoted themselves to husbandry. They are slowly becoming an agricultural people, and their characters have been much improved. We met at the foot of one of the mountains of the Alps, a chamois hunter who in speaking of his hair-breadth escapes and the perilous life of privation he chose to lead offered as a reason that the passion was born with him, having descended through the blood of many generations. A good friend of ours has often related to us an anecdote of the son of an Indian chief whose education had been entrusted to him; after the indulgence of an occasional paroxysm of anger and revenge, he was in the habit of excusing himself by saying that "it was the

Indian that was in him." If these were hereditary propensities might we not hope that as reason and conscience are given to all men, these dispositions might in time become modified and then entirely eradicated and a different train of propensities be transmitted to their posterity? In this way the pursuits and habits of nations have become changed, their aspirations have been directed into other channels, and finally all their ancient characteristics have disappeared, and they have been converted into new psychological races.

Defense of Luther and the Reformation

Bachman wrote, "I enter on the defense of Luther and the Reformation not so much on account of my being a Lutheran clergyman professing to hold the fundamental doctrines he taught, but because the principles he inculcated form the key-stones of Protestantism—the right of private judgment—of free enquiry into all matters of religion—and of the great privilege we enjoy in being permitted to read the Scriptures in their vernacular tongue.... I enter, therefore, on this defence, not as a Lutheran but as a Protestant" (58).

The principles that had been adopted by all Protestants succeeded "in breaking the chains that had so long fettered the mind" (58–59). Luther initially fought "a single-handed warfare against prejudices, errors, human power, and malignant hate," and he "lived to see established the freedom of thought" (147, 92). Bachman quoted what Frederick Schlegel wrote about Luther in his *Philosophy of History*: "'the boldness of his speculations and the vigour of his eloquence will be found to form an epoch, not only (as is universally acknowledged) in the history of the German language, but in the progress of European science and European Culture'" (272).

Bachman's goal initially was to defend Luther against vague and untrue charges, but the continuing unfairness of his opponents provided the opportunity to deal specifically with the reforms that Luther had considered essential: "The Romanists who attack Luther cannot expect him to be defended without mention of the reasons for his withdrawal from Papacy" (246). To achieve these goals required great erudition, strong powers of reasoning, a wide knowledge of history, intellectual honesty, and courage.

Bachman defended Luther's dictum that every individual needs sufficient education to be able to interpret the Bible for himself. Luther wrote, "'we should let every one believe what he thinks right.... With Scripture and reason, we should try to convince them—but not with fire and sword'" (394). Luther's purpose in translating the Bible was to provide complete and "unrestricted access" (117). Bachman asserted, "hitherto, men had blindly submitted their faith and their souls to the guidance of others. Luther had taught them to take the word of God for their guide, to choose Christ instead of Peter as the head of the church, and to think for themselves" (119). Luther accepted only Scripture as the basis for religion, and he rejected all aspects of Catholicism that had no biblical basis.

The controversy that led to the creation of this book began in March 1852 when John Bellinger, a Charleston alderman and physician, warned city council that riots would ensue if a former Catholic gave a publicly announced lecture on questions that women were asked in the confessional. Citing the right of free speech, the council refused to prohibit the lecture, but the lecturer was so intimidated that he left Charleston.

The renowned classicist Basil Gildersleeve, a Charleston native, asked Bellinger either to retract his remarks about Luther or to support them with evidence from Luther's own writings. Bellinger responded by repeating vague charges that had been made against Luther by various Catholic writers since the Reformation, by paraphrasing and misquoting some of Luther's most controversial writings, and by offering statements that were incorrectly attributed to Luther. Bachman responded with quotations from Luther's collected works to refute fifteen separate charges (230–32). Bellinger refused to continue the debate, but the *United States Catholic Miscellany*, a newspaper published with the approbation of the Bishop of Charleston and edited by a priest under his control, made further attacks on Luther and ridiculed Bachman, calling him malignant, senile, an infamous liar, and an imbecile (331, 435, 439, 501).

Bachman emphasized that Luther's personal failings did not invalidate the principles of Protestantism. Luther had never claimed infallibility, and he asked no one to adopt "his opinions without the exercise of our reason and judgment" (52). By contrast, "the Roman Catholic is taught to render implicit obedience to the church. That church he believes infallible" (55). Bachman noted that in *A Doctrinal Catechism* published in 1849 with the approval of the Bishop of New York, American Catholics were instructed "'that Protestantism cannot be the religion of Christ, because, if the church of Christ required reformation, a God of purity and holiness would never have chosen such an immoral character—an apostate—a wholesale vow-breaker and sacrilegious seducer for that purpose'" (237–38). This catechism stated that Luther "'by his own account, received his training and instruction in the school of satan'" (240). As Bachman lamented, "prejudiced men can distort the plainest facts in history" (115).

After completing his defense of Luther, Bachman dealt in detail with the doctrines of the Catholic Church that Luther had opposed (357–510). He presented specific evidence to support his contention that the Church is "the most arbitrary and absolute despotism that has ever been devised to crush the freedom of the human mind" (369). He rejected Catholic assertions of its right to be intolerant and to persecute. He warned that everywhere in its control the Catholic Church had been completely intolerant of dissent and had always persecuted dissidents as fully as possible; "in our own country, they now openly teach that they have the right of compelling us to adopt their unscriptural creed" (378). He quoted the editor of the Catholic newspaper the *Rambler* as stating that "religious liberty, in the sense of a liberty possessed by every man to choose his religion, is one of the most wicked delusions ever foisted upon this age by the father of all deceit. . . . None but an Atheist can uphold the principles of religious liberty" (378).

Bachman dealt with the confessional as "the fulcrum of their power." He stated that it was the ability of the Catholic Church to interrogate any Catholic on any subject that gave the Church its "possession of the minds and consciences of all its members. . . . Here they gain all their information" (417).

Bachman concluded by addressing a letter to the Bishop of Charleston: "I have not concealed the fact that I am the opponent of the Romish Church. I regard its whole system as opposed to the plain teachings of the word of God. I regard her priests as engaged in perpetuating error—and by the doctrines they teach enslaving the human mind and bending it to a creed which leads to intolerance, to persecution, and the grossest immoralities. . . . I have endeavoured to show that the evils were inherent in the system" (434). He considered Catholic doctrine to be "subversive of our liberties, civil and religious" (436).

When an anonymous Catholic wrote fictitious letters purportedly by Luther, Bachman provided a sample of his own ability to write satire: "The archbishops and cardinals go to the pope: 'May it please your infallibility, you first burnt the books of this man and then declared them to contain immoralities and blasphemies, which unpardonable sins it has been proved are not in them. What are we now to do?' 'Depart in peace, my children; what we did was for the good of the church. It was a lie once, but the decision of the church has made it a truth, and I command all men to declare it a truth on pain of excommunication.'"

Bachman wrote that he personally knew many priests and lay Catholics "as good citizens and as virtuous members of society" (243), but he was "fully warranted in pointing out all the iniquities of their unscriptural, corrupt, and persecuting hierarchy" (250). He had often befriended Catholics, and he had found many jobs for recently arrived Irish immigrants. Of them, he wrote, "They have not only been faithful but grateful, and I doubt very much whether your editor or . . . [an anonymous critic] could ever persuade them to get up a mob for my special benefit. . . . I believe I am a better friend to the Irish Catholic than are his priests" (496).

This series of exchanges between Bachman and others was largely published in the *Evening News*, a Charleston newspaper, from March through October 1852. The book based on the series was published by the newspaper's editor.

A / Defence / of / Luther and the Reformation. / by / John Bachman, D.D., LL.D., / Against the Charges of / John Bellinger, M.D., and Others. / To Which Are Appended / Various Communications of Other Protestant / and Roman Catholic Writers Who En- / gaged in the Controversy. Charleston: / Published by William Y. Paxton. / 1853.

The Roman Catholic Doctrine of Oral Confession. The doctrine stated as laid down by the Council of Trent.—Decree of the Lateran Council.—Romish bishops secret spies of Rome.—The Papacy the enemy of liberty, civil and religious.—The confessional leads to immortality.—Extracts from the "Garden of the Soul," the Roman Catholic prayer book.—Dr. Lynch's defence of the prayer book.—Effects of celibacy on the morals of the priesthood.—Experience of Blanco White.—Answer to Newman's charges against the Protestant clergy, and to Dr. Lynch's homily on christian charity.

By the doctrine of the Romish Church established by the Council of Trent, confession of sins was enjoined as a part of their sacrament of penance. Penitents are bound to disclose to the priest every sin, even the most secret thoughts. The following is the language of a decree of a Lateran Council, confirmed by the Council of Trent, Sess. 14, C. 5. 8, and which is recited among the five precepts of the church: "Every one of the faithful of both sexes, after he (or she) shall have reached the years of discretion, must faithfully confess all his (or her) sins, alone, at least once a year to the proper priest," &c. Children are obliged to confess, and even the deaf and dumb persons must do this by signs. The punishment for neglect is: "living, let him be sequestered from the threshold of the church; and dying, let him be deprived of christian burial." The punishment of one who does not commune Easter is the same. One of the heinous sins to be confessed is that of having entered a Protestant church.

The confessor is bound to keep the secrets of the confes[sional] so faithfully that he is allowed even by an oath to deny that he knows any thing about it. The following is the language of Dens (No. 159):

"What, therefore, must the confessor reply, who is asked concerning the truth, which he has learned through sacramental confession alone?

"He must reply that he does not know it, and if it is necessary he must confirm the same with an oath."

"Objection. In no case is it lawful to lie; but this confessor would lie because he knows the truth, &c.

"Answer. I deny the minor: because such a confessor is interrogated as a man and answers as a man; *but now he does not know this truth as a man, although he may know it as God*," &c.

May the King of Kings, and the Lord of Lords, save us from such would-be Gods.

Thus, the Romish Church has possession of the minds and consciences of all its members and is acquainted with the secrets of the lives of male, female and children, from the monarch on the throne, to the baker and lazaroni in the streets. Here is the secret of her omnipotence. The priesthood hold the key by which the thoughts of every bosom and the secrets of every family are unlocked. Here they gain all their information. Here they find the δος πovστώ of Aristotle, the fulcrum of their power; and with this lever they are able to move families and nations and govern empires.

The effects are seen and felt, both in a political and moral point of view.

In a political, because every bishop in our land is the secret spy of Rome, to which he has sworn obedience. Here, then, is the cause why in all purely Roman Catholic countries such as Spain, Portugal, Italy, Central and South America, Mexico, Cuba, &c., no toleration is extended to the Protestant. If there is observed in the monarch a spark of liberality, the priests extinguish it. If he acts in opposition to the demands of the church, the archbishop orders the bell to be tolled and arrays the church against the government, and it is compelled to succumb. Rome cares not what governments are established provided she rules supreme. The forms of government have frequently changed from monarchy to elective governments; this has been particularly the case in South America and Mexico; and the priests in the United States have held up to us these examples of the republicanism of our southern neighbors to show that they are the advocates of constitutional liberty. But they have omitted to state that one of the most essential ingredients of true liberty was entirely left out of this code of freedom. The first act of all these Republican priests was to establish the Roman Catholic system as a *sine qua non* in the Republic and to deny all Protestants the right of worshiping God according to the dictates of conscience and the principles of civil and religious liberty. Without this clause in their constitution they would have opposed the resolution. If there is a single exception in the world where the priests possessed the power and did not enforce the decree, I have yet to be made acquainted with it. The infidels of France after their terrible revolution wrung from the government a toleration for the persecuted Protestants, which had been for ages denied them by the Romish Church. She had deluged her fields with the blood of her Protestant subjects; now the voices of Atheists and Deists were united in urging the claims of justice to the consciences of their fellow man. When the power of the priesthood had departed, then only was there a hope even under the most adverse circumstances for some relief from the intolerable oppression.

If at the commencement of the Revolution of these United States, ours as well as the mother country had been Roman Catholic, where would there have been the slightest hope for the establishment of such a constitution and government as that under which we live? The priests would have been in the possession of every secret movement in the land. Every wife would have been compelled to become an informer against her husband, and the first dawnings of the light of liberty would have been extinguished by the spies and widely and unprincipled agents of intolerance and despotism appointed to look into men's secret thoughts. If, indeed, civil liberty had been won by the sword, there was a power stronger than the sword that would, for ever, have extinguished the hope of religious liberty.

In a moral point of view, the evils of Romanism are deplorable. I have already alluded to the training of the priests in such a manner as can scarcely fail to sully the purity of the female mind and compel her to reveal to a man thoughts at which innocence recoils. Whoever will wade through the instructions recorded by Sanchez, Ligouri, Dens, or any of their saints and doctors to train their priests as confessors must, if he possesses any knowledge of human nature, plainly perceive that it is a system which, in the nature of things, must lead to the grossest immoralities. At the tender age of five years, the little girl is dragged to the confessional to have sentences suggested to her which have never entered into her mind; and as she advances in life, she is questioned and cross-questioned in order to ascertain whether she has, as yet, had any carnal desires (Dens, vol. 6, p. 204). The blushing virgin must explain every thing to her bachelor priest. The secrets ("circa actum conjugalem") of the nuptial couch dare not be withheld. Even the disconsolate widow is not spared in her weeds; she is told that if she only calls to mind "the joys once sanctioned by her marriage vow, it is a mortal sin, and must, therefore, be specially confessed." I cannot conceive how a father can permit his lovely daughter or a husband allow the wife of his affections to be dragged to the confessional of an unmarried priest to submit to such a torturing as is calculated to degrade if not to destroy. I invite the Roman Catholic father or brother who doubts the truth of my assertions to go to the library of his priest at the Cathedral or if he is denied admission there, then to go to the Charleston Library and enquire for Den's Theology. It is a book that has been in use for ages for the instruction of young priests. It was recommended by all the bishops of Ireland and was dedicated to and highly approved by Archbishop Murray. It is not, therefore, to them like the writings of Luther, a prohibited book. Dens has outrun Basil on the stadium of blackguardism. The filthiness contained in his Theology cannot be surpassed even in a brothel, and the saint who studies it ought to be protected by the shield of Origen.

But it is not only the priests that are disciplined in Latin in the mode of ferreting out all the secret thoughts of women. The virtuous and refined female has placed into her hands "a manual of fervent prayers and pious reflections" by which means she is enabled to prepare her mind for the confessional. The following is a sample:

"Have you procured or thought to procure a miscarriage? or given any counsel, aid or assistance thereunto? How often? p. 212.

"VI. Have you been guilty of fornication, or adultery, or incest, or any sin against nature, either with a person of the same sex, or with any other creature? How often? Or have you designed or attempted any such sin, or sought to induce others to it? How often?

"Have you been guilty of self-pollution, or of immodest touches of yourself? How often?

"Have you touched others, or permitted yourself to be touched by others immodestly, or given or taken wanton kisses, or embraces, or any such liberties? How often?

"Have you looked at immodest objects with pleasure or danger? read immodest books or songs to yourself or others? kept indecent pictures? willingly give an ear to, or taken pleasure in hearing loose discourse, &c., or sought to see or hear any thing that was immodest? How often? p. 214.

"Have you exposed yourself to wanton company? or played at any indecent play? or frequented masquerades, balls, comedies, &c., with danger to your chastity? How often?

"Have you been guilty of any immodest discourses, wanton stories, jests or songs, or words of double meaning? How often? and before how many? and were the persons before whom you spoke or sung, married or single? For all this you are obliged to confess, by reason of the evil thoughts these things are apt to create in the hearers.

"Have you abused to the marriage bed by actions contrary to the order of nature? or by any pollutions? or been guilty of any irregularity in order to hinder your having children? How often?

"Have you without a just cause refused the marriage debt? and what sins may have followed from it? How often?

"Have you debauched any person that was innocent before? have you forced any person, or deluded any one by deceitful promises, &c., or designed or desired so to do? How often? You are a obliged to make satisfaction for what you have done.

"Have you taught any one evil which he knew not before? or carried any one to lewd houses, &c.? How often? p. 214.

"Have you taken pleasure in the irregular motions of the flesh? or not endeavored to resist them? How often? p. 216."

Extract from the "Garden of the Soul," &c., to which is prefixed an historical explanation of the vestments, ceremonies, &c., appertaining to the holy sacrifice of the Mass, by the Rt. Rev. Dr. England, late Bishop of Charleston. With the approbation of the Rt. Rev. Dr. Hughes, Bishop of New-York.

When Luther's writings were paraphrased and regarded as too impure to be translated into English, I translated them and gave them to the public to show that he had been grossly misrepresented. In return, I asked Dr. Bellinger to give and defend the language in the Garden of the Soul contained in a designated page of the book. Dr. Lynch, in answer to some of my charges, resorted to criticism, proving to the satisfaction of all concerned the fact, in copying from an author I had, "through the hurry of writing," changed semicolons into commas, etc. I certainly had not the slightest objection to this criticism, inasmuch as it neither altered the sense or removed the objections to the book.

I had further stated that the Garden of the Soul had been partly written by Dr. England. It was the book of the church composed by various authors of whom Dr. England was one. This he admits. I had inadvertently stated from the title page, which had been copied for me by another hand, for I could not at that time retain the book, that Bishop England had also recommended it. This he denies, but adds that it was in existence before Bishop England was born. He acknowledges that it was the prayer book of the Roman Catholic Church and admits that Archbishop Hughes had recommended it. He goes on to say, "the work has been published more than once in England."

I perceive that Sir Henry Parnel in 1817 in a speech before the British

Parliament in praise of his, the Roman Catholic, church asserts that the annual sale of the "Garden of the Soul," by a single bookseller in Dublin amounted to 2,000 copies. This was in Bishop England's native Ireland.

Dr. Lynch, in correcting what he calls my mistake, says, "this is a trifling matter, in itself scarce worth the ink and paper I am using." I believe the public thought so, likewise, and only wondered why this trifling matter was so seriously dealt with. It looked like dodging the real question—like throwing down his armour and waking up a flock of geese to save Rome in her hour of danger; since, if Bishop England had this book in use during so many years in his Diocese, he by this very act recommended it. Dr. Lynch says of it, "As to the real character of that pious prayer book, I need not say a word in its defense." He goes on to state that the fact that Dr. Bellinger "did not deem it unfit to be presented to an excellent and esteemed lady, will go far in the minds of those who know his personal character, to satisfy them that the book is not 'infamous.'" The extracts given above are in plain English—they are in the hands of ladies and of the children who go to the confessional. Inasmuch as part of that on the Mass was written by Dr. England, as Archbishop Hughes recommends the whole work, as Dr. Bellinger presents it to ladies, and Dr. Lynch calls it "a pious prayer book," they will be gratified, no doubt, at seeing it quoted by a Protestant to convince the world how pure and pious a book it is.

To my mind, these extracts from the confessional are calculated to produce the reverse of purity and piety. If it be said it is a part of our faith to recall every particular temptation to sin, in order that we may confess and receive absolution—then I would answer that a faith that takes such an erroneous view of man's nature can not have been derived from the pure and holy precepts of the Saviour of sinners. There are some things that should not be spoken of or alluded to. The virtuous mind has no desire to become familiarized even to such thoughts. He would be a very unwise parent who would permit his son to look on at the gaming table and to visit scenes of intemperance and debauchery for the purpose of being thereby enabled to receive a moral lecture on the evils of dissipation.

The effects of this training of priests and of penitents to think and to speak of the temptations to impurity have been most disastrous. Volumes have been written, even by Roman Catholics themselves, to betray the evils that have flowed from the confessional.

In the proceedings of the Council of Trent as given us by Romanists, we have an account of the ineffectual labours of that body in reforming the priesthood. The 15th chap., Sess. 25, is headed contra "clerici filios," which throws some light on the licentiousness of a celibate clergy, about whose sins this infallible council was now engaged in consultation. In a preliminary address of Benedict VIII, to the Council of Ticine, he speaks of a large number of the sons of the clergy under the name of "filii concubinarii." In the Quarterly Review of June 1851, the writer goes through a long list of the ancient Catholic Bishops in Scotland and shows how one after another they lived openly with concubines. He says, "the most cultivated, the most amiable among them were, in this respect, not a whit purer than the others. Such of them as were contented with one woman were esteemed virtuous," etc. Whose son was the infamous Cæsar Borgia, who caused his brother and several other persons to be assassinated? and was he not made Archbishop and Cardinal by his father, Pope Alexander VI.? and what right had Pope Alexander VI. to a son? Look into the old annals of Spain, of Flanders, of Ireland, and of Norway, and see if you cannot trace the descendents of cardinals and bishops, who were legitimized by their several governments. Look at the annals of Scotland, and you'll find a history of Cardinal Beaton, of Archbishop Hamilton, of Bishop Chisholm of Dumblane, of Bishops Lesley, Morey and Argyle, and then inquire why it was that the legitimation of nearly all of the families stand on record "for the satisfaction of all concerned."

Laws have been enacted by Catholic legislators to prevent the illegitimate children of priests from holding offices in the state. The statement given by Luther of the immorality in his day, which resulted from the celibacy of the clergy, has been confirmed by every priest who left their communion from that day until the time of Blanco White and Conelly of our own times. I was myself while in Europe witness to one priest accusing another to his face of his gross immoralities. They supposed that I did not understand their language. The accused did not deny it, but excused himself on account of his constitution and gave as a salvo to his conscience, the admitted fact that others of the fraternity were as guilty as himself. Blanco White, a man of unquestionable veracity who had been a priest of great celebrity in the Romish Church in Spain, gives us the following testimony, in 1835:

"I cannot think on the wanderings of the friends of my youth, without heart-rending pain. One, now no more, whose talents raised him to one of the highest dignities of the church of Spain, was, for many years, a model of christian purity. When, by the powerful influence of his mind and the warmth of his devotion, this man had drawn many into the clerical and the religious life (my youngest sister among the latter), he sunk at once into the grossest and most daring profligacy. I heard him boast that the night before the procession of Corpus Christi, where he appeared nearly at the head of this chapter, one of two children had been born, which is to concubines brought to light within a few days of each other, etc.

"Such, more or less, has been the fate of my early friends, whose minds and hearts were much above the common standard of the Spanish clergy. What, then, need I say of the vulgar crowd of priests, who coming as the Spanish phrase has it from coarse swaddling clothes, and raised by ordination to a rank of life for which they have not been prepared, mingle vice and superstition, grossness of feeling and pride of office, and their character? I have known the best among them; I have heard their confessions; I have heard the confessions of young persons, of both sexes, who fell under the influence of their suggestions and example; and I do declare that nothing can be more dangerous to youthful virtue than their company. How many souls would be saved from crime, but for the vain display of pretended superior virtue, which Rome demands of her clergy, etc.

"If a priest, in spite of his proverbial cunning, is discovered, and if he has been denounced to the bishop by public opinion, he will be removed to silence the scandal, and sent to another village, where he will be unknown, and where, by and by, he will begin again the same mode of life," etc.—(Practical and Internal Evidences, pp. 138–43.)

Newman, the recent proselyte from Pusyism to Romanism, has stated that the violation of these laws of purity was as common among the Protestant as Catholic clergy. This is very far from being true. Where rare instances have occurred among Protestants, they were excommunicated and with difficulty restored. They become public because these transactions are published in the proceedings of their various ecclesiastical bodies. On the other hand, the sins of the priests in this particular are studiously concealed by their brethren of the same faith; and what injured woman would reveal her own shame and thereby draw down upon herself the censures of her church and her whole communion? She is degraded and has no other redress or consolation than the solemn mockery of a priest, who insults the religion and the power of his Saviour by prostituting his language — "Thy sins are forgiven thee."

At the same time it must be recollected the priests cannot enter into lawful wedlock. Although believing themselves members of an infallible church, they feel that they are fallible men. Most of them, if we judge by their conduct, are unconverted, and possess not that regenerate heart which is calculated to raise them above temptation. The great majority have sprung from the lower walks of life. They now live luxuriantly and generally indulge in stimulating drinks. They have much leisure and have none of the cares or joys of a family to occupy their time or wed them to a domestic life. They are surrounded by temptations at the confessional and know the weakness of every female who resorts thither to pour into his ear a detail of her temptations and her sins. From the very nature of their situation they are infinitely more liable to fall than the Protestant clergy, and we know from history and experience that this is the fact. The doctrines of the celibacy of priests, of oral confession, and the power of the priests to forgive sin have been the ruin of the souls of millions of our race. If the Romish priests in Protestant countries are either more guarded or more pure and if, where the two churches are placed in juxtaposition, they do not exhibit the foul spot that rests on their order, it is an evidence of the incalculable benefit they have derived from the Reformation.

When Dr. Lynch came out in defense of the "Garden of the Soul," the confessional part of which I had characterized as "infamous," he declined quoting or defending the chapter which I had censured, but gave a quotation from another portion of the book, which recommended at least one quarter of an hour daily in devotions. The Scribes and Pharisees did more than this, and yet the Saviour characterizes them as "hypocrites," "blind guides," "whitened sepulchers," and "the children of them which killed the prophets." When the immoralities of the priesthood, the effects of the celibacy enjoined on them by the church, as proved by the Romanists themselves who enacted ineffectual laws to check these horrid crimes, were referred to, Dr. Lynch brought forward as a test, their conduct during the prevalence of pestilential diseases. Were all this true, it would be no evidence of their purity. We have seen thousands of the most depraved men fearless in pestilence and in battle. Besides, as Bishop England once informed me, the priests of that church are compelled by the rules which govern them to be present with the dying to administer extreme unction. This is not required of the Protestant minister. He visits, instructs and prays with the sick that he does not feel himself empowered to give the dying man a passport to heaven; and regards it as arrogance and a simple mockery and those who pretend to a power that belongs only to God. He performs his duty without parade or show.

In regard to the charity of the priests to the famishing poor in Ireland, Dr. Lynch has given us the Popish version. I have no doubt this is true in regard to some of them. There is, however, another phase in this movable panorama. I have lying before me several publications in which I place the most implicit confidence, which go to prove that it was to a considerable extent the Protestants who supported hospitals for Roman Catholic inmates. In 1834 Mr. Inglis found 2,145 persons on the books of the Dublin mendicity house, of whom only 200 were Protestant; and although the vast majority of the inhabitants are Roman Catholics, he states, "that for every pound subscribed by them, there were fifty pounds subscribed by Protestants." It is so everywhere. The Church of Rome impoverished those its people by exactions to build showy cathedrals and colleges and leaves the Protestant public to contribute to very largely to the support of their paupers. It is stated by Dr. Dill (p. 161) that, "while the Protestant labourers of Ulster and Britain were sending their wages to the famished south, it was the topic of public remark that balls and entertainments were going on as usual in several southern towns though paupers were nightly dying in their streets." He says, "even Ulster we have often seen the small farmers driving to the priest's haggard their half load of corn, under the name of the 'priest's stock.' In Connaught a similar tax is levied called 'the priest's bart.' At the various stations for confession, which are held in different parts of each parish, 'a treat,' as it is termed, must be given to the priest, and the poor people have been known to sell their pig thus to entertain him. During the late famine, when the Hindoos of Calcutta and the Copts of Alexandria were sending relief to Ireland, its own priests in many cases not only left the people to perish, but robbed them of the almonds bestowed by heathen and Mahomedan charity. One priest made large sums by selling holy salts to cure the potato disease, and many gave their last six pence to purchase this 'specific.' When the potato failed in 1847, there was a poor man in the neighbourhood of Westport who had nothing left but a patch of oates, amounting when reaped to sixty sheaves. The priest came around for his 'bart.' The wretched man pointed imploringly to his wife and family. But deaf to every entreaty, dead to every feeling, he commanded his servant to count into his cart twenty sheaves of the sixty, and then he marched off with his booty."

If, then, we take Dr. Lynch's "practical test" of true Christian piety, viz.—charity and mercy to the poor and afflicted—we will discover that even here the priesthood is not infallible.

He has administered to me a homily on Christian meekness and charity. He feeds with the milk of human kindness, and annoints with holy oil his "separated brethren;" yet at the same time, he who holds as he informed me the office of "vicar general" and is therefore possessed of ample power, inculcates through his priests and the books of his church the doctrine that there is no salvation for those "separated brethren" out of the pale of the Papacy.

The Priest replies, this is my doctrine; why censure me for acting up to its principles? I answer, if that intolerant doctrine is calculated to deprive me and my fellow men of our rights, then is he amenable to society for an abuse of that reason and judgment which God has given him. Protestants say we have a right to read the Bible—it is God's book—we are commanded to "read the Scriptures," and are assured that "they will make us wise unto salvation." The Papist says, "you shall not read it;" and even in America have frequently burnt our Bibles. The helpless Madiai are now prisoners and in the gallies for no other offense than that of reading it. Many of the Romish doctrines, such as transubstantiation, the works of supererogation, extreme unction and the like, are simple absurdities; they do not, however, infringe on our rights; but when the Romish church tells us if we do not swallow these absurdities, she will confiscate our estates and either banish us from the land or consign us to the Inquisition if she has the power, then we say, you have no right so to prostitute your reason as to embrace the doctrines that would destroy individual rights and inflict an intolerable oppression on society. The robber may work himself up to the belief that he has a conscientious right to my purse, but if he wrests it from me, the law will consign him to the penitentiary in spite of his imagined rights.

He distributes from the depository at his own residence a catechism which represents all these "separated brethren" as having received their religion not from the holy teachings of the word of God, but "from a man by the name of Martin Luther," who derived "his doctrines from the devil" and was the most depraved of men. It further teaches that "all Protestants are bound to obey the decisions of the Council of Trent, which consigns every one of these "separated brethren" to the flames of hell.

The question now forces itself on our minds, how can Dr. Lynch, who professes such sentiments of charity and speaks of his "separated brethren" in language soft as "the dews of Hermon," cling to a church that has in every age exhibited not the shadow of a claim either to charity or to mercy towards these "separated brethren" and has at all times and under every form of government when it possessed the power and forced by its "holy severity," as Bossuet designates it, its unscriptural dogmas by the most unrelenting persecutions? How can he, even for a single day, remain in a church that admits of prevarication and fraud—that dispenses from oaths—that enjoins the confessional and arrogates itself to the power of absolving from sin for the purpose of building up its own ambitious and worldly hierarchy? How can he consent to become a fellow labourer with the Jesuits, whose whole history is one of intrigue, corruption and bloody crimes? How with such sentiments of charity and mercy can he continue to advocate the claims of the Romish church, which for thirteen centuries has shut up its bowels of compassion and closed the fountains of mercy and has by a moderate calculation shed the blood of more than sixty millions of the human race?

Address on Education

As president of the board of trustees, Bachman gave *An Address on Education, Delivered on the Day of the Laying of the Corner-stone of Newberry College, July 15, 1857*. He began by indicating the increasing need for higher education as societies progressed, and he stated his intention to "point out briefly, 1st, the nature of the studies to be pursued in the college; 2nd, explain the principles on which the institution is to be conducted; and 3rd, the benefits it is calculated to confer."

Lutherans had taken the initiative to create a college in a predominantly Lutheran part of South Carolina, and they bore most of the expense. The board was mainly Lutheran, but its membership included other faiths as well. The new school would be a Christian institution, but not a sectarian one like many other colleges that had been created throughout South Carolina following the presidency of Thomas Cooper at South Carolina College. Cooper had outraged so many parents by his "heresies" that they withdrew their sons, and as a direct result most denominations in the State soon afterward created separate colleges for the education of their youth. Bachman stated, "To denominational colleges—in the usual meaning of the term—that is literary institutions conducted on the sole principle of teaching a peculiar set of religious dogmas—we have ever been opposed." He believed that this was the work of churches and seminaries and that "our halls of learning and science should be open to all."

Of the advantages to be expected from an institution of higher learning, Bachman said, "a college always creates a town and then a city, wherever it may be located. Families of wealth, education, and influence will take up their residences there to enjoy the advantages of education, and the benefits of society." Another great benefit would be to "supply suitable teachers for our common schools and thus elevate the standard of education among our people.... The students will carry home with them to their families and various neighborhoods the seeds of knowledge—the love of study and the ambition to excel."

The desire to continue learning on one's own was the greatest benefit of education. Bachman wrote, "By reading and study, the vast storehouse of nature, the mysteries of art, and the histories of the past and present generations of the world are all brought familiarly to his [an educated individual's] mind.... We are suddenly introduced into a high and lofty society which we cannot find among living men."

An / Address on Education, Delivered on the Day of the / Laying of the Corner-stone / of / Newberry College, / July 15, 1857. / by / John Bachman, D.D. / President of the Board of Trustees. / Charleston, S.C. / James & Williams, Printers / 1C State-Street. / 1857.

Correspondence.

Pomaria, S.C., July 17th, 1857.

Rev. J. Bachman, D.D.

Dear Sir—at a meeting of the Board of Trustees of Newberry College, held yesterday, the undersigned were appointed a committee to solicit for publication, a copy of the very able and interesting address delivered by you on the occasion of laying the corner-stone of the college building.

Sincerely hoping you may be able to comply with our request, we are

Yours very respectfully,

T. S. Boinest, / O. B. Mayer, / P. Tod,} *Committee.*

Rev. T. S. Boinest, Dr. O. B. Mayer, and Dr. P. Todd.

Gentlemen—The interest alluded to, in too flattering terms, in your note was hastily written and without an idea that its publication would be called for. Under the hope, however, that it may awaken an additional interest in favor of the institution we are all desirous of fostering, I will waive all private considerations and cheerfully place the manuscript at your disposal.

Yours respectfully,
Jno. Bachman.

Address on Education.

Friends and fellow citizens:

Believing that the event we are commemorating this day is the commencement, in this portion of our State at least, of a series of efforts and labours in the promotion of knowledge—the foundation of enlarged means of usefulness and the increase of human happiness, you will indulge us in inviting your attention to a subject which should be prominent in our minds on an occasion like this, namely that of *Education*—the rearing up of the intellectual and moral man, which is to prepare him for his struggles through life, for his labors and efforts in the various duties which are before him, including his moral and religious training; this latter will impress on his mind the sentiments of truth, justice, honour, benevolence, and purity of life, fitting him for that higher destination to which the Christian aspires—the hopes of immortality and the bliss of heaven.

Education in the general sense of the term may be defined as the art of training and instructing the mind and forming the character of the young.

According to this definition, the education of youth not only embraces the instruction given for the regulation of his manners and his improvements in literature, science, and morals, but every opinion he has imbibed and every habit he has acquired, either from his associations, from the reflections of his own mind, or from reading the thoughts and sentiments of others. It further includes the regulation of his propensities and passions and that self-government which will preserve him from the contagion of examples of evil and will enable him to profit by the wisdom and virtue of the good.

The importance of a well directed education in this comprehensive sense of the term is so evident and so generally admitted that it would appear to be almost superfluous to enlarge on the subject before this enlightened audience.

There are however some well disposed persons who do not readily admit the importance of any studies that are not practically of importance to our own immediate necessities—in other words, that nothing is worth noting that does not supply us with food and raiment. If time would permit, it could easily be shown in what manner education increases the facilities of labour and adds to those productions which are necessary to the sustenance and comfort of the world.

We may be told that our forefathers, possessed of very limited attainments, were enabled to convert the forest into fertile fields, that they raised their own products and were as contented, as virtuous, and as happy as any of their successors; where then, they will ask, is the advantage of contributing so largely to the cause of education if so little is apparently gained by the change? An answer to this, we will observe, that if we for the sake of argument admit that this was the case in the generation which is now fast disappearing, the important fact must not be overlooked, that they were surrounded by men of their own pursuits—habits of thought, education, and manner of life. They had therefore few rivals, and being on a general equality, they were in a measure freed from the mortifications attendant on a consciousness of inferiority.

But let us not overlook the changed circumstances under which the rising generation is summoned to engage in labors and efforts that are required of them not only in their social capacities, but as men and citizens of our common country. The progress of all civilized nations in every department of knowledge and especially in scientific attainments has been unprecedented in any former period of the world. Our own population has increased since the organization of our government from three to twenty-six millions. The number of our States has been multiplied from thirteen feeble sovereignties to thirty-one powerful independent States and many territories. The barriers presented by the Alleghenies have succumbed to the science and indomitable perseverance of our race. The wilderness of the far West has yielded and fallen before the axe of the sturdy and independent husbandman, and the once solitary desert has been made to "rejoice and blosom as the rose." California has yielded its auriferous treasures. The mighty rivers, among which the father of waters in the west, the lakes, the inland seas of the North, the wide Pacific Ocean that now rejoices in rolling its billows on our own shores are all white and with the sails of our commerce; and the screams of the loon, the tern, and the seagull have been interrupted by the puffing of the huge steamer and all the other appurtenances of an enterprising and progressive nation. The arts have advanced, and manufactures have multiplied the articles of clothing a thousand fold through the agency of steam. Our railroads are daily increasing the facilities of travel and commerce, and our telegraphic wires seem destined to encircle the globe and invite to rapid and familiar converse all the nations of men so that the ear can catch the sound almost at the moment it is uttered on the opposite sides of our hemisphere. Science and the arts have combined to supply the necessities, comforts, and luxuries of the increased population of our teeming earth.

From what sources have all these wonderful improvements been derived? Is it not self-evident that they were solely the result of education? Ignorance can never become the mother of invention. Unenlightened Africa has stood still for ages and centuries shrouded in barbarism and gloom, whilst the educated nations of the world have carried the lights of knowledge, the treasures of commerce, the aides of civilization and the blessings of religion to the farthest earth.

Under these improved circumstances, when knowledge is advancing with such rapid strides, it is impossible for you not to see and to feel that unless you and your children follow in this march of improvement, they will be left at an infinite distance behind; and instead of being associated as our fathers were with a band of equals, they will be compelled for want of education to fall back into the lower ranks of life and have the mortification of witnessing those who in many circumstances may have been their inferiors, now by their improvements in education rising above them. It is in vain for us to expect that our children can maintain their position and prospects of usefulness in society without being entered into the ranks in the march of human knowledge and progress. The world will go forward without any regard to our indifference. Men have felt the pleasure which is derived from information and knowledge. They have experienced its good effects, and one acquisition has brought another within their reach. "Knowledge," says Johnson, "always desires increase; it is like fire which must first be kindled by some external agent, but will afterwards propagate itself." We may feel assured that in this march of improvement men will not take a single step backwards as soon might we expect that the travellers who have been accustomed to the comforts and rapidity of the rail road car, would prefer going back to the old-fashioned party-based stage-coach or the community be willing to relinquish the fine and cheap products of the steam factory for the spinning wheel and the hand loom as to believe that in the present day we could do well enough without schools or seminaries of learning and that a simple cross with "his mark" written above and beneath will confer any additional respectability to the individual who is compelled to make it.

It is almost impossible to conceive to what an extent the principles and conduct of every man, his successes or his misfortunes, the happiness or misery of his life, depend on his early education and training in knowledge, morals, and religion. He is left by nature a weak and helpless creature; he is dependent on the care of others; he cannot provide for his own sustenance or safety. But how wonderful is the difference between what he is at his birth and what he may become at his maturity. God has given him the privilege of enlarging and forming his various powers by his own diligence and skill, so that with a considerable force of truth, it may be said, he enjoys the proud pre-eminence of being his own maker.

We do not desire to be understood as supporting the doctrine that every thing in the intellectual and moral system is the result of education. It is admitted that the rudiments of disposition and capacity are very different as beheld even in children. In some, the sensitive powers are quick and lively whilst in others they are dull and sluggish. The external structure of the organs of the body and the mind differ widely in different individuals, and it is therefore reasonable to suppose that the internal structure and the more concealed corporeal system on which the offices of the mind depend must also be essentially different.

Although it is admitted that education cannot elevate all men to the same high standard, it can improve the minds of all and greatly increase their capacities for usefulness and enjoyment. Having presented these views of education in general, let us proceed to the subject which is more immediately connected with the object of our present assemblage.

We meet together to-day, to lay the corner stone of the first college ever erected in the district of Newberry. Its inhabitants have set an example to the neighboring districts, of their devotion to the cause of education and of their determination to open the halls of learning and science to their children and their posterity.

A college is an institution endowed with certain revenues, with competent professors, where the several parts of learning are taught in halls and classes arranged for that purpose.

A university is an assemblage of several of these colleges. Thus, in the English universities of Oxford and Cambridge there are in each upwards of twenty colleges. In the universities the different professions, such as theology, law, and medicine are taught. The individuals who attend are men who had previously received a collegiate education. They simply attend the courses of lectures and are not subject to the rich strains, the daily tuition, examination, and discipline of the college.

In America, universities of this character are less needed inasmuch as the various religious denominations have theological seminaries supported by themselves, and our schools of law and medicine are found to prosper most where they are unconnected with the classical, the mathematical, and literary studies of the common college.

What is most needed in our country are colleges conducted on the plan of the German gymnasiums, whose youths are thoroughly grounded in those studies pertaining to our colleges, where their lessons are daily recited to competent professors, where they are stimulated to industry by the honors that await the most distinguished, where their moral and religious duties are faithfully instilled into their minds, and where a course of rigid discipline is observed by which they will be preserved from the contagious examples of vice, imbibe the principles of integrity and honor, be qualified to fill important stations in life and become the ornaments of Society, the pride of their families and a blessing to their country.

Such an institution we have resolved, under the favor of heaven, to rear up in your midst. We have met this day to remind each other of the arduous work we have undertaken to accomplish, to solicit in behalf of our labors and immense expenditures, to countenance and support of patriotic and good men, and to invoke the blessings of Almighty God.

We will endeavor to point out briefly, 1st, the nature of the studies to be pursued in the college; 2nd, explain the principles on which the institution is to be conducted; and 3rd, the benefits it is calculated to confer.

1. The studies to be pursued are those usually taught in all of our colleges. Your sons will be instructed in the classics, the mathematics, philosophy, history—in a word enjoying all the advantages of other colleges in the United States. Fortunately there appears no difficulty in obtaining suitable men as Professors. Men of sound learning, of unimpeachable characters, and attached to the peculiar institutions of our Southern country. Without this latter essential qualification they could not under any circumstances be received or countenanced among us.

We yesterday elected as President of this college a gentleman of education, of high principles, of honor and integrity, polished in manners, eloquent and pious, and a southern man by birth and education.* We regard this election as most fortunate for the best interests of the institution.

We feel confident that we will be able to establish such a discipline in the college under the direction of intelligent, firm, and able men, that our young men will know that they have entered into our institution for the purposes of study and not to be indulged in idleness, riot, and dissipation and in those rebellions which have so frequently thrown our colleges into chaos. It is intended when their course of studies shall have been completed that they shall receive their diplomas and graduate with all the honors of the college. Owing to the deficiencies in the grammar schools in this and the surrounding districts, it will be necessary to attach a preparatory grammar school or academy to the college, which, although attended with considerable labor and an additional expense will be of great advantage to those who are preparing to enter the college, and will be of especial benefit to the inhabitants of this town, who will be able to have their children educated without the necessity of removing them from the control and discipline of their families.

2. In this part of our address it may be necessary to explain how far this is intended to be a denominational college. To denominational colleges in the usual meaning of the term—that is literary institutions

---

* Rev. F. R. Anspach, of Virginia.

conducted on the sole principle of teaching a peculiar set of religious dogmas—we have ever been opposed. Our idea is that whilst students intended for the ministry should be thoroughly grounded in the doctrines of their peculiar faith in their own theological seminaries and that our people should be instructed in our churches and Sunday schools in those articles of faith which are the characteristics of the several bodies of Christians, our halls of learning and science should be open to all. Our young men of every religious profession are destined to mingle together in all the walks of public and private life, and they will be prepared to live and labor in greater harmony if they have associated together in the same schools and colleges. It is even a matter of regret that in our religious views and our doctrines and forms of worship we could not all harmonize.

It is true the church of which he who addresses you is a humble representative is firmly and devotedly attached to the sentiments of the reformation and venerates the name of Luther as the father of Protestantism and the successful advocate of the freedom of religious thought and the holy Scriptures as our guide in doctrine and in duty; yet that church inculcates an extensive charity and liberality in all its opinions and feelings. Its pulpits and its communions are open to all Christians. It has never been a proselyting church. Satisfied that our brothers in other denominations have adopted from our creed all the doctrines that are essential to salvation, we welcome them as of the same Christian fold and devoutly pray that we may all "endeavor to keep the unity of the Spirit in the bond of peace."

Whilst however this is not designed to become a sectarian college, it must not for a moment be supposed that in this institution the great truths of our common Christianity will not be prominently acknowledged and faithfully inculcated. Sentiments of piety should be impressed on the minds of the young and should form a part of all our instructions. Religion forms the relation between man and his God, not only as the Creator and creature as governor and subject, but has the support of the relation between man and man as the foundation in principle of social and moral duties. Religion is equally the basis of private virtue and public faith, of the happiness of the individual, and the prosperity of the nation. Thus far we intend to go, and we feel assured that every

Christian parent to whatever denomination he may be attached will second us in these resolutions.

Whilst however the members of the Lutheran church are desirous of throwing open these halls of learning for the benefit of all, we expect to derive no small advantage from the institution. Our theological students who are to succeed their elder brethren in the ministry will with few exceptions be educated here before they enter the theological seminary. Few of them have the means of meeting the increased expenditures of an education at our State college, and they would naturally prefer being associated with professors who they felt assured would take an interest in their improvement. Parents of our own faith whose children may be destined for other professions than those of the ministry will feel greater security in sending their sons here than to more distant colleges in whose discipline they have less confidence. But beyond the advantages we expect to derive from the education of our theological students, we have no interest but that which all other denominations will enjoy in common with us. It is true we have stipulated for a majority in the Board of Trustees, but it will be borne in mind that some of your most influential and intelligent men who are not identified with our church are also trustees and co-laborers with us, and their very names will be the guarantees that the affairs of the college will be conducted on liberal principles. We have voluntarily assumed a great proportion of the labor; we must make provision to meet the heavy expenditures, and we take a large share of the responsibility in conducting the affairs of the college. Thus it will be perceived that all denominations enjoy equal advantages with us as far as the education of their sons is concerned. In a word, we voluntarily assumed the labor and the responsibility, and they will enjoy equal privileges with us without either labor or responsibility.

It is difficult to conceive in what other mode a college under our peculiar circumstances could be sustained with any prospect of success. The State supports its own institution very liberally but will not render aid to any other. Our college has not as yet been endowed with gifts or legacies, and we have no funded capital. If we were to depend upon having the college endowed by all denominations and have an equal number of trustees among the various societies of Christians, it would soon be discovered that no denomination in particular would take an interest in the institution. It would be difficult to convene a Board of Trustees; sectarian feelings would be generated, and the best interests of the institution would be jeopardized. This subject has for many years engaged our earnest and prayerful attention. We could not conscientiously support an institution but on the broadest principles of Christian liberality. It is on these principles that we intended this college to be conducted. We have called it Newberry College. The name of this growing and flourishing town and this fertile district will be a rallying point to the lovers of learning and science not only in this district, but in those by which you are surrounded.

Inhabitants of Newberry, it is your college, named after your town and district. Cherish her as the young daughter of your love and training. Be proud of her for the fair promises she holds out to you in the years of her maturity when she will become the mother of many sons whose voices will be heard at the forum, the bar, in the Senate, and from the sacred desk and who when duty shall require it will become the defenders of the time-honored institutions of our Southern land. Thus, "she shall give to thine head an ornament of grace, a crown of glory shall she deliver to thee." Throw over her the mantle of your protection and bestow on her the fond and benevolent smiles of a parent; then, when in other years, men effeminated by luxury and grown giddy by the pride of life shall display their ornaments, their gay equipages and their trappings of silver and gold, she, the alma mater, like Cornelia, the mother of the Gracchi, will point with proud exultation to her sons, and proclaim, "These—these are my jewels."

3. Let us now proceed to point out some of the advantages which we may reasonably hope will be derived from the institution whose foundation is this day laid. So full of interest is the subject that it is difficult to decide where we ought to begin and equally difficult to be restrained at the point where we ought to conclude.

It must be left to men of more experience in pecuniary affairs to point out to you the advantages which this town will derive from an increase in the value of property in consequence of the vicinity of the college. A college always creates a town and then a city wherever it may be located. Families of wealth, education, and influence will take up their residences there to enjoy the advantages of education and the benefits of society. By this means cities have sprung up where before nothing but a solitary farm house existed. This is the result of our experience in regard to every college both in Europe and America; we are warranted, therefore, in believing that the same results will attend our present efforts.

But whilst these temporalities are not to be disregarded, we should look for far higher and infinitely more beneficial results which will flow from our present efforts. Our schools of learning in our halls of science are intended to build up the inner man and entitle him to the honor bestowed on him by his Maker, who has described his high mission and exalted destination in these emphatic words: "Thou hast made him a little lower than the Angels, and hast crowned him with glory and honor. Thou madest him to have dominion over the works of thy hand; thou hast put all things under his feet."

One of the most important advantages which a well conducted college confers on a community is, not the simple rearing up of a class of intelligent men who will graduate at the institution, but the effects which in the silent progress of time will be produced as the result of the education of the few for the ultimate benefit of the whole. What is most needed in our southern country is an intelligible and practicable system of popular instruction—and that the business of teaching should be better understood, more highly appreciated, and more liberally remunerated. The education of the people is the hope of our very existence. Such institutions as ours can have no permanent standing but on the basis of knowledge and virtue. Our nation is passing through a great trial. Let luxury and excess be permitted to grow in our cities; let vice stalk abroad fearlessly in our villages; let our hearty yeomanry become indolent and inefficient; let our noble youth lose the principles of virtuous education and indulge in extravagance and revelry, then farewell to our country's hope. Though the semblance may remain for a while, the spirit will have fled forever.

Another of the great benefits expected to be derived from this institution is that it will supply suitable teachers for our common schools and thus elevate the standard of education among our people. A body of intelligent, laborious, virtuous, and pious professors will exert a most salutary influence on the students and the community around them. The students will carry home with them to their families and various neighborhoods the seeds of knowledge, the love of study, and the ambition to excel. Parents will be convinced that their money and efforts in behalf of the education of their sons have been doubly remunerated. These young men will enter on their various professions; many of them will become teachers in our common schools. They will from their knowledge and experience be admirably qualified for the work before them. A desire and a taste for knowledge will be widely diffused among the masses. A well educated yeomanry is a blessing to any community. At present at your various agricultural meetings, who are they that address the assembled crowds? They are either lawyers, clergymen, or politicians, and we need not be surprised if men without experience should advance wild and speculative theories. Why does not the farmer often address these meetings? He has more experience and knowledge on these subjects than the combined wisdom of all the professional men on the ground. He is now silent because he has not been educated. When, however, he shall have received the benefits of an education, he will hold up his head among his equals and will save others the trouble of making speeches for him, either on agricultural or political subjects.

Where men are well educated, their wives, sisters, and daughters will not consent to remain far in the rear. Woman is the companion and the equal of man, and it will soon be perceived, that although she is not destined to occupy the posts designed for the more rugged sex, yet that in all that is valuable in education—in all that can inform the mind, regulate the affections, and adorn her character as a Christian woman, she is fully capable of qualifying herself for her high destination.

Thus the college exerts its silent but progressive influence. Like the light of the morning, its rays penetrate everywhere. In time it changes the aspect of society, and if its teaching of knowledge impress on the hearts of men that higher knowledge of duty that leads to salvation, then has she fulfilled her mission in rendering man wiser, better, and happier, qualified for usefulness on earth, and fitted for the society of angels in heaven.

If we are told that many valuable men rose to eminence and usefulness without a collegiate education—that Washington was a wise statesman, a heroic leader of armies, and the best of men—that Franklin pierced to the clouds and rendered the lightning submissive to his call—that Rittenhouse carried the knowledge of astronomy beyond that at his age—and that our records contain the biographies of thousands of other great and good men, who are entitled to the gratitude of posterity for their discoveries, or their invaluable services to mankind—and yet, that none of these entered within the walls of a college. The answer is at hand; these great and good men were not even the exceptions to the general rule that education is necessary to success in every department of life. True, they did not receive collegiate educations, but they educated themselves. The work was more laborious, but they accomplished it. They acquired knowledge by the slow process of study—of thought and self-discipline. The college did not make them, but knowledge—that knowledge which is taught at the college, many branches of which they pursued with the intensity of thought—of reading and study, made them great men. Thus, although in one sense they were self-made men, yet they drank from the same fountain that gave pre-eminence to other men, and they are entitled to the additional credit of having accomplished great ends by surmounting the difficulties that, in their cases, obstructed the paths of knowledge.

In speaking of the advantages of education, it is fully admitted that there are dissolute and bad men among the educated, since all men have inherent propensities to evil, and that in these cases their adroitness and skill in the commission of crime render them the more dangerous to society. But it cannot be denied that crime is more common among the uneducated classes. From the statistics of criminals in the penitentiaries in the United States, it has been ascertained that five-sixths at least are unable to read or write. It should be further remarked that the man of education and polished manners seldom indulges in brutal violence or unpardonable asperity of language; on the other hand the ignorant savage has immediate recourse to the firebrand or the knife.

Reading and intellectual pursuits supply those resources to the mind which will render it independent of meaner excitements. The man who flies to the intoxicating bowl is led to this degrading habit generally because he has no resources in his own mind; he is unaccustomed to find pleasure in books—his evenings are dull to him—he goes abroad for relief and generally finds that relief, which is his ruin. Let such a man be educated to the love of knowledge—let him have some acquaintance with the laws of nature—let him have access to books, and leisure to him will not be a burden, nor will his home become irksome. He will find new resources and a new impulse to life, and he will be raised above sense and matter to intellect and virtue.

By reading and study, the vast storehouse of nature, the mysteries of art, and the histories of the past and present generations of the world are all brought home familiarly to his mind. Within the last few years a visionary sect calling themselves spiritualists have greatly startled weak minds by a pretended power to recall to the earth the spirits of departed worthies—of holding converse with them and extracting a variety of opinions from this intercourse. From these conversations it would appear that the intellects of these ancient sages have become considerably blunted since their long absence from this earth. The student, however, need not resort to these necromances or to any system of jugglery and fraud to be indulged in the privilege of holding intercourse with the wise and talented of past ages. We cannot enter a well-selected library without feeling an inward sensation of reverence and without being excited to emulation by the mass of mind scattered around us. We are suddenly introduced into a high and lofty society which we cannot find among the living men. We associate with the men of the past and find the human mind displayed in its highest flights in all its walks through science and the cycle of its thousand intelligences. We are permitted to ransack all the stores of learning and knowledge and revel in the mysteries of thought. Thus we become associated with men whose works have outlived monuments and pyramids and still survive in unspent and undiminished youth. Why, in man's folly, would he call back the fossil remains of departed greatness when we have in their works before us, their minds in their fullest development and when they in their best attire and kindest manner will come to us at our bidding. The pleasure of intercourse with minds of the highest stamp—in their works—especially when they are presented wearing the garb of hoary antiquity can scarcely be surpassed.

Have we a taste for classical learning—do we delight in going back to the days of ancient Greece and desire to know the thoughts and habits of men before the Christian era? We have access to the thoughts, clothed in their own words, of Homer, Plato, Demosthenes, Xenophon, Plutarch,

Sophocles, Pindar, Aristophanes, and a host of others; or do we wish to be introduced into the families of the Romans, their Latin contemporaries of the same ages, we may turn to Cicero, Livy, Cæsar, Horace, Virgil, Juvenal, Ovid, Tacitus, and others whose works have immortalized their names. Do we delight in philosophical studies? We may summon Bacon from his closet, and he will give us the conceptions of his mighty mind; with him are ready to come Locke and Reid and Stewart, Condillac, Berkeley, Hartley, or Paley. Are our minds thirsting for the knowledge that is derived from the higher mathematical studies? We may at any time call upon Plato, Aristotle, Copernicus, Leibnitz, Newton, Kepler, or Herschel. In history and the arts we have vast libraries at our command. Are we devoted to the natural sciences? Buffon, Linnæus, Cuvier, and an army of naturalists will wake up at our invitation and tell us the history of the earth we tread on, of the birds of the air, the beast of the field, the fishes of the sea, and every creeping thing; they will also discourse to us of the trees and plants "from the cedar that is in Lebanon even unto the hyssop that spring up out of the wall." The poets that sung in every age are here also to commence their songs anew, and Shakespeare and Milton, Pope, and Dryden, Klopstock, Schiller and Goethe, Tasso, Racine and Corneille invite us to leave the busy haunts of living men for a season and partake of the rich festival, which these departed worthies have prepared for all the world.

But we are compelled to break away from the indulgence of these fascinating reminiscences; not however without recalling Milton's lamentation of the mother of the human family when driven from Paradise or the lingering desires of the wife of Lot when she looked back upon Sodom. We are reminded that there is danger of relinquishing the duties of life in the luxurious leisure of study. The men of letters must resolutely counteract their propensities to indolence and too great a love for retirement.

The fact must not be overlooked that languages and literature are far from being the only studies of the college. The greater proportion of those studies are of a nature adapted to the practical duties of life, and there is no department either in agriculture, in architecture, in mining, in the manufactures, in surveying, in the construction of rail roads and canals, in composition, and keeping accounts, and in all the ministers to wealth and comfort that is not aided by those instructions derived from a collegiate course. Men of reflection and foresight can scarcely doubt that in the course of the next half century the value of property in this and the surrounding districts will be increased four-fold in consequence of improved modes of agriculture, manufactures, mechanics, etc. In fact you have all the resources within yourselves so that if driven to the necessity you could render yourselves independent of the world, not even excepting the production of tea and sugar. All this can be accomplished in no other way than by a general diffusion of knowledge and its judicious application in those industrial pursuits that contribute to man's wealth and comforts. How far a well conducted college will aid you in arriving at these desirable results, you are now preparing to ascertain, and the problem will be solved by the success or failure of your institution.

But why wait on the tardy footsteps of time? The problem has been already solved. Look at Scotland, with its barren soil and ungenial climate, once trodden down and plundered by robber chieftains. The seeds of knowledge were sown broadcast among the people, and gradually the nation became regenerated, and they have carried their knowledge, industry, and enterprise to every land. Look at Switzerland, romantic from its towering Alpine mountains and its deep, but fertile and blooming vallies—rearing its mighty glaciers above the clouds of heaven—the land of Tell and of freedom—shut out from the commerce of the world, and without the command of a navigable river leading to the ocean. What must such a people do to preserve their independence? A solitary republic surrounded by jealous, powerful, and warlike monarchies. They discovered the secret of human power. In their cities, they reared gymnasiums and universities, and in every nook and corner of their vallies and on the slopes of their mountains, the school house is seen, and the church not far distant. A sound and practical education enabled them to excel in the arts, and many of their manufactured articles have taken precedence of the world. It may here be added that the finer works of nearly every time piece that we carry in our pockets have originated from the workshops of that ingenious, free, and independent nation. Such a people, whose knowledge has enabled them to find resources within themselves, are invincible. Their confederacy of free and independent states has already lasted five and a half centuries. Italy, Austria, France, and Bavaria successively strove for ages and centuries, to conquer them. They invaded their land with fierce warriors, scaled their mountains, and carried fire and sword into the villages of their peaceful vallies; but they were all compelled to retire in discomfiture and disgrace. Look at Protestant Germany with her unrivalled schools of learning. Select for instance Upper Saxony; she has no river of commerce, and her natural soil is less productive than that of Austria, from which it is separated by no other land mark than a pillar of stone. Every child in the kingdom is taught to read and write and keep accounts. Her University at Dresden is an ornament, an honour and a blessing to the country. No traveller passing from Saxony into Bohemia, the neighbouring Territory of Austria, can fail to observe the vast difference in all that constitutes an intelligent, prosperous, and happy people, between an educated and an uneducated nation.

In conclusion, you will yet indulge us in briefly relating an anecdote which we trust is not inappropriate to the occasion and the objects which have brought us together.

Nineteen years ago, in one of our visits to the University of Berlin—the most eminent in the world—we were kindly conducted through the various halls of learning by the prime minister of the aged king, who has since deceased. In the course of a conversation in reference to the value of institutions of learning to a nation, he related the following very striking incidents.

When Napoleon with his armies had overrun Prussia and all Germany was lying prostrate at his feet, the king summoned his political ministers to his side. He inquired, in the look and language of despair, what in this emergency could be done? After a long pause, one of his counsellors said: "We have tried all that physical power could effect; we filled our ranks with strong, brave and well-disciplined men, but our armies have been conquered—even our tall grenadiers from Potsdam have been prostrated—and now the heel of the oppressor is on our necks. I would advise that, as a last resort, we try the effect of intellectual and moral power. Let us educate the people of all ranks. Let us begin here at Berlin and establish a university that will give a tone to every gymnasium and people's school in the kingdom. Let us give to all our people that knowledge which will enable them to build up the resources of their country and that courage which will make them ready to defend it. Let religion, which teaches the love of country and the duty we owe to God and man, be inculcated in all our schools and seminaries of learning." That very day the erection of a university was decided on. Every child in Prussia was compelled by a law of the kingdom to attend school. Education was widely diffused among the people, and the intellectual man from the highest nobleman to the poorest peasant became educated. All protestant Germany became animated by the same zeal in the cause of education. In Prussia, education was compulsory by the laws of the land; in the adjoining kingdoms it became at least the law of custom.

In the silent progress of time, a new arm of power was bestowed on the nation. Science and the arts gave a stimulus to agriculture. Manufacturers of all descriptions were carried into successful operation, and all the sinews of war became strengthened. By the general diffusion of knowledge, writings and speeches now emanated from the most intelligent of the common people; patriotic songs were composed and became national songs, a volume of which entitled "The Lyre and the Sword," was written by Koerner, who was originally a volunteer soldier in the army. He, like Burns, a ploughman and he like Hogg, the Ettrick sheppard, sprung from the lower ranks of society, and was a poet by nature. He sung of the wrongs and oppressions of Germany, his native land—of patriotism and of the duty of sacrificing life for the good of our country. Through these instrumentalities, the whole nation of Germany was roused up to a burning desire to free their country from foreign rule. An enthusiasm was awakened by these patriotic writings and discourses, and these touching and soul stirring melodies, scarcely equalled by the effect of the Marsailles hymn on the French, or the *Ranz de vaches* on the Swiss. Koerner fell on the field of battle, thus sealing his devotion to his fatherland with his blood.

Let us now look at the sequel and mark the effect of education on the security and prosperity of the nation. Two thirds of a generation had scarcely passed away when that very king with his army of heroes lived to become one of the conquerors at the battle of Waterloo and to unite his victorious legions to those of the allies in their entrance into the streets of conquered Paris; and that now aged counsellor who had given the advice was before us.

Fellow citizens and especially our Lutheran brethren: Your forefathers were long engaged in cultivating the physical and moral man. By the former they were enabled to fell the forest and render their fields productive and by the latter their characters as men of integrity were established. Whilst you are enjoying the fruits of their labors in the light of their religious example, resolve that you will now bring to your aid another and an additional power—that power which can create new resources and surmount all difficulties—a power that gives a lever to move the world—*the power of knowledge*. Let that knowledge be regulated and controlled by the pure precepts of that gospel which deters from the evil by a consciousness of accountability, and stimulates to goodness by the smiles of conscience and the approbation of God; then will you have fulfilled your mission as intelligent beings and through the mercies of heaven may hope for the rewards of a blissful immortality.

Vindication

In 1865 a Philadelphia clergyman named E. W. Hutter accused Bachman of refusing to aid Union soldiers who had been hospitalized in the Confederate city of Charleston. He asserted that "no man in Charleston gloated so openly over the barbarities inflicted on our prisoners." In fact, Bachman had remained in Charleston throughout the period from 1861 to 1865 while Union troops blockaded and besieged the city, and he continually visited hospitals to minister to soldiers on both sides.

As Sherman's army approached in 1865, Bachman was urged to leave the city to avoid reprisals for having given the opening prayer at the Secession Convention. He went to a junction near Cheraw, where he was thought to be the owner of a plantation and was brutally beaten by a federal officer in an attempt to get him to reveal the location of valuables. He was then seventy-five years old, and his left arm was permanently injured (Stephens 1999: 832). When Confederate troops later captured the federal officer, Bachman recognized but did not identify the man to his captors. In this essay Bachman recorded numerous other acts of unmilitary and unlawful behavior. In his history of the Confederate government, Jefferson Davis quoted the entire portion of Bachman's account that relates to Sherman's army (Davis 1881: vol. 2, 710–17).

When Bachman returned to Charleston, Union troops controlled the city and its government (as they continued to do for a decade). Everything of value had been plundered, and a vivid account of the postwar conditions also forms part of his vindication.

Vindication / of / Rev. Dr. John Bachman, / of Charleston, S.C. / in Answer to / Rev. E. W. Hutter, / in Regard to an Article Published in the "Lutheran / and Missionary," of the 27th of July, 1865. / Published by a Personal Friend. / 1868.

Vindication / of / Rev. Dr. John Bachman.

Rev. E. W. Hutter:

Rev. Sir—In the *Lutheran and Missionary* of the 27th of July, 1865, I perceive an article headed "Southern Lutheran Church," under your signature, dated Philadelphia, July 20th, 1865, one paragraph of which demands some notice from me. The bad taste betrayed in the whole article, drawing its illustrations from the "barn-yard," the "kitchen," and "finny-prize floundering in the net" is not the greatest objection, neither am I disposed to consume time by criticizing your prejudiced comparison between your Northern and Southern synods. It is the spirit and manner of the whole article—the narrow, one-sided views, the censorious, illiberal remarks, and the bitter personalities betrayed that characterizes the temper of the writer.

I would here just remark that your discussions whether the Northern General Synod will, or will not, receive the Southern Churches into their body are premature inasmuch as it appears to me it would be a wiser policy first to ascertain whether they have evinced any disposition to be re-united. As far as I am acquainted with the sentiments of the Southern Lutheran clergy and the people, there is not one in a thousand who would for a moment entertain the slightest idea of a re-union with the Northern General Synod, more especially as long as it retains such a mouth-peace as Rev. E. W. Hutter. All, with unexampled unanimity, are in favor of retaining our present organization and as early as possible continuing the publication of our Book of Worship and highly valuable Southern Lutheran [which Bachman coedited, 1860–62 (Stephens 1990: 832)].

I am quite sure that I would not have noticed this offensive article if you had stopped here. The following paragraph, however, as it refers to me personally, calls upon me for something more than a passing notice:

"By one of the most eminent citizens of Charleston, a native and life-long resident of that city, we recently were favored with an item of intelligence concerning Dr. Bachman, the first received by us since the fall of 1860, when he so profanely invoked the divine blessing on the South Carolina ordinance of secession. To show what a melancholy change has been wrought in the doctor's spirit, only about two months ago, although besought by prayers and tears, he refused to administer the holy supper to a dying Lutheran soldier. Rather than not receive it at all, the expiring hero received it from a Catholic priest! We have the same authority, too, for stating that, than this same Dr. Bachman, no man in Charleston gloated so openly over the barbarities inflicted on our prisoners."

Here we have a charge penned by your self and printed by the editors of the *Lutheran and Missionary* reflecting on my character as a clergyman, which if true would destroy my usefulness and render me to scorn and contempt of all Christian men. I am accused, 1st, of withholding the communion from a dying man on political grounds. 2d, "of openly gloating over the barbarities inflicted on our prisoners."

Now, Rev. Sir, I pronounce these charges made and published by your self the addictive, malicious, and unmitigated falsehoods. I never was besought "by prayers and tears to administer the holy communion to a dying Union soldier." I never heard of "the expiring hero," or of the "Roman Catholic priest" to whom this highest office was assigned. The paper in which these vindictive charges were made and which you ought in justice to have sent to me through the mail, having been sent by a friend. You might have easily satisfied yourself and saved yourself much trouble by writing a line to me. You would then have had both statements before you. You say this occurred only two months ago. Your article was dated July 20th; consequently, it must have been sometime in May when you state that I refused the communion to this imaginary "dying hero" and when, to render the episode more impressive, according to your pathetic statement, a Catholic priest was sent for to lighten the load of his sentence on his passage to eternity. My church had long been in the "shelling district," my people were scattered throughout the country, and I followed them on the 13th of February and did not return until the middle of May.

Since then, I have had one public communion in my church. It was largely attended by all denominations. The community had heard of the barbarities inflicted on me by the officers in General Sherman's army and had for several weeks yielded to the current report that I had died of my wounds. They now crowded around their blessed Master's table with feelings of love to the aged man who had been spared to minister at the altar and of gratitude to God for His mercies. Among the congregation were several United States officers. From that day to this I frequently administered the communion in private to the sick and saw no reason to deny the ordinance to those who desired to partake of it. My rule in my whole ministerial life has been never to administer the communion to the sick without an examination into their state of preparation to receive it. They must have penitence, faith in the Savior, and resolve to live Christian lives. Hence, I have always regarded the old German practice of depending on communion on a deathbed as savoring of superstition—looking on it as a salve for sin.

During the war, I administered the communion to several hundred sick soldiers of both armies. I naturally saw more of those who belonged to the Confederate than the United States Army. In no case were their political opinions allowed to sway my judgment; but in every case, the requisites to the worthy communicant were carefully examined. Whilst during the long period of four years I either postponed or rejected a few applicants for the communion among the sick in the Confederate Army, I postponed but one among the United States soldiers.

The day after the battle on Morris Island, in my usual rounds of visits to the hospitals, I was asked by a German to administer the communion to him. On inquiring into his life and conduct, he informed me that he had been engaged in breaking the locks and rifling the drawers of rebel ladies on the islands, had taken a considerable quantity of children's clothes and silver spoons, and that he had stolen some from his fellow-soldiers, and that his Colonel had sent all to New York, whence they would go to his wife and children in New Hampshire. He thought he had sent enough to last for several years. I asked him whether he was willing to make restitution for his robberies, particularly the articles stolen from his fellow soldiers. He said no; his Colonel had told him that he had a right to take anything from the rebel ladies, and that he had grabbed from them as much as he could get, and that the soldiers all stole from one another. He said he had, in common with the officers and soldiers, up to the time of that battle, lived in criminal intercourse with the negro women in the camp; and he was, moreover, a terrific swearer, even on what he feared was his dying bed. I perceived, and was so informed by a surgeon, that his wound was not mortal; and I allowed him to consider himself in danger of death and of hell, hoping that his very terrors might lead him to repentance and amendment. When I arrived on the next day, he had been sent away as convalescent, and I did not see him again. If this was the dying Union Lutheran soldier, the expiring hero whom you refer to, you may still have an opportunity of administering to him the "holy supper" in your own way. The Army is now disbanded, and you may find him at home in New Hampshire. Please ask him whether he did not say to me, before I left him, that "he felt that he was not fit for the communion then, but would try to be better prepared for it." The only other case where the communion was referred to was that of a German who had been shot through the lungs and believing his wound to be mortal, I in my daily visits to the hospital apprized him of his danger and the almost certainty of his death. He said that he had not been inside of a church for seven years, and if he was to die, then "by got," he must have the sacrament; but if he was not to die, he swore he would not take it from any "Pfaff" in the land. He did not ask me to administer it. His comrades informed me that five of the seven years in which he had been in this country had been spent in a western penitentiary, where he had been enlisted for the war. His companions-in-arms represented him as the most quarrelsome, profane, and thieving villain they had ever known. I asked him if he wished to be prayed for. He said he did not understand English well enough. "Will you have a German prayer?" He shook his head. The next morning I visited the hospital again when a most revolting spectacle was presented. A wounded Lieutenant who had been in command of a black company (his captain having been killed) was lying in a cot opposite to that of the German. They had just had a quarrel and a fight, the German insisting that he had been fighting for the Union whilst his opponent had been fighting for the negro. The Lieutenant was unable to rise; but the German had crawled out of his bed and beaten the officer unmercifully, and the German had been forced back to his bed, growling and cursing. It was at this moment I entered the ward. I was told that in his rage the German had enlarged the rupture of the wounded blood vessel. He was in too great a passion to speak with me, and I left him cursing. On the subsequent morning the Lieutenant was dead, and I performed his funeral service. On the same afternoon, the German died and was buried. If this "jail bird" is the "expiring hero" you refer to, you may canonize him in the overflowing of your patriotism and the bitterness of your fanatical fury; but be assured that he called for neither Protestant nor Catholic priest and died without a sign.

Up to this day, I have never refused to visit any United States soldiers, etc., and am still engaged in administering the instructions and comforts of religion to all who sent for me. 'Tis true, I cannot discharge these duties as quickly and with as much comfort to myself as I once did. I am compelled to travel miles on foot to visit the hospitals; all my means of conveyance have been taken from me. In my large congregation, all the carriages and horses, including those of the aged widow and non-combatant were seized by the government; there is but one left, which was saved by being claimed as British property; it has no horses and therefore is of no service. President Lincoln by his proclamation tendered free pardon with restoration of the right of property except that of slaves to all who would take the oath to support the United States Government. That oath has been taken by all of us. But what has been the result? We were told to identify our property. My carriage, buggy, and the barouche of a benevolent widow were by an order from General Hatch taken from my premises, which were occupied by an English family with the protection of the Consul and were not, in any sense of the word, what could be construed as abandoned property. When I inquired about the buggy, which I needed most, I was sent from one office to another—from post to pillar—for a few days until time was afforded to send them to Hilton Head, but was informed that it was shipped to New York. My carriage I found in a depot in the city; but when the men placed as guard ascertained that it was mine, they ordered me away and locked the door. That night they removed the pole, and the cushions and the wheels; and by these manœuvers, I am left without any conveyance. Pictures, bedding, and clock, etc., were taken from my house by the Rev. Mr. French, who has speculated largely and profitably among the poor negroes in urging them to be married over again at only a dollar and two candles a pair. Many had no objection to the change and in the state of utter demoralization of the negro have been married several times since, enjoying their freedom *ad libitum*. I was sent from one office to another. Whilst thus amused, my articles, which I had detected in my neighborhood in the house of the United States officers were removed to the Pavilion Hotel. I followed them there and was told to write to the Treasury Department, and my goods would be restored. I wrote accordingly, but received no answer until a month had elapsed. I then went for the articles, but was refused even to enter the room where they were stored. The women of the officers had selected what they wanted; the remainder, which was of but little value, was sold at auction. What became of the proceeds, let the heads of our government inquire. Certain it is that of the ten thousand persons deprived of their property in Charleston, not a thousandth part has been recovered. We are in the situation of a certain man in the Gospel fell into strange company (Luke ii, 30). When these officials and the ladies under their protection return to their homes in the North (God speed them on their way!), they will be much richer than when they came here, and, alas! the poor will be poorer still. Watches, ladies' ornaments, silver spoons, and all manner of household furniture, etc., must by this time be at a discount in the North. The Rev. Mr. French, who made a clean sweep from the houses in my neighborhood, must by this time be a man of wealth, and General Hatch and another officer cannot be far behind. The elegant carriage of Miss Annaly could not be retained here, but was sent to the North to accommodate Mrs. Martel. Our carriages have not all disappeared from Charleston since. Although the owners cannot get them, they can see them perambulating the streets, not only with the officers, but carrying negro soldiers and women of all colors. Many of our horses are still here, as may be seen any afternoon tearing through the streets on their way to the race course, their riders making the welkin ring with screams and blasphemies. When these horses are worn down in flesh, they are sent to the auction and sold to the highest bidder. Who pockets the cash?

We were invited by a proclamation to pay a great tax by a certain day; the parties knew that they had destroyed the railroads and bridges and captured all the carriages and horses. They refused to have these taxes paid by agents. Before we could arrive here, they closed the offices, leaving no one to attend to the business. Thus we have to pay additional taxes to hire our own houses from the government officers. Thus the last dollar is taken and the citizens reduced to beggary. We now and then see the pictures in Harper's, etc., representing the North feeding the South. The representation would have been more true to nature if the cause had been stated, viz.: the previous plunder. Our situation, however, is not altogether peculiar. There is a parallel case of the Israelites under Pharaoh's iron rule. Exodus v, 7.

As an evidence of the different feelings towards me, which exists in your mind and those of the other abolitionists of the North compared with those in this community, your own soldiers included, I would just remark that a number of officers and soldiers in the United States army have asked the privilege of uniting with us on the solemn occasion of our communion if they remain here and that some of Pennsylvania and Ohio have offered themselves for confirmation. I would just add that the slanders which you have so extensively circulated are sooner believed anywhere else than in Charleston.

I now proceed to your 2d charge:

"We have the same authority, too, for stating that than this same Mr. Bachman, no man in Charleston gloated so openly over the barbarities inflicted on our prisoners."

Here I am held to pause and gaze at the picture presented by the fanatic. He is narrow-minded, stern, indefensible, vindictive and cruel. He appears never to have read the chapter on charity taught us by St. Paul. He knows nothing of the law of human kindness and the sweet charities of life. The angel of mercy seems never to have visited his cold, pulseless heart, and he becomes the slanderer of his neighbor, believing that he may thereby promote the cause which prejudice and malignity have induced him to espouse.

I have been the pastor of the same church and people for nearly fifty-one years. During that long period, when five generations have been under my ministry, the harmony that existed among us has been disturbed by no discordant. When the handful of persons with which we began had increased into three large congregations, I was under the hope that I had not been a useless laborer in advancing the interests of the Church in the South and strove to unite discordant material which composed the old General Synod in the Northern and Middle States. I certainly did not expect that the voice of slander would reach me in the advanced period of my life, being in the 86th [76th] year of my age. Here I have lived and labored, and here I expect my remains to rest with those that loved and cherished and clung around me from youth to age.

I defy you or your contemptible informer to produce a single case of my inhumanity—and when you publish to the world "that no man in Charleston gloated so openly over the barbarities inflicted on our prisoners as this same Dr. Bachman," you certainly do not place yourself in the position of a meek and lowly servant of Christ. You do not regard the command which enjoins us not to bear false witness; you drop the lamb and assumed the attitude, the growl, and the malignity of the tiger. I appeal to every virtuous citizen of Charleston if I have not devoted my life to mitigate evils of yellow fever, cholera, and civil war. I was in Charleston during all the seasons of yellow fever, but one (when I was in Europe on account of health). I will venture to affirm that I have seen more cases of that disease than any man in America—having on one occasion buried forty-one victims in one day. Hundreds of times I could not find time for an hour's repose during many long and weary nights. My own congregation as natives were exempt from this fever, and therefore my services were not required to bury them. The sufferers were in most cases the people of the North, to whom I sacrificed my days and nights—the very people over whose inflicted barbarities I am accused of gloating. During the war, I will venture to say I have visited, succored, and attended at the bed-sides of more United States prisoners than you have done to the sick and wounded, including both armies. Allow me here to give you a few specimens of my "gloating over the barbarities inflicted" on your prisoners. You will be able to judge what were the causes of my resentment and how I sought revenge when it was in my power.

When Sherman's army came sweeping through Carolina, leaving a broad track of desolation for hundreds of miles, whose steps were accompanied with fire, and sword, and blood, reminding us of the tender mercies of the Duke of Ayla, I happened to be at Cash's Depot, six miles from Cheraw. The owner was a widow, Mrs. Ellerbe, 71 years of age. Her son, Colonel Cash, was absent. I witnessed the barbarities inflicted on the aged, the widow, and young and delicate females. Officers high in command were engaged tearing from the ladies at their watches, their ear and wedding rings, the daguerreotypes of those they loved and cherished. A lady of delicacy and refinement, a personal friend, was compelled to strip before them, that they might find concealed watches and other valuables under her dress. A system of torture was practiced towards the weak, unarmed, and defenseless, which, as far as I know and believe, was universal throughout the whole course of that invading army. Before they arrived at a plantation, they inquired the names of the most faithful and trustworthy family servants; these were immediately siezed, pistols were presented at their heads, with the most terrific curses they were threatened to be shot if they did not assist them in finding buried treasures. If this did not succeed, they were tied up and cruelly beaten. Several poor creatures died under the infliction. The last resort was that of hanging, and the officers and men of the triumphant army of Gen. Sherman were engaged in erecting gallows and hanging up these faithful and devoted servants. They were strung up until life was nearly extinct, when they were let down, suffered to rest awhile, then threatened and hung up again. It is not surprising that some should have been left hanging so long that they were taken down dead. Coolly and deliberately these hardened men proceeded on their way as if they had perpetrated no crime and as if the God of Heaven would not pursue them with His vengeance. But it was not alone the poor blacks (to which they professed to come as liberators) that they thus subjected to torture and death. Gentleman of high character, pure and honorable and gray-headed, unconnected with the military, were dragged from their fields or their beds and subjected to this process of threats, beating, and hanging. Along the whole track of Sherman's army, traces remain of the cruelty and inhumanity practiced on the aged and the defenseless. Some of those who were hung up died under the rope, while their cruel murderers have not only been left unreproached and unhung, but have been hailed as heroes and patriots. The list of those martyrs whom the cupidity of the officers and men of Sherman's army sacrificed to their thirst for gold and silver is large and most revolting. If the editors of this paper will give their consent to publish it, I will give it in full, attested to by the names of the purest and best men and women of our southern land.

I, who have been a witness to these acts of barbarity that are revolting to every feeling of humanity and mercy, was doomed to feel in my own person the effects of the avarice, cruelty and despotism which characterized the men of that army. I was the only male guardian of the refined and delicate females who had fled there for shelter and protection. I soon ascertained that the plan that was adopted in this wholesale system of plunder, insult and blasphemy and brutality. The first party that came was headed by officers, from a colonel to a lieutenant, who acted with seeming politeness and told me that they only came to secure our fire-arms, and when these were delivered up, nothing in the house should be touched. Out of the house, they said that they were authorized to press forage for their large army. I told them that along the whole line of the march of Sherman's army, from Columbia to Cheraw, it had been ascertained that ladies had been robbed and personally insulted. I asked for a guard to protect the females. They said that there was no necessity for this as the men dare not act contrary to orders. If any did not treat the ladies with proper respect, I might blow their brains out. "But," said I, "you have taken away our arms, and we are defenceless." They did not blush much and made no reply. Shortly after this came the second party, before the first had left. They demanded the keys of the ladies' drawers, took away such articles as they wanted, then locked the drawers and put the keys in their pockets. In the meantime, they gathered up the spoons, knives, forks, towels, table cloths, &c. As they were carrying them off, I appealed to the officers of the first party. They ordered the men to put back the things; the officer of the second party said he would see them d——d first; and without further ado, picked them up, and they glanced at each other and smiled. The elegant carriage, and all the vehicles on the premises were seized and filled with bacon and other plunder. The smoke-houses were emptied of their contents and carried off. Every head of poultry was seized and hung over their mules, and they presented the hideous picture and some of the scenes in "Forty Thieves." Every article of harness they did not wish was cut in pieces. By this time, the first and second parties had left, and a third appeared on the field; they demanded the keys of the drawers, and on being informed that they had been carried off, coolly and deliberately proceeded to break open the locks, took what they wanted, and when we uttered words of complaint were cursed. Every horse, mule and carriage, even to the carts were taken away, and for hundreds of miles, the last animal that cultivated the widow's corn-field and the vehicles that once bore them to the house of worship were carried off or broken in pieces and burned.

The first party that came promised to leave ten days' provisions, the rest they carried off. An hour afterwards, other hordes of marauders from the same army came and demanded the last pound of bacon and the last quart of meal. On Sunday, when the negroes were dressed in their best suits, they were kicked and knocked down and robbed of all their clothing, and they came to us in their shirt sleeves, having lost their hats, clothes, and shoes. Most of our own clothes had been hid in the woods; the negroes who had assisted in removing them were beaten and threatened with death and compelled to show them where they were concealed. They cut open the trunks, threw my manuscripts and devotional books into a mudhole, stole the ladies' jewelry, hair ornaments, etc.; tore many garments into tatters, gave the rest to the negro women to bribe them into criminal intercourse. These women afterwards returned to us those articles that, after the mutilations, were scarcely worth preserving. The plantation of one hundred and sixty negroes was some distance from the house and to this place successive parties of fifty at a time resorted for three long days and nights, the husbands and fathers being fired at and compelled to fly to the woods.

Now commenced scenes of licentiousness, brutality, and ravishment, that have scarcely had an equal in the ages of heathen barbarity. I conversed with aged men and women, who were witnesses of these infamous acts of Sherman's unbridled soldiery, and several of them from the cruel treatment they had received were confined to their beds for weeks afterwards. The time will come when the judgment of Heaven will await these libidinous, beastly barbarians. During this time, the fourth party, who I was informed by others we had the most reason to dread, had made their appearance. They came, as they said, in the name of the great General Sherman, who was next to God Almighty. They came to burn and lay in ashes all that was left. They had burned bridges and depots, cotton-gins, mills, barns and stables. They swore they would make the d——d rebel women pound their corn with rocks and eat their raw meal without cooking. They succeeded in thousands of instances. I walked out at night, and the innumerable fires that were burning as far as the eye could reach in hundreds of places illuminated the whole heavens and testified to the vindictive barbarity of the foe. I presume they had orders not to burn occupied houses, but they strove all in their power to compel families to fly from their houses that they might afterwards burn them. The neighborhood was filled with refugees who had been compelled to fly from their plantations on the sea-board. As soon as they had fled, the torch was applied, and for hundreds of miles those elegant mansions, once the ornament and pride of our island country, were burned to the ground.

All manner of expedients were now adopted to make the residents to leave their homes for the second time. I heard them saying, "This is too large a house to be left standing; we must contrive to burn it." Canisters of powder were placed all around the house, and an expedient resorted to that promised almost certain success. The house was to be burned down by firing the outbuilding. These were sitting near each other at the firing of the wall and would lead to the destruction of all. I had already succeeded in having a few bales of cotton rolled out of the building and hoped if they had to be burned, the rest would also be rolled out, which could have been done in ten minutes, by several hundred men who were looking on, gloating over the prospect of another elegant mansion in South Carolina being left in ashes. The torch was applied, and soon the large store house was on fire: this communicated to several other buildings in the vicinity. At length the fire reached the smokehouse, where they had already taken off the bacon of 250 hogs; this was burnt, and the fire was now rapidly approaching the kitchen, which was so near the dwelling house that should the former burn, the destruction of the large and noble edifice would be inevitable.

A captain of the United States service—a native of England—whose name I would like to mention here if I did not fear to bring down upon him the censure of the abolitionists as a friend to the rebels mounted the roof, and the wet blankets we sent up to him prevented the now smoking roof from bursting into flames. I called for help to assist in procuring water from a deep well; a young lieutenant stepped up, condemned the infamous conduct of the burners and called on his company for aid; a portion of them came cheerfully to our assistance; the wind seemed almost by a miracle to subside; the house was saved, and the trembling females thanked God for their deliverance. All this time, about one hundred mounted men were looking on, refusing to raise a hand to help us; laughing at the idea that no efforts of ours could save the house from the flames.

Mr. Hutter, allow me to ask who are the most criminal, the men who were rejoicing that a house was to be burned and women and children be deprived of a shelter and home and driven into the woods or he who slanders an aged clergyman of his own Church and would bring down upon him the odium of all good men when he had it fully in his power to ascertain that the whole invention was an infamous slander, concocted by the mean, the worthless, and the malicious for the purpose of getting offices and money?

My trials, however, were not over yet. I had already suffered much in a pecuniary point of view. I had been collecting a library on Natural History during a long life. The most valuable of these books had been presented by various Societies in England, France, Germany, Russia, &c., who had honored me with membership, and they or the authors presented me with these works, which had never been for sale and could not be purchased. My herbarium, the labor of myself and the ladies of my house for many years was also among these books. I had left them as a legacy to the library of the Newbury College and concluded to send them at once. They were detained in Columbia, and there the torch was applied, and all were burned. The stealing and burning of books appear to be one of the programmes on which the Army acted. I had assisted in laying the foundation and dedicating the Lutheran Church at Columbia, and there, near its walls, had recently been laid the remains of one who was dearer to me than life itself. To set that brick church on fire from below was impossible. The building stood by itself on a square but little build up. One of Sherman's burners was sent up to the roof. He was seen applying the torch to the cupola. The church was burned to the ground, and the grave of my loved one desecrated. The story circulated that the citizens had set their own city on fire is utterly untrue and only reflects dishonor on those who vilely perpetrated it. General Sherman had his army under control. The burning was by his orders and ceased when he gave the command.

I was now doomed to experience in person the effects of avarice and barbarous cruelty. The robbers had been informed in the neighborhood that the family which I was protecting had buried $100,000 in gold and silver. They first demanded my watch, which I had effectually secured from their grasp. They then asked me where the money had been hid. I told them I knew nothing about it and did not believe there was a thousand dollars worth in all, and what there was had been carried off by the owner, Colonel Cash. All this was literally true. They then concluded to try an experiment on me which had proved so unsuccessful in hundreds of other instances. Coolly and deliberately they prepared to inflict torture on a defenceless, grey-headed old man. They carried me behind a stable and once again demanded where the money was buried, or "I should be sent to hell in five minutes." They cocked their pistols and held them to my head. I told them to fire away. One of them, a square-built, broad-faced, large-mouthed, clumsy lieutenant, who had the face of a demon and who did not utter the five words without an awful blasphemy, now kicked me in the stomach until I felt breathless and prostrate. As soon as I was able I arose again. He once more asked me where the silver was. I answered, as before, "I do not know." With his heavy elephant foot he now kicked me on my back until I fell again. Once more I arose, and he put the same question to me. I was nearly breathless, but answered as before. Thus was I either kicked or knocked down seven or eight times. I then told him it was useless for him to continue his threats or his blows. He might shoot me if you chose; I was ready and would not budge an inch, but requested him not to bruise and batter an unarmed, defenceless old man. "Now," said he, "I'll try a new plan. How would you like to have both your arms cut off?" He did not wait for an answer, but with his heavy sheathed sword struck me on my left arm near the shoulder. I heard it crack; it hung powerless by my side, and I supposed it was broken. He then repeated the blow on the other arm. The pain was most excruciating, and it was several days before I could carve my food or take my arm out of a sling—and it was black and blue for weeks. (I refer to Dr. Kollack of Cheraw.) At that moment the ladies headed by my daughter, who had only then been made aware of the brutality practiced upon me, rushed from the house and came flying to my rescue. "You dare not murder my father," said my child; "he has been a minister in the same church for fifty years, and God has always protected and will protect him." "Do you believe in a God, Miss?" asked one of the brutal wretches; "I don't believe in a God, a heaven, nor a hell." "Carry me," said I, "to your General." I did not intend to go to General Sherman, who was at Cheraw, from whom, I was informed, no redress would be obtained, but to a General in the neighborhood, said to be a religious man. Our horses and carriages had all been taken away, and I was too much bruised to be able to walk. The other young officers came crowding around me very officiously, telling me that they would represent the case to the General and that they would have him shot like ten o'clock the next morning. I saw the winks and glances that were interchanged between them. Every one gave a different name to the officers. The brute remained unpunished, as I saw him on the following morning as insolent and as profane as he had been on the preceding day.

As yet no punishment had fallen on the brutal hyena, and I strove to nurse my bruised body and heal my wounds and forget the insults and injuries of the past. A few weeks after this, I was sent for to perform a parochial duty at Mars Bluff, some twenty miles distant. Arriving at Florence, in the vicinity I was met by a crowd of young men connected with the militia. They were excited to the highest pitch of rage and thirsted for revenge. They believe that among the prisoners that had just arrived on the railroad car on their way to Sumter were the very men who committed such horrible outrages in the neighborhood. Many of their houses had been laid in ashes. They had been robbed of every means of support. Their horses had been seized; their cattle and hogs bayoneted; their mothers and sisters had been insulted and robbed of their watches, ear and wedding rings. Some of their parents had been murdered in cold blood. The aged pastor to whose voice they had so often listened had been kicked and knocked down by repeated blows, and his hoary head had been dragged about in the sand. They entreated me to examine the prisoners and see whether I could identify the men that have inflicted such barbarities on me. I told them I would do so provided they would remain where they were and not follow me. The prisoners saw me at a distance—held down their guilty heads and trembled like aspen leaves. All cruel men are cowards. One of my arms was still in a sling. With the other I raised some of their hats. They begged for mercy. I said to them, "The other day you were tigers—you are sheep now." But a hideous object soon arrested my attention. There sat my brutal enemy—the vulgar, swaggering lieutenant, who had rode up to the steps of the house, insulted the ladies, and beaten me most unmercifully. I approached him slowly, and in a whisper asked him, "Do you know me, sir—the old man whose pockets you first searched to see whether he might not have a penknife to defend himself and then kicked and knocked him down with your fist and heavy scabbard?" He presented the picture of an arrant coward and in a trembling voice implored me to have mercy. "Don't let me be shot; have pity! Old man, beg for me! I won't do it again! For God sake, save me! Oh, God, help me!" "Did you not tell my daughter there was no God? Why call on him now?" "Oh, I have changed my mind; I believe in a God now!" I turned and saw the impatient, flushed and indignant crowd approaching. "What are they going to do with me?" said he. "Do you hear that sound, click, click?"—"Yes," said he, "they are cocking their pistols." "True," said I; "and if I raise a finger you will have a dozen bullets through your brain." "Then I will go to hell; don't let them kill me. Oh, Lord, have mercy!" "Speak low," said I, "and don't open your lips." The men advanced. Already one had pulled me by the coat. "Show us the men." I gave no clue by which the guilty could be identified. I walked slowly through the car, sprang into the waiting carriage and drove off.

Rev. E. W. Hutter, this is the way in which I have "gloated over the barbarities inflicted on the prisoners." This is the man whom you have wantonly and cruelly traduced. I defy you or any one else to produce a single instance to the contrary in my whole conduct, from the beginning to the close of the war.

I claim, as an act of justice, that you send me the name of your author, whom you call one of the most eminent citizens of Charleston—a native and a life-long resident of that city, whom you have given us as authority for the slanders which you have perpetrated against me. I defy you to produce the name of a single "eminent citizen" who will dare, in the face of this community, to make the assertion which you have in such a cowardly and unchristian manner published to the world. When that name shall be ferreted out, I will venture to predict that this "eminent citizen of Charleston—a native and a life-long resident," will be proved to be an unprincipled, time-serving demagogue—a spy, a political turn-coat, a defamer of the reputation of others, to obtain notoriety, power and money—not many degrees removed from a drunkard—a man without credit or character and who has never had either.

It is scarcely necessary to add that I have not sought this controversy and only defend myself when grossly and unprovokedly traduced. It should be here remembered that we are writing under surveillance and are at the tender mercies of a Provost Marshal. The time may come when men can speak freely. Under present circumstances, it is but a contemptible, cowardly act to drag men into a discussion, where freedom of the pen is restricted to one party, and given with unbridled license to another.

Yours, &c.,
John Bachman
Charleston, Sept. 14th, 1865.

Selected Letters, 1831–1871

Approximately two hundred of Bachman's letters are known to have survived largely in the correspondence of recipients and principally in his correspondence with members of the Audubon family. The following sixteen letters are addressed to seven correspondents, and they have been selected to represent the best of Bachman's correspondence, his principal interests, and the various periods of his life.

Bachman's correspondence with John James Audubon (six letters) and with Audubon's son Victor Gifford Audubon (five letters) indicates how they collaborated to produce the *Viviparous Quadrupeds of North America*. Audubon and his other son, John Woodhouse Audubon, painted the 150 folio lithographic plates that were issued serially from 1842 to 1848. With research assistance from Victor, Bachman began preparing the text that was issued in three volumes in 1846, 1851, and 1854. The eleven letters about the *Quadrupeds* cover nearly the entire period of its creation. They reveal a great deal about Bachman and Audubon's working methods and friendship.

The included letters were selected from a nearly complete set of photocopies of all of Bachman's known correspondence in the Jay Shuler Collection, which is part of Special Collections in the College of Charleston Libraries. Shuler assembled copies for a dual biography, *Had I Wings: The Friendship of Bachman and Audubon* (1995). The largest group of original letters is in the archives of the Charleston Museum, and except for one missing letter, I have compared photocopies of the selected letters with the originals. One of these meticulous transcripts has a note signed "E. B. C.," indicating that the curator E. Burnham Chamberlain probably prepared them.

A large part of the Charleston Museum's collection of Bachman's correspondence was published in *John Bachman, D.D., LL.D, Ph.D.; the Pastor of St. John's Lutheran Church, Charleston* (1888). This is the biography that allowed Bachman to "speak for himself" (6), that was begun at Bachman's request by the Rev. John Bachman Haskell, his grandson, and that following Haskell's death was edited for publication by C. L. Bachman, Bachman's daughter. However, most letters were published only in part, and some were paraphrased. Most of the following letters are being published in full for the first time.

(1) 1831 Nov. 15: to Lucy Bakewell Audubon (Mrs. John James Audubon) on meeting Audubon (Bachman 1888: 95–96).

Charleston, 15th November, 1831

Dear Madam—I comply with a request of your kind and worthy husband, who laid an injunction on me this morning, that I should write to you. He, together with Mr. [George] Lehman [view painter] and Henry Ward [taxidermist], left for St. Augustine. They were all in good health and spirits, and enthusiastically bent on the accomplishment of the object of their expedition to the fullest extent.

The last has been one of the happiest months of my life. I was an enthusiastic admirer of nature from my boyhood, and fond of every branch of Natural History. Ornithology is, as a science, pursued by very few persons—and by no one in this city. How gratifying was it, then, to become acquainted with a man, who knew more about birds than any man now living—and who, at the same time, was communicative, intelligent, and amiable, to an extent seldom found associated in the same individual. He has convinced me that I was but a novice in the study; and besides receiving many lessons from him in Ornithology, he has taught me how much can be accomplished by a single individual, who will unite enthusiasm with industry. For the short month he remained with my family, we were inseparable. We were engaged in talking about Ornithology—in collecting birds—in seeing them prepared, and in laying plans for the accomplishment of that great work which he has undertaken. Time passed rapidly away, and it seems but as yesterday since we met, and now, alas! he is already separated from me—and in all human probability we shall never meet again.

I am well aware of all the difficulties your husband will have to encounter in a wild and, in some respects, an unexplored country. He proposes traversing the swamps of Florida—the wilds of Missouri—the snows of the Rocky Mountains—and, if possible, to reach the Pacific. He will have to encounter not only the climate, but the animals—the savages—the parched deserts of the Southwest—and the snows of the North. But I depend much on his hardy constitution, on his knowledge of the countries through which he has to pass, and on his admirable tact in avoiding and extricating himself from difficulties. But, above all, I have a firm reliance on the goodness of Providence that he will spare his useful life, and enable him to answer the high expectations of his friends and his country.

Mr. Audubon has promised frequently to write to me, and I shall feel as much interested in all of his movements, as if he were a brother, or the dearest friend I have on earth.

I need not inform you that Mr. Audubon was a general favorite in our city. His gentlemanly deportment, his travels and experience, his information and general talents, caused him to be sought after by all. But your husband knew that the great objects before him required his unremitted attention, and he was obliged to deny himself to his friends, on many occasions, and devoted to them only his evenings.

There seems quite a blank, in my house, since he has gone, for we looked on him as one of our family. He taught my sister[-in-law], Maria [Martin], to draw birds; and she has now such a passion for it, that, whilst I am writing, she is drawing a Bittern, put up for her at daylight by Mr. Audubon.

I hope that Charleston may be able to give a few subscribers to your husband's work, and I wish that she was richer, and had taste, and a knowledge of Natural History, to encourage her to do more.

I shall always be glad to hear from one so intimately associated with my friend, and, with the best wishes for his and your prosperity and happiness, I beg leave to subscribe myself,

Yours, in great sincerity,
John Bachman.

[Addressed to: "Mrs. J. J. Audubon"]

(2) 1832 Dec. 27: to John James Audubon with information for the *Birds of America* (Bachman 1929: 180–82).

Charleston Decem. 27th 1832

Dear Audubon

I propose filling this letter with remarks on Birds & when I have done I will send it to you. It will at least prove that although [the] nullification [controversy] may for a time dampen, it cannot altogether strangle my zeal for the Science.

I have this moment shot in my Garden, the Regulus about which we have so often disputed and although your mind is made up against me, yet you must listen like a reasonable man, we both have a spice of obstinacy. Now hear my reasons for believing it to be a new bird.

I received recently from Gibbs [Lewis Reeve Gibbes?] the old & young of the Golden crowned Wren, on comparing these Birds with several specimens I have of the present Bird, I cannot doubt their being different Birds. They are this moment before me. The Bill of the Gold crest, is thick & black, that of the present bird is at least one third less, thin Brown, Nostrils not covered with feathers as the Golden crown and its legs are brown instead of black, its bill partakes more of the Genus Sylvia whilst the other partakes of the Muscicapa. Make further observations my friend, the above are permanent differences [species] setting aside the difference of colour [varieties]. It cannot be the young of the ruby crowned wren, it is considerably smaller & differs in all its habits. The present Bird loves thick bushes, I never saw it high from the ground & never on pine trees like the other, its voice differs also very much. I am preparing by the help of Ward skins of each. Let us both resolve that we will not resist the evidence of our senses. I question much whether either the Golden or Ruby crowned wren are found in the Autumn without some colour on the head either red or yellow.

May not the Northern Marsh Hen, be the Bird which we here call the Fresh Water M. Hen & our Ash coloured one that keeps in the Marsh be peculiar to the South? I should like to have this matter ascertained. The greatest mystery in Ornithology to me is the circumstance you mention of the young of Peale's Egrett being when young some brown & others white. My bird is improving. It was white originally, is so still, which I take to be the colour of the old bird, & I hope to preserve it through the winter.

The new Frigilla of which you have a drawing has not entirely disappeared. It departs in November & returns in April. I have never found a nest, but saw three & 4 young ones with the old. It undoubtedly breeds on, or very near the ground. This bird although not very rare is seldom seen, in consequence of its darting toward the earth as soon as it is approached when it runs along the bushes & in the broom grass of the pine woods & is with difficulty made to fly. It is somewhat strange that it leaves us about the time the other of the numerous sparrow family visit us & it comes to take their place as they retire to the North in the spring. It is the only high land sparrow that breeds in the lower country of South Carolina, the three salt marsh sparrows, the sea side, sharp tailed, & the new sparrow I intend to send you, breed in near the situations which they most frequent. The present species I suspect breeds but once, of this however I am not quite certain. Next summer I shall be able to say something about the habits of the new Finch that you have not seen, & which I take to be the long lost F. Caudacuta of Latham.

The Ardea Cristata, your large White Crane's, were in—tolerable but not in the perfect plumage when I had them destroyed. Till then I kept them in my Garden, but they became dangerous, killing a Duck & on another occasion a half grown cat at a single blow & on the day I had them killed they made a pass at a young negro, this was when they were hungry. They became somewhat clamorous at night uttering the usual harsh guttural note of their species. They managed to mount the highest arbours to sleep. Were remarkably expert flycatchers & I frequently remarked at their making a point equal to a setter before they darted on the butter-flies & Sphynxes, which they never failed to sieze. Indeed this was their amusement every afternoon & evening. I question whether these birds would ever in confinement have attained to the growth of those in their wild state. I tossed a dozen Mulletts of a tolerable size to one in succession one Morning, he seized them before they reached the ground & they disappeared in an instant. In fact they were enormous gormandizers. I think I speak within bounds when I state that one could easily consume a peck of Fish in a day.

I sat down to make a note or two & here is a whole letter. It may contain information which you may desire & so I send it. I will keep on making remarks & hope by the first vessel to send you the long promised drawings & Birds. I write that I may not be forgotten. A letter from you, always acceptable is now a cordial. We are all well & all beg to be remembered to you & yours. Our political horizon remains dark & gloomy. The Winter Wren has been sent me from Columbia.

Yr frd
Jno. Bachman.

[Addressed to: "J. J. Audubon Esq. / F.R.S. etc. etc. / Boston."]

(3) 1837 Feb. 17: to Samuel George Morton, M.D., on a species named unethically by Richard Harlan, M.D. (ALS [autographed letter signed] and typescript, American Philosophical Society Library).

Charleston Feby 17th 1837

Dear Sir,

I addressed a letter to you about a month ago—the answer to which being of some importance to me I have looked for with some anxiety. Believing it possible from my not hearing any thing of you that either my letter or your answer may have miscarried, I take the liberty of addressing you again.

In a letter to Dr [Charles] Pickering some time ago I gave a description in a hurried manner of a new Hare (Lepus [*Sylvilagus*] palustris [Marsh Rabbit]) the engraving of which you had the kindness to say in a letter to Audubon was finished. The object of my letter to you was to inquire whether my description was already printed. It was written very hurriedly & not at that time intended for publication & I have kept no copy. Since then I have given a careful description of its character & habits & my object was to say that if the article was not published I should like to substitute the present in its room. [Illegible name] however obtained a new and large species from Alabama—and a specimen of Lupus [*Lagomys*] princeps from Mr Nuttall—the latter animal was so much overlooked by Dr. [John] Richardson—who described it without a tail & placed it in a wrong sub-genus that it is necessary to describe it as new. In short I have written a history as well as considerable opportunities for observations afforded me of 6 species of Hares inhabiting the U. S. Now if your society has not published this last number & you should think my remarks on this genus worth having I would send them for their perusal & disposal. If however I am too late I will send it elsewhere.

As Dr Harlan has recently published one of my quadrupeds in Sillimans Journal—an animal which I had submitted to his inspection through Dr Pickering & which he Dr H[arlan] did not know & sent it back with the name of a well known species—and as I cannot answer for some one taking the same liberty, with my new Hare, (a specimen of which he has also seen) I do not mean to wait so patiently or confide so implicitly as I have done heretofore. With these little bickerings you have however nothing to do & fortunately for my peace they do not much incommode me. As my paper is finished & the drawings made I would take it as a favour if you will give me an early answer to this letter.

Since I am troubling you let me ask you another favour. My friend Mr Casper of N.Y. has requested me to permit him to examine the bats I sent to your society & Dr. Pickering wrote to ask me for the privilege of his doing so. When I sent the specimens I considered them the property of the society & beyond my control. He has any permission that I have a right to give. I have however yesterday sent him all my specimens on that difficult genera, to enable him to make his article complete & when he receives them I presume he will not think it necessary to apply for yours. In the mean time as I am trying to give a monograph of our Shrews Dr P[ickering] stated that you had recd. a specimen of a Sorex differing from those I was describing & advised me to ask the loan of it for a short time to examine it. Could you do this without infringing on any of the rules of your society you would do me a favour. I am receiving those from N.Y. & Boston & my article as far as the species I am acquainted with is finished.

There are several things, which I said in my former letter that I would repeat here did I not think it probable that you may have received it.

Audubon & son & Mr. Edwd. Harris left this [place] for Mobile this morning on the rail road. They hope to find a suitable vessel there or at

N. Orleans to carry them along the Mexican Gulph—up the Sabine & enable them to examine the nat[ural] history of the apalusas & some of the prairies in that vicinity they are in.

With great regard health & full of zeal in the cause.

Yrs truly / J Bachman

[addressed to: "Samuel George Morton MD / Corresponding Secty. / Academy Nat Sciences / Philadelphia"]

(4) 1839 Jul. 5: to John James Audubon on research needed in Europe about quadrupeds (typescript, Charleston Museum).

New York July 5th 1839

Dear Audubon

Victor to day read to me a letter from you requesting me to give you a more particular list of American Quadrupeds than I sent you. I do not know what Dr [John] Richardson has written, but he is a very good judge & withal an honest & worthy man, & it would be well for you to spend a couple of days with him at Portsmouth. Take notes & write down every thing. One of the most valuable assistants you will find is [G. R.] Waterhouse at the Zool[ogical] Museum. It would be wise for you to study the skulls & teeth a little. You take these things by intuition. Although I would not go as far as others in the determination of species from skulls you will find that they go very far towards it. Waterhouse will tell you what Books and other transactions of societies have been published that you ought to consult & bring with you. Grey [John Edward Gray] of the British Museum may be consulted *cautiously*. The animals that you want are there but in a bad state of preservation. However you must keep a book as you did of the birds leaving room for 100 species. I doubt whether we will go beyond that unless we include the whales & mammalia of that character. The Quadrupeds in the following list (I have unfortunately no book to refresh my memory) might as well be painted in England as you will scarcely find them in this country. In all cases where the animals can be preserved in the flesh it must be done. There will then be about 18 or 20 that you must draw from skins. In taking descriptions mark the different annulations in the hair. It was from the above that I ascertained that Richardsons Lepus virginianus was a new animal. I have taken a description of the skin in his possession—the only one in the world that I know of. By having an English Hare before you, you may be able to make a good figure of the winter colour. There is in the Zool Gard[en] a Canadian marmot that you ought to figure & the moose at Earl Derbys & if I mistake not also at the Zool. Gard. that you ought to figure & measure. The prong horned antelope—mountain sheep, (of which it is conjectured there may be 2 species) musk ox a white bear & several of Richardsons Deer are not to be obtained here. The Wolves (confound the Dogs) will give us plenty of trouble. Look at the one of Richardsons from the N. The majority are I think mere varieties but so strongly & permanently marked that they ought to be given figured. Nearly every thing from beyond the rocky mts. will prove new. Look to this & lay your hand on every thing you can beg buy or borrow (which means stealing) from the far North & West.

Leave nothing in England that you may be obliged to send for hereafter. I promised Harris & other that I would give a full synopsis of American Quadrupeds. I have done no more than make pretty full notes. These and all the information I have to give are at your services But the Book must be original & creditable—no compilation & no humbug. I am not often far out of the way in predictions & I assure you it will, if managed with your usual zeal and the boys industry & attention to the commercial part, be the most profitable speculation into which you have ever entered. It will amuse & occupy you in your old days & keep you from snuff & grog. Look out for the stones on which you are to make your Lithographs, you may obtain them cheaper & better in England than here. We find great difficulty in obtaining good engravers in this Art. I have some Books such as Richardsons Fauna, Harlan, Godman, Catesby, D'Azara, Lewis & Clarke, Langs expedition, Trans. Zool Society, the Works of Lichtenstein & of Ehrenbach of Dresden—But there are various scattered papers in Journals that might be of use to us. There is also the Zool. & Bot[anical] Transactions, by Jardin & others that I have not carefully examined but one or two things I saw there struck me as worthy of note. The figures may I think be given without reference to any scale—those of the size of a skunk full size those above as taste or space will dictate. The work should not come at more than $100 but not much less when completed. It had perhaps best be done at Boston. Philad[elphia] might be more convenient but I know your love to that *city of Brotherly love*. I am not sure but you might more conveniently make a drawing of the Buffaloe in England than here. It will do not harm to take a figure or two of such Indian Dogs obtained far N. & West as you may see in the Zool. Gard. & we will talk the matter over afterwards. The porcupine & Wolverine too may be figured in England & if you can find a real skin from America of the Sable do [not] let it pass—also the Fisher (penants martin) the common mar[tin] & the rascally grizzly bear. There is some doubt about the ol Brown bear—ask Richardsons opinion. All [Richard] Harlans names but one perhaps will have to go by the board. [Thomas] Says will hang on—& [George] Ord will if I am not mistaken be entitled to one. [Constantine Raphael] Raffinesque may as well be left in his Glory, for no one can know what he intended to describe. The Bats will not give us much trouble except as to their arrangement as to Genera & in this they are puzzlers. The shrews I have given, but they still require over hauling. There are some Hares & Squirrels in the British & Zool. Museums particularly the latter that you ought to figure—Ask Waterhouse. Your range should be from the farthest N. & on the West the pacific—south Texas & Californias a wide range & few species. I possess good specimens of the British Ermine & Weasel that will do to compare with ours. Bring with you the English field mouse, it will do to compare with our Mus leucopus. The other quadrupeds differ from ours very materially. Look out for every thing from our country in the shape of meadow or field mice & shrews. Bring the skin of the common water rat (not the Musquash). Figure the Canada Lynx. Write to [James] Trudeau about the wild cats of Louisiana—it is said they have one with a long tail—also the Louis. muskrat. Indeed I fear you will have to spend a few months in Louisiana. Ask Waterhouse whether he thinks you ought to give the whales porpoises &c in your Book & the antediluvian animals. If so it will be very troublesome but I am afraid we cannot otherwise call it American Mammalia. I meant to have added a list, but I am shockingly nervous & cannot lay my hands on a Book. Richardsons Fauna will however be sufficient for you. I have given you work enough to do for the present. I am afraid I may be too late for the mail. Toronto not yet arrived. Love to all.

Yr frd
Jno Bachman

[addressed to: "John J. Audubon Esqr / Care of Messrs Wright & Co / 6 Henrietta St. / Covent Garden / *London*"]

(5) 1840 Jun 25: to Victor Audubon on his father's drinking (typescript, Tyler Collection).

Charleston June 25, 1840

Dear Victor,

Yours of the 21st making some inquiries respecting the cause of difference between your Father & myself came to hand just now & I will try to answer it before the close of the mail, although as I have but a quarter of an hour, I fear I may be too late.

I will now state to you what I intended after the lapse of a few months to say to your Father & whenever you think it is seasonable & proper you may show him this letter. I have no desire to irritate. A sincere wish to render him a benefit is my only motive in disclosing to you what I intended at a suitable time to say to him.

Shortly after your Father's arrival I discovered that he used liquor too freely. In the afternoons especially he was almost unfit for conversation. He became garrulous—dictatorial & profane. He stated as facts things which I knew were not such & which he seemed not ever to recollect on the foll'g day. The ladies in my family all remarked it, & begged me on those occasions to let him have his own way. I bore with it for many weeks. I have a perfect abhorrence of intemperance & when I perceived he had taken too much I had only one resort, to go into my front room & lock the door. Thus time passed away—in the midst of other heavy trials & afflictions. But things grew worse. There were scarcely three days during the last month of his stay in which he was not the worst for liquor in the afternoon & on several occasions before dinner. His friends Dr[s]. [Elias] Horlbeck, [Samuel] Wilson and [Eli] Geddings who were almost the only gentlemen who ventured to call on us in our afflictions had observed this growing miserable habit. They called on me separately to say that something shd. be devised to save him from degradation & unhappiness. They supposed the habit might be increased from trouble & want of occupation. They therefore advised that he shd. be urged to return home under a hope that change of scene might make an alteration in his habits. I cd. not well propose it to him & got John [Woodhouse Audubon] to do it. Maria [John Woodhouse's wife and Bachman's daughter, who was then critically ill and who died on 15 Sept. 1840] however grew worse at that time and he still remained. But things cont'd in the old way. I saw one whom I esteemed beyond almost any other living man ruining his health & his intellect, setting a bad example to my children & rendering me perfectly miserable. I wished the infernal whiskey jug at the bottom of the sea.

The immed. cause of the quarrel was a trifle. I said at the dinner table in reference to our writing about Maria's [health] that I placed no confidence either in his or John's statements—he representing her worse than she was & John as always better—meaning that one was too sanguine & the other too desponding. He then told me that I was equally bad—as a few days before when some message came from Maria's room [he claimed] I had jumped up exclaiming—My God Maria is dead. I remembered the message well. I had heard it but did not utter a word—& no one wd. have known my sensations, had I not afterwards told Maria that she had frightened me with her message. I told yr. Father that I had not on that occasion even opened my lips & when he contradicted me I was so vexed at this miserable forgetfulness that I told him he must have lost his brains. He had taken too much whiskey & I was angry. He went immediately to make complaints to Dr. Wilson who guessed at the secret & begged me to humour him. It had one admirable effect—he indulged no more to excess during his stay & although he avoided conversation with me, except a few words yet his conduct was that of a perfect gentleman. There has been no difference bet[ween] John & myself—believing that I ought not to influence him against his father's only fault, I have avoided conversing with him on the subject. John is positive, but a good fellow. I cd. agree with him, believe me in [this rather] than weaken his attachment to his father. We are I think very good friends.

I feel pretty confident that yr. Father is not sensible of his fault, & his folly when [he] indulges in it. But alas, others see it. I have not spoken even to my friends on the subject since he left us, but if he wishes for the opinions of others, let him ask Drs. Wilson, Horlbeck or his kind friend Maria [Martin], who has shed many a tear on that account. In our world I am a clergyman. I cannot countenance intemperance. I owe much to my family & much to society & whilst on all occasions my house is open to your father, whilst he uses liquor in moderation, I cannot consent to welcome him here under habits such as myself & family were recently made the painful witnesses of. I have written this with reluctance & regret. I am willing to admit that I am hasty. I am willing to take all the blame on myself provided this miserable habit is abandoned. I wish to save him from degradation. I am willing to do any thing & every thing to effect it. His name & memory are very dear to me, therefore have I taken this stand. If I knew any other mode of effecting my object I wd. adopt it. I pray God to direct us both in the right way. I am afraid I shall be too late for the mail.

Affectionately yours, / Jn. Bachman

[Addressed to: "Victor G. Audubon, Esq., / 86 White St., New York"]

(6) 1841 Aug. 5: to John James Audubon on the quadrupeds project (ALS and typescript, Charleston Museum)

Charleston 5th Aug. 1841

Dear Audubon,

Yours of the 1st inst was recd. to day in company with that of Victor to Maria [Martin]. It found those of us that are at home all well. Mrs B[achman] I fear has not gained very much in the country & as she is anxious to return. I shall go for her on Mond. & return with her at the end of the Week. Harriet [Bachman Haskell] is well & has gained about 21 pounds in 5 weeks. As I shall be occupied tomorrow & shall be preparing for my duties on the day after I write you this evening although I am fatigued & have Mrs. Davis, Mrs Strobel, & some other old Ladies in the adjoining room.

Your letter & Johns few lines were perfectly satisfactory & it is best all round to say no more on that subject. Let us act & labour in all particulars.

As I shall have to write you very often & you will do the same to me, let us make some arrangement that our writing may be profitable to the Book. Once a fortnight you will be sending to Hart—let your long letters go in the Box. Occasionally I will be sending you specimens & other matters by water—then I will write you on Nat. Hist. I have not time to keep copies, but you may as well copy such parts as are important—so that you may refer to them. Your memory as well as mine is none of the best. Send me in Harts box a few sheets of your large letter paper. We cannot get it here.

By our friend J. D. Legare who left this aft. I sent you a farther list of the quadrupeds that are known bringing up the list to 190 true species & 23 doubtful ones about which inquiries are to be made. I had sent you the previous one before I had ransacked my notes which I have now done thoroughly. I have I think all that has been done in this department up to 1838 & my correspondents are keeping me informed if any thing new occurs. They have not I think any collectors at present in the country. Legare will be with you in a few days after this letter arrives. He appears to be much attached to you all. I am sorry to say his health is very delicate.

But although I can see my way pretty clearly & think we will be able to give more new species even than you did in the Birds. (I include those I have already published which scarcely any one know any thing about.) I have some fears lest [John Edward] Gray of the British museum may try to diddle us out of half a doz species. You are aware that there is a rivalry between him and his Museum & that of the Zool. Soc. In the latter place I found some of my own Animals Brewers Shrew Mile—my dwarf Mouse & several others. I showed them my manuscripts & they immediately labeled the species accordingly. They also allowed me to describe those of their own as far as they belonged to our Country. Within a few weeks they sent me their last catalogue—in which they give my names marked Bachmans catalogue—manuscript. All this is generous, but [G. R.] Waterhouse says Gray came to inquire where my descriptions were printed. He put him off by saying—they were published in America. He however advises that they be published in some societys papers with short descriptions at once to save the names & prevent piracy. Gray is of the same Genus with Dr [Richard] Har[lan]. At the same time the Frenchman whom you thought dead is publishing in Paris. Might it not be as well to publis[h] [t]hese species at once—at least those that there is danger about? The Frenchman (I cant th[ink] of his name) has some new things, as [Charles] Pickering told me so. A few of these species have already been read at Philad. but I stopt the publication as I intended at that time to publish by genera. I will wait your opinion, but wish to save [Thomas Mayo] Brewers species—as it is called after him. All I know about the squirrels you will find in my paper on the Genus—the magnicaudatus if it exists [not included], is in the Philad. Museum. You are not so far out of the way about the Mink— but I rather think we will find the fellow growing lighter & larger as he grows older—certainly however [in] our up country one is black & our salt water one much paler. Nothing but specimens & a plenty of them will settle the matter. We have the Muskrat in the upper country. It is smaller than in N.Y., but so is also the Deer. I have not been able to find a specific difference. I have not seen that of Louisiana. Dr Dekay [James E. DeKay] may be right about new species—certainly there are in N.Y. one new arvivola [mouse]—& at least one Sorex [shrew]—but when I last saw the Dr he was a great a blunderer as his friend H[arlan] of Philad. [John Kirk] Townsends small Lepus artemesia [Worm-wood Hare] was returned to him. By the way have you his dusky wolf in spirits? I would only give the skulls of one in each genus—unless the species differs very much. I will send you in the Box a drawing of Lichtensteins Sea otter. The Skull comes up to my notion. The Mountains of Texas & the plains too are to be our grand resort for new species—so will be the plain for a thousand miles this side of the Rocky Mts—Missouri Territory &c. You will see what I have said of Sciurus Macrourius [not included] in my list. We will have the honour of christening this child. Spec. Philad Museum. You may let some of the Editors say our Work is to be published—but I would not yet issue a prospectus, but lay up a better stock of materials & decide on the Work. I am afraid a Big book will be a heavy business. A work like [John Edwards] Holbrooks 8 or 10 Species in a number or half bound vol at $10 per Vol would run like wild fire & with your drawings would I think be a most profitable concern. In the mean time your large drawings would always be valuable. Tamias [ground squirrel] or Scirus lateralis [squirrel] of Say—I find by my notes is half way between

Tamias & Spermophilus [marmot-squirrel]. I have set it down for the latter Genus & in this you must correct the catalogue—so also for the Genus Geomys [*Pseudostoma*]—which is one of Rafinesques. All our pouched rats that I know of are of Says Genus Pseudostoma [rat]—with cheek pouches opening outside of the mouth. [John] Richardson in this made an awful blunder. The pouched rat I shall certainly obtain from Georgia—in the flesh—if not alive. The Neotoma Floridana [Florida Rat] is abundant here. In cold weather I can get one any day. You had better draw it from the flesh.

This species as well as the cotton Rat—which the Philadelphians made so much fuss about as coming from Florida was all the while under their noses within 5 miles of the City. The Ermine & Weasel are different—but hard to distinguish in summer. Cooper has some specimens but perhaps not the one I want. It never becomes white. I kept it alive through the winter. I fear it does not come out in Winter. Here we never get it at that time.

I will send you a number of things at once to be shipped by the first vessel—also a few things that I may pack up in spirits—our black Rat [Mus Rattus]—Molossus fuliginosus [Jamacan Bat] all of our Bats in skins—Cooper has some.

This is a bad time to get specimens. It is hot—the country sickly—the Animals young—the fur bad.

Make some experiments with West Indian rum about preserving Quadrupeds. I have suffered sadly in my specimens. They were all dissolved.

Now is the time to open correspondence with every part of the world. We cannot do without specimens—[and] with them we have nothing to apprehend.

Our distant military posts are the places to write to. Ask Townsend to write to a missionary of his acquaintance near the Columbia river.

The animals must be skinned over the skull—up to the nose & lower jaw—& well arsenicked & the skull may be taken out at once. The who[le] skull must be preserved.

Have a few cages made & this will answer to keep living Rodentia & study their habits. Now is the time to get the Ermine & Northern Hare also to study the changes of colour. The Maryland March Hare, the Tamias Lysteri [ground squirrel] & the Mus leucopis [American White-footed Mouse]—common animals should be kept to ascertain their state when lying dormant.

I have specimens of both the common weasel & Ermine of England to compare with ours.

I have no specimen of our Otter. I left one with a good skull at the Zool. Soc. for comparison. I see they have set it down Braziliensis with a query? They think it may be intermediate between that & canadensis.

Remember in observing from Museums that the colour fades dreadfully. You may correct this in some animals at the furriers with fresh skins.

The skins of polecats from different states are much wanted.

You have two months labour at Philad. Museum & Academy—but not yet. I want to see LeContes specimen. He ought to have one of two new species—tell me what he has given you? He knows a good deal. Cooper—cold—cautious & lazy ought to be visited & won over. In your interviews with them listen rather than talk just now. In a year you will know a great deal about our species. Make a [manuscript note]Book as with the Birds & set down all you hear. But try to put the habits under the right species lest we bark up the wrong tree. Adieu G. bless & prosper you.

Yr frd / J B—

May not your Tamias lateralis [not included] prove to be the palm Squirrel of the Eastern continent?

[Addressed to: "J. J. Audubon Esqr / 86 White St / New York"]

(7) 1841 Dec. 7: to John James Audubon on the quadrupeds project (typescript, Charleston Museum)

Charleston Decm. 7th 1841

Dear Audubon

On my return from Georgia much fatigued & not very well, I found yours of the 21st Nov. waiting my arrival & will endeavour to answer it this evening if I am not interrupted. It is amazing that you can walk 20 miles before dinner. Were I to do that, they would soon write Hic jacet [here lies] over my grave.

The expression you use about "Sunny England" might be well enough for a staunch John Bull, but I am an American found the sun rather a scarce article in any part of that vast kingdom. I am glad we can give [John Eatton] LeContes name to the Bat. I think him rather a decent Gentleman—no great beauty indeed but genteel. As I feared I have been interrupted—as Dr Wilson & another Dutchman came to talk mosses & Brooms all the afternoon & it is now pitch dark & I will have to write the remainder of this letter amid the chattering of the children. Harriett [Bachman Haskell] went in the Steam Boat this morning to spend a month or 6 weeks at Judge [Benjamin Faneuil] Dunkins Plantation. She seems quite well but I think a recreation in the country will be of service. My wife has I am sorry to say a return (although slight) of Tic douloureux. The rest are as well as usual. The black Fox squirrel, as you have learnt by Marias [Martin's] letter is on the way. If it does not reach you in good order—never mind, I will send you more. I promised to send you all our Carolina specimens in the flesh—*& it shall be done*— Cats, foxes & all—but positively until now it would have been a waste & an imposition on the olfactories to have sent them. Next Monday I go to [C. L.] Desels. Then look out for a cat if there is one in Goose creek. Johns Dog is a roarer at Cats & Foxes. We have a bell on his neck as he hunts too widely. I wish I had his blunderbuss for a few days to help me astonish the old bucks. I am sorry you gave so bad an account of friend [John Edwards] Holbrook, but I will give you a short episode which may possibly induce you to think more favourably of his *professorship*. Some weeks ago he called during my absence & borrowed my Gun from Mrs. B. He did not return it for some days when he brought it back with the stock broke in two. He never said a word about the accident. When I charged him with it, he laughed & said some of the visitors at his farm had broken it; when I asked him who they were he said he could not in honour tell me. So I have my 18 Dollars for my wifes good nature, to pay. I tell the story every where & my friends repeat it to him & he laughs most heartily. Never mind I shall mark him yet. I think it quite likely that the cat squirrel, which is found in Pennsylvania, may also be found in the adjoining states of Maryland & Virginia. You seem rather stiff about looking at [Richard] Owens dissections & somewhat unwilling to understand my plain words. His manner in dissecting & his anatomical investigations in deciding on closely allied genera—I most heartily approve of. There is very little use in measuring the diameter of the stomach & length of intestines Trachaea &c unless we draw practical conclusions, an error which appears to me you have sometimes fallen into in the dissections of your Birds. Owen has given very few of our American Quadrupeds. Lichtenstein has examined the anatomy of the Sea Otter. More original work remains to be done in this way than you or I or Dr Goldsmith will ever finish I fear. I do not wish the dissections of others to be copied. We may however learn something from their manner which would do us I think no harm.

In regard to the oppossum I have I think written you before. I long since read [Benjamin Smith] Bartons [1806] paper, which then appeared to me to come nearer the truth, than the experiments of others. I do not however just now recollect it very distinctly. I have kept the female through all its stages & have only one point to satisfy myself with—that is—the finding of the young embryo in the womb, which evidently remains there but a short time before its exclusion—for they are seen on the teats when but a little larger than pease. The perfert formation of the parts of generation, prove that they copulate like other Quadrupeds. They remain in the womb but a short time—then they are excluded into the pouch which the female can reach with the parts & then fasten to the teats. Then they are protected by the pouch & receive further development & growth & nourishment. The Florida rat & the white footed mouse—approach the opossum in this. They fasten their young on the teats when very minute—& for several weeks drag them along wherever they go. On the contrary the wood rat like the rabbit suckles its young & when they are satisfied let go their hold like the kitten & the puppy. I have seen the Kangaroo—the half brother of the opossum dissected & we found the minute young in the womb. It is quite impossible that there can be any other process. The notion that the young are born on the teats is about as reasonable as the Girls story—but I am too modest to relate it. I am glad the cages come to you at last—you only told me the Neotomas [rats] were alive. I hope the young were living also—& also the two other species in the other cage. One was LeContes pine mouse which you must colour after the skins you have in possession—the other the wood rat. But I suppose I shall hear all about this matter soon. The Neotoma of Florida is like ours, as Dr [Edward Frederick] Leitner brought me a specimen—LeConte's opinion to the contrary notwithstanding. Query, have you not by this time found out that he is oftener wrong than right? You have his specimens of sublimate media—& the Dickens knows what—you are now able to say what they are. Our motto must be *nature—truth & no humbug.*

I told you sometime ago that I would yet have to send you a few specimens in rum. I will now try your olfactories once more. I obtained in Georgia after infinite labour a pouched rat—caught in a rat trap—with a broken leg. The weather was warm & I chucked him in whiskey. I opened him to day & find him in fine order. I will send the fellow in 3 days—& you will figure him & smack your lips for more specimens. I also have a Hoary Bat in spirits—a rare species every where—you will figure this too & say thank you. We had a roaring Bear too of 400 lbs but I had no bottle large enough to put him in so we ate him & that is next to making a figure. Say has the opossum ever been found East of the Hudson, without having been carried there? I think not.

As I shall now send you a good specimen I hope in spirits of the Georgia Hamster—which I believe old Mitchell first described & must be called Pseudostoma [*Arvicola*] pinetorum [LeConte's Pine Mouse]. It would be well to figure those also from Townsend, as you are not likely to go to the Columbia river to get them in the flesh. Have you a specimen of the Canada Hamster P[seudostoma] bursarius [Canada Pouched Rat]? I think there is a specimen or two in the Museums at N.Y.

I believe you have not yet figured the Carolina Squirrel. I will send you a pair on Thursday—if I can find them in the Market—as I think this cold change will enable us in future to send specimens in the flesh. I have a splendid pair of Curassoe birds, sent me by [Joel Robert] Poinsett, who recd them from Mexico. They are very tame & eat out of the hand. Your number 43 has come to hand but I hear nothing more of the Rabbit skins & other matters you promised to send. I have only just returned & am resting. As soon as I get a little better I will go to work. At present I am sadly out of sorts.

I will just mention to you a matter which I don't wish you to speak of farther. The presidency of the So Carolina College was tendered to me by the Trustees a month ago—Salary $3,000 a fine house &c—But

the state of my health & the fatigue attendant on the duties would soon finish me & I promptly declined. I did not speak about it to my family for I think, Maria [Martin] at least, who does not like a city, over much, would have persuaded me to go. Such a step would moreover have put an end to my amusements in Nat. History.

During my absence Mr Bee, Rhett, [Christopher] Happoldt & others sent a considerable number of all sorts of field rats & mice alive. Unfortunately I had only one cage the other having been sent to Georgia for pouched rates. My folks did not know what to do with them—so they put them in a cage containing wood rats—& these carnivorous scoundrels had a glorious feast of them. Nothing but the tails & legs are left to tell what they were. All desire to be remembered to all.

Yr frd / Jno Bachman

[Addressed to: "J J Audubon Esqr / 86 White St / New York"]

(8) 1843 Mar. 12: to John James Audubon, St. Louis, Missouri, on research needed in the western United States on quadrupeds (ALS and typescript, Beinecke Rare Book and Manuscript Library, Yale University).

Charleston March 12th 1843

Dear Audubon

By this time I presume you are on your way westwardly. Now may God bless & prosper your great undertaking in this great cause of natural science. I hope and pray that you & your companions & assistants may be preserved in health & in spirits. I do very much rejoice that you have been so fortunate in your associates & assistants. [John G.] Bell [taxidermist] I know to be a first rate fellow & not slow as a naturalist & Friend [Edward] Harris, too modest by half, knows a vast deal more in Nat. history than those who have bragged & scribbled a great deal too much. Now friend, whenever you get high up the yellowstone—where the prairies end—in sight of the Rocky Mountains turning toward heaven—where the wide waste of prairie lies to the East covered with thousands of Buffalo and Deer—where you breathe the free pure air of heaven uncontaminated by the breath of our cunning, treacherous race & are warmed by Gods own bright sun,—there on some Indian mound let Harris stand up—with head uncovered & his cap on the ground—

place your right hand on his head and dub him—not a Knight (for in our day this is a comparatively worthless appellation) but by a much nobler and more dignified title—namely that of Naturalist. He deserves it all for his love of science & for this last fearless adventure where I trust he may reap an abundant share of health knowledge & fame.

I hope to work hard this summer at the quadrupeds. My health has been slowly mending for some time. My wife is much better & all the rest around me are well. We this moment recd. your letter to Maria [Martin]—I will look over it presently & see whether it contains any thing for me to answer. I will try to do my duty toward you whilst you are absent & although you to all appearances will outlive me, yet in case of accident I will take care of your well merited fame. And now let me say a few words about your contemplated collections.

1. In the first place, do not entirely overlook the birds. My word for it you will see several new species & those two which you have heretofore thought were South American. Figure them & preserve skins.

2. It will not be much trouble to collect seeds and plants, but this will of course be a secondary consideration.

3. There will not be many snakes or lizards, but they are so easily preserved that they should not be thrown away.

4. After all you must attend to your business which is looking for quadrupeds & knowing all about their history. I need not say I know nothing of western animals, but what the Books tell me & the few I saw in confinement.

Every thing therefore, you write about them will be new. Of course Bell will prepare skins—he is first rate at this—but I am under the impression that in many cases he will have to take out the stuffing after they are dried—to save bulk. Mark the sexes & localities—date when killed—dimensions & weight—save all skulls. Write up your notes every evening—use plenty of arsenic. Whenever you can sketch your animals in the flesh—note habits time of breeding—number of young—Holes and nests. Be careful to keep a specimen of every species you may procure although you may think them similar to those in our Atlantic states. You know my theory—that every species—except a few Bats—& a single species each of Bear & Deer is new in the far west.

I begin with the Bats—you will, of course, save some of each species.

The shrews if you could get at them may present something new & I do not doubt it. Look out for the condilura [moles] and scallops [shrew moles] especially the latter. The Black bear you have not as yet figured & the Grizzly Bear if he does not eat you will give you a chance of telling some thundering yarns about him. You will find the Badger in abundance & you must study the beast's habits. Look at the weasels, martins, mink, otter & Beaver, the prairie & common wolf—see whether there are more than two & whether the Dusky Wolf is only a variety or a bonafide good species—even if varieties, they should be figured. Satisfy yourself about the Black Wolf. He presents this variety of colour all over the World—in fact colour is no very good guide in the wolves, squirrels, Foxes & a few other Genera. Paint the Indian Dog and the Kit Fox (canis velox). Don't forget the cats. Note the range of the different quadrupeds. In the meadow mice (aruicola) you may find something new. Try to ascertain whether there are any true Lemmings at the West. You will hardly do much in the Genera Neotoma [rats] & Mus [mice], Arctomys [marmots] & spermaphilus [pouched squirrels] will be genera, especially the latter that will demand all your attention. I doubt whether you will find all the species of the latter. In mountainous regions you must look for Tamias. The squirrels must be most carefully studied—especially as they differ so much in colour—we must try to separate species from varieties. Flying Squirrels. The pouched rats much be carefully examined—my word for it Richardson is wrong in Geomys [*Pseudostomo*; pouched rats]—you will find all the cheek pouches entering on the outside. The Sewellel (Aplodontia of Richardson—arctomys rupa of Harlan who never saw it). I have never seen it is undoubtedly a good species. The Hares and Lepus [*Ochtona*] princeps must be hammered at. The Horse of the prairies is as much naturalized as the Norway rat & may as well be figured. The Elk and Buffaloe you will shoot & skin and figure. Cervus macrotis and C[ervus] leucures [mule deer] of Douglass. I need not say any thing about the Rocky Mt Goat, Sheep, and Antelope.

I don't know what more I can add. Your own good sense will add more than I can at this distance. Don't overlook Hunter's skins—sometimes these bring something new to light—especially when they come from localities where our visits do not extend. Write down the accounts of trappers & Hunters when you are sure you can depend on their veracity. In a word, give a true history of every species that inhabits the plains & mountains—the earth, rocks & trees. This is the duty of the Naturalist. I hope to hear from you frequently.

Please remember me to your party, although I am not acquainted with only the one half of them—but I seem already to know them. And once again God bless & prosper you all.

Affectionately, / yr friend / Jno Bachman
[Annotated: "Recd 31st March / Answered 1st April / 1843." Addressed: "John J Audubon Esqr / F.R.S. / St Louis / Missouri"]

(9) 1846 Jan. 17: to Victor Audubon on the insufficiency of materials being provided for the quadruped project (typescript, Charleston Museum; the original of this long and important letter could not be located, but the typescript is reliable.)

Charleston 17th Jan '46

Dear Victor:—

I always try to answer all your letters and those of your Father, although I sometimes kill two birds with one stone after waiting till there is a slight accumulation of epistles. The day before yesterday I received your German account of Lepus Nanus by Schreber and this morning your Father's of the 12th came to hand. I have just finished my Saturday's work and will devote an hour to you on a subject which I could wish you to lay to heart.

You may perhaps have sometimes conjectured that there is a disposition in me to shrink from the heavy duty and responsibility of the labour thrown on me in writing these descriptions and as I have at last found in preparing the whole work. No Victor it is not this. I have all my life been accustomed to hard work and if I could have the materials essential for such a work it would be pleasant to me. I have also confidence in myself and feel that I can accomplish what I undertake provided I have health. I was under the hope that some of you would just have read enough to know what belongs to such a work and then you would be able to see what I wanted and would be able to give judicious answers to my inquiries and supply me with materials necessary to fulfill our obligation to that public that expects much of us.

When at last I became aware that I could obtain no literary or scientific aid from any of you I was not discouraged but proposed a method by which all could yet go well. It is true the mistakes in the names attached to the plates were terrible owing to the want of knowledge of your never looking at books where these animals are given etc. but even this might be in some measure remedied. I then proposed to you to do what I required. To substitute my knowledge in these matters for yours and to get you just to attend to my plain written requests. In this I confess I have been sadly disappointed. My letters are either not answered at all or they are not answered with the necessary precision. I will give you a few instances and then you will be able to ascertain whether I am a grumbler without a cause.

I wrote you most pressingly to go to [John] Bell and ask him whether he had stuffed a beautiful striped Tamias or not. I urged you not to wait a day. The only answer I ever received was on the following up to send it on—you had not as yet seen Bell. To all my inquiries whether Bell knew anything about it I have received not a solitary line of answer. To me such conduct is very strange. I presume it is from thoughtlessness but I wish you to consider that it worries and mortifies me since I know how easy it would be to satisfy my mind.

About this long looked for account of Lepus names of Schreber, I told you I wanted the dates when the particular volume was published in order to ascertain what author had the precedence. You understood me for you wrote begging me not to use the Lepus names. Now your answer has come and although I can make out your German quite well you have not given me even a clue to the time the species was described or the book published. Luckily I have obtained the information elsewhere and so have done with Schreber.

I see both you and your father are hinting at the old worthless books you have the trouble of copying. I wish I could save you the trouble and if I had them I could by running my eye over them ascertain in a moment whether they are worth copying—but so it is with some of them—the first description of Genera and species for instance I cannot do without—to ascertain whether it is the real species referred to. I ask what describer of species can do without them? Just oblige me by turning over to the end of Richardson's preface p. 37 and ask whether he could do without books. To give a compilation such as you had commenced would have been a reflection upon all our pretensions as naturalists. I thought to save you a thousand dollars in books by getting you to take copies. Some of these are no doubt worthless but I must see them to ascertain whether they are of any use. These books are necessary to enable me to pursue an independent course and to show that I am untrammeled.

I feel no desire to turn critic on your Father's drawings and if it gives offense (as I perceive it does) I shall not say another word about it. But then I feel as if something was wrong and although I shall be silent about not giving characters I beg to say my opinion remains unchanged. For instance the last plat in the first Vol[ume] Sp[ermophilus] Richardsonii a figure is given without the long hair sin the tail. Now you may say as your Father says in his letter "I cannot help copying nature." The fault I find is that nature is presented in a distorted way. Suppose he had given his Wild Turkey with its tail pulled out. That [ground squirrel's] tail was not as God made it, but as Townsend let the hair drop out. That tremendous scrotum of the California Squirrel was not given to it by its creator whose works are all natural but was stuffed out of character by Bell. The Carolina Gray Squirrel is nearly always grey. The varieties are given, three of them, but the one which we want the light grey one is omitted. The Northern Grey Squirrel has three figures, why not give our black Squirrel two. But I will not multiply cases as it may be unpleasant and I will desist by just remarking that the plate of the Mus leucopsus seems to me to be a true representation of what I call a character—two brown and one blue with every characteristic quite visible.

In your Father's letter about Townsend's Hare he says "I do not mention in my account of Townsend's Hare that they turned white in winter because I thought Townsend had told you so." Now turn to Vol. 8th part 1st Journ. Acad. N. Sciences for the description p. 92 where I give Townsend's own words. "I made particular inquiries etc etc they all agreed that it never was lighter coloured." Is this not a proof of what I have said that all you think of is to get up the drawings. If Townsend's Hare turns white and there is no other white hair on the Prairies then the name must be Lepus Campestrii. If my anxious written instructions had been attended to this would not have occurred. If your Father can show me in a letter that I advised three pale figures of the black rat to one black one to be made I will make him a present of a pair of horses and shallow the carriage. I wished a figure of this as well as the young Neotomas where LeConte made the mistake, but I am incapable of advising such an absurdity as making a variety more prominent than the species.

I am exceedingly troubled about one of the animals I am quite sure will be named on the plate—I mean the prairie dog. I have never seen a specimen. Sometimes I think that a parcel of legs and a head that came in the box may be that species. There is however as usual nothing labeled. Now I warn you not to make Spermophilus Ludocivianus—unless you either make it out a spermophilus—or send it to me to decide on the genus. It has never been examined by any naturalist. Do tell me whether you have a specimen.

There is a skin of a deer in a box—what is it? In the box are some weasels marked Lemstela erminea are they American or English? Your Father writes about a specimen he sends of sciurus longicaudatus. I don't know such a species. He asks whether he shall figure arvicola without a skull. I can't tell what he refers to. LeConte's species had a skull.

Dear Victor it is this want of precision—this loose rambling way of answering my questions that renders my situation in this so unpleasant. Sometimes I have to write three or four times before I can be understood. As regards the common [Northern] grey squirrels you certainly don't understand me. I have specimens enough but I want to see how many are tufted in winter. DeKay has made a new species of the Tufted one. So has Emmons (whose [1842] Report is a work you could easily get). I think the old ones get tufts after a year old—in the winter is the time to make the inquiry. The New York market is the place to find it out. I can't settle the species without it. If I were with you I could settle this matter in a single week. Your Father sent me a dozen letters to prove that it was a different species. Now he seems to have forgotten all about the matter.

Maria [Martin] sends a box next week and I will send one of the tufted ones and one without tufts so you may see what I mean. None of your Father's on the plates have the tufts I refer to. I am sure he drew his figures either in summer or from animals under a year old. Your

Father thinks it curious that the large wooley flying squirrel is marked P[teromys] Sabrinus. If he looks at Richardson's P[teromys] Sabrinus or Alpinus he will find it described there. I wanted to compare the three specimens. So far all is right *except my old friend's memory* which sometimes goes wool gathering. By the way don't send me any copies from the transactions of The Zoological Society. I have them all.

The European fox skin I would like to compare. We have not the European fox in this country but I wish to show how they differ. Your Father writes about my sending him Lichtenstein's work—why he had it already and criticized it most unmercifully. I am now describing the skunk from it, when I have done I will send it again. He asked why I did not send him Lichtenstein's work when he figured the skunk—why it was figured you know before he sent the specimen. The squirrel of Capt. [John C.] Fremont I will send in the box to be figured—also a shrew from Oregon. The striped Tamias I will not send as a punishment for that outrageous trick that was played on me with sciurus rubicaudatus. In the manner of making the figures etc. your Father has a right to use his own description in spite of my criticisms but in naming species with my name my consent must be obtained.

Now I have told you what is unpleasant under the hope 1st that you will have my letters before you when you write answers. 2 that you will use precision which is everything in Nat. Hist. 3 that as you have not studied the species you will just attend to my advice. Now my good friend I will not magnify my offence in this matter but would just ask on your own account—if you had not my poor services at your elbow would you not make a fine kettle of fish of the Quadrupeds. Who in America could you get to aid you. You ought never to have commenced the work when you did. I urged your father to study—to make drawings and to wait. So you had better make good use of me whilst you have me for I am now perfectly convinced that you cannot move a step by yourselves. I have the outlines of a preface written but this we are in no hurry about as it is always finished last.

I have now 40 species finished with the exception of the gaps I must fill up when I receive your answers to my letter in which I summoned up all I wanted to complete the book. In asking for information on the muskrat's foot your Father always talks to me of the webbed feet of the mink. But I know the mink like a book. But my muskrat skins are almost an eighth webbed on the feet—are they so in nature. Answer me yourself—my old friend's gun goes off half cocked—a world too fast for a slow sober naturalist. [William E.] Haskell is copying the articles beautifully. I have done [my] best—if any man can do better let him undertake it. Look here—have you nothing about the peccary—skunk—Pennant's Martin—Beaver—if not say so—and I will do what I can but answer me yes or no.

This is a long letter—I began it—then went and dined with the Judge [Benjamin Faneuil] Dunkin—now late at night I am writing still.

You cannot think how much of my time is taken up in writing about Quadrupeds. Today's mail (thanks to cheap postage) brought me 13 letters. Some are worth nothing but others are valuable—but Dr. [John] Wright [of Troy, New York] is a man—*the man*. He knows with what precision to answer a question in Nat. Hist. We have the Puma—Jaguar—Peccary—American Badger—Grizzly Bear—Black Bear—Fox—Raccoon—Prairie Wolf etc. in a menagerie alive here. I work away at them every two days and find them very serviceable.

Love to all. My poor wife is still very unwell with Tic doulourex.

                         Affectionately your friend / J Bachman

P. S. Rodentia collected in the far west—what became of the skins. What did Bell do? Nothing but hunt Buffalos. I am putting you to [a] little expense just now of 50¢ a week. I have engaged two assistants per week till April (females) for dissection. In this way I hope to find the [opossum's] embryo young in the matrix before it finds its way into the pouch. The little girls trifling portion is attending to etc I will inform you shortly. I have not heard from Mrs. B since I left her 3 days ago. She was no better. Love to Jane.

(10) 1846 Mar. 6: to John James Audubon on his western journal (ALS and typescript, Charleston Museum)

                                   Charleston March 6th 1846

Dear Audubon,

I have returned from seeing poor Dr [John] Wright. He is evidently no better, although I do not think he has fallen off as much as my friends represented him to have done. His cough has increased a little & his hectic symptoms are still hanging around him, but he keeps up his strength & was able to ride out once. I think however we will have to abandon all hope of his recovery. My poor wife still continues labouring under her old disease with very little mitigation. All the rest are well. I go up on Monday to bring down my Daughter [Harriett Eva] Mrs [William E.] Haskell—after that for a while I shall be stationary.

For the last four nights I have been working at your [western] Journal & not till now am I able to say any thing about its merits.

To me it has been a very interesting one although here is less in it of quadrupeds than I had expected to find. The narratives however are particularly spirited—often amusing & instructive. On what you write on the spot I can depend—but I never trust to peoples memories. I admired the remark of Dr Wright on this head. I wished him to give me an account of the glands of the Skunk—but he answered "I must write for my notes. I cannot depend for these particulars on a fading memory."

The following has struck me as matters rather to be lamented than to be found fault with for it is very probable that you could do no better than was done under the circumstances of your party.

1. Your journal terminates abruptly. Have you nothing more?

2. You do not go far enough. You had only got to the mouth of the Yellow Stone. Did you not go up any farther. If not then you had left a most interesting country unexplored.

3. I cannot find that you ever set a trap or looked at the smaller rodentia. This was a terrible error. The man who collected all the fifty species of rats &c in Beecheys voyage told me that the people told him there was only one species—& he knew not of their existence till he set his traps & caught them.

4. I think you all spent too much time among the Buffaloes & hunting to feed other people.

As for the rest you all had many hardships & certainly seemed to hunt well—although the birds sometimes took you off.

Lewis & Clark however give a much larger account of the Quadrupeds on the Missouri than you have mentioned. I am afraid the broad shadows of the Elk Buffaloe & big horn hid all the little marmot squirrels, Jumping mice rats & Shrews. I wish very much you would write to Culbertson & find out some way of getting—not his princess, brain eating-horse straddling squaw for I suspect neither of us would like to be bothered with such a specimen from the black foot country—but for

1. The Skunk—better than the black foot princess. Lichtenstein figures it as new.

2. Hares in winter colours. You will yet see when they come.

3. The rabbit that led you so many chases. It may be the L[epus] Artemesia—but that species was on the Columbia river—or it may be something quite new. Why man what poor trappers you proved yourselves to be. Do you know what I would have done? Why make a brush fence a foot or 18 inches high in the thick brush wood 100 yards long. Set a snare at every gap ten feet apart & I would have had a couple every night. I never have failed—rabbits—Northern Hare—marsh hare—I care not—in they go & by the neck they hang. The wolf could scarcely have found the small creature.

4. The large red fox—you will I think find that to be a new species & if I am not mistaken the big cross fox for skin is a cross of that same red fellow—I predict this. If you had only brought a hunters skin it would have helped us. I would have given up three Buffaloe hunts & dug out the red fox.

5. You do not say a word of the Mountain goat Capra Americana—but if you did not go farther up the Yellow Stone you would scarcely have heard any thing about it.

6. You speak of a mouse like the M[us] [leu]copus—what is it—did you send it to me? Has it a mark? It may prove a new species. I know nothing of Mus leucopus with a short tail. *Hear me* tell me about this little fellow? In my opinion from your locality if you had brought a dozen species—they all would be new.

It just occurs to me that you will not quarrel with me any more about William Penn & the noble generous race of Indians. Why friend from your description they must be the most blood thirsty, lazy, lying, thieving, murdering, carrion eating, lousy set of rascals that ever disgraced this habitable globe. The wolf & grizzly bear are civilized gentlemen compared to them. By this time I hope you have the manuscripts. Add at once to the grey rabbit that it was found up the Missouri. The papers were gone before your last acct. of habits were received. Is the Pennsylvania Meadow Mouse found west of the Mississippi? Please remember us all to Jane & all your good family.

Yr friend / Jno Bachman

Why not paint the Jaguar & Cougar from living specimens you may get no other. Who has heard from John?

There must have been storms & bad weather at the north as we have had no Mail for 5 days—owing it is said to snow in Virginia. My eyes are well again. I have now only the pouched rat to describe—but don't like the article on the Beaver. I will have to write it over again. There are in your Journal some Buffalo hunts that were first rate. But after all your expenses were greater than the knowledge was worth.

[Addressed to: "J. J. Audubon Esqr / 78 John St / N. York"]

(11) 1847 Sept. 8: to Victor Audubon on the death of Julia Bachman (ALS and typescript, Charleston Museum; partially published, Bachman 1888: 237–238)

Red Sulphur Springs [West Virginia] Sept. 8th 1847

Dear Friends,

The event which we for sometime past were reluctantly & sorrowfully anticipating has at last become a stern reality. Our dear Julia—the object once of our pride & more recently of our sympathy & unwearied watching & care has at last been called to her peaceful rest. She was taken away from us yesterday at 2 O'C. & is to be interred this afternoon.

The day before her death she called me to her bed side & said: "Tell all the Audubons from the oldest to the youngest child farewell. They were all kind to me & I was grateful to them & I now feel that I ought to have been more grateful still for all their indulgence. Tell them not to mourn for me, for though my body is racked with excruciating pains my soul is happy—more happy than I can find words to express. Implore them to bring up their children in the fear & love of God, for if they possess not religion say to them for me that they yet cannot know what true happiness is." I have endeavoured to use her own words but the fervour, the love that shone in her countenance at the moment & the soft & gentle tones of her voice I possess no language to convey.

It appears that for some time before her death she had anticipated the event & sought by the help of God to prepare for it, although she concealed it from us from a fear of giving us pain. When at length she called me to her aid & disclosed her whole thoughts to me I found that she had already advanced far in the Christian life. From that day her hopes became brighter & brighter. She lost every fear of death & said she was waiting in hope for that joyful hour when her Saviour would call her to his blessed arms. I never witnessed so surprising a change of character. [Your] Aunt Maria [Bachman] likened it to inspiration. Her natural reserve & timidity were thrown off. Her judgment was strong. Her mind was clear. There was a purity of thought & a propriety in her words that indicated that we had never sufficiently appreciated as we should have done the powers of her mind. Her whole character had undergone a change & she seemed, although yet on earth to partake of the character of some superior angelic being. I have witnessed the decease of hundreds, yea thousands but in the whole course of my long ministry I have never witnessed so triumphant a death.

She had set apart the last Sabbath for meditation for prayer & the holy communion. Oh it was a comforting day to us all. Monday was devoted to her spiritual duties & sending fond remembrances & kind messages to all her beloved friends. Among the rest she laid the request on me "Father when I am going to my heavenly rest pray at my pillow that I may ascend to my saviour's home on the wings of my earthly Father's prayers." When about to die & we were every moment looking for her last—I thought she had forgotten her request & I wished her to breathe out her life in peace. She suddenly raised up her hands—clasped them in prayer—opened her eyes & Fervently looking up to heaven—said "my time has come Father pray." I commended her soul to God—in the midst of the prayer she ceased to live—& that prayer which was commenced for the dying was concluded in behalf of the poor sorrowful heart broken mourners that knealt around her insensible remains.

She wore the face of an angel even in death but her short spiritual life was one of the brightest & loveliest I have ever been privileged to witness. Her kind Physician who was with her every hour & watched with her many a night on closing her eyes exclaimed "If ever a soul has gone to heaven it is hers."

We are as you may well suppose all worn out with watching anxiety & sorrow. Maria & Lynch [Bachman] send their love to your all, the latter is in bed & I fear will not escape a fit of sickness as the parting between her & Julia was one of the most touching scenes I have ever witnessed. It was I fear too much for the delicate nerves of Lynch. Maria will write to you all when she is a little stronger.

Thank you for the newspapers which you may now discontinue sending. We rest a week at the blue sulphur & then go to perform our duties at home.

Think not that sorrow has unmanned me. No I am at my post of duty. I trust in God & will not repine. My energies will soon be restored & I will go on to perform the duties that are yet enjoyned on me.

Affectionately yr fr'd / Jno Bachman

Write to me at Richmond.

[Addressed to: Victor G. Audubon Esqr. / 78 John Street / New York]

(12) 1847 Dec. 13: to Victor Audubon on nearly losing his sight (typescript; Charleston Museum; partially published, Bachman 1888: 250).

Charleston 13th Decm. 1847

[My dear] Victor,

Maria [Bachman] had promised me for 10 days past to [write to you] all as I was unable to do so. In looking ever over her letter this morning she has I find wholly omitted what I was anxious she should have communicated to you saying she thought I had written it myself. However I am able thank God now to do it for myself. I had but returned from the Synod in Georgia for two days when a sad accident befell me which but for Gods providence might have rendered the rest of my days like Milton's blind & wretched. I had prepared a mixture of Gunpowder, sulphur & lard to bathe a mangy dog—& gave it to Sam our little servant to be carried to the yard. I was intently engaged in writing on my chair with my feet on the [fire] fender. In his wisdom he supposed the fat should be melted & clapped it on the fire about 18 Inches from my nose. An explosion took place like that of a cannon. It was ½ pound of powder. I was knocked over—saved 25 cents in the hair cutting lost my eye lashes & eyelids & lay on my back for 10 days with grated irish potatoe poultices. Nothing but my spectacles (bless them) saved my eyes from total blindness. I was not able till yesterday to use them. I have now a new skin from chin to forehead & my eyes are so much better that you see I have been able to show you the scratch of my pen & have left the dark room & look again on the light of heaven.

18th Decem.

I thought the above was sent to you—but having once more been compelled to go to a dark room for 5 days I only saw this morning that my letter with Marias enclosure had not been sent. I am now quite better save eyelashes & eyebrows—will preach tomorrow I hope. I see as well as ever although I cannot bear a glare of light. This is Saty. Morng. & I am preparing for Sund—have several extra duties—confirmation—communion & c. On Mond. day after tomorrow I am to have a long conversation in my study with [Louis] Agassis. That evening I will write to you in full about all your queries. I am now ready to resume my work although I know not what I will do at night. It is a hard job, but I do [everything about it with] pleasure. I find Agassiz opinion which I prize more than [any other naturalist] in America, most favourable both to the Book & the engrav[ings. He says] it has not its equal in Europe in this department. I know he is sincere [for he is candid.] But alas—alas! we are sadly in want of materials. The [materials are needed] much earlier than I wished—but you know it all. Lynch [Bachman] is only tolerable gains nothing—all the rest well. Lynch & Jane [Bachman] are going to [William E.] Haskells plantation.

Ever Yr frd / Jno Bachman

[Addressed to: "Victor G Audubon Esqr / 78 John St / NY"]

(13) 1848 May 11: to Maria Martin on the senility of Audubon (typescript, Charleston Museum; partially published, Bachman 1888: 255–56).

New York 11th May 1848

My Dear Maria,

The Girls say I have been snoring loud & strong in an arm chair with my feet on the hot fender this chilly evening & I am half inclined to think they were in part right for I feel a little drowsy just now & believe I had best try to shake off lethargy by writing a few lines to you. But how shall I collect my thoughts amid the din & confusion that prevails around me.

The old Gentleman [John James Audubon] has just gone to bed after having eaten his eleventh meal handed his [snuff] box all round kissed all the Ladies & heard his little evening song in French after having seen that John [Woodhouse Audubon] had fed all the dogs. Jenny, Lynch [Bachman] & Wm Rhett are keeping up an incessant chatter — talking of Miss Mac. & her mother & other jokes which I cannot fully understand & Mrs. John [Woodhouse Audubon] as much of a girl as any of them is joining in the fun & such a breeze as they are kicking up would almost drive a steam car. They are popping all manner of questions to me which I pretend not to hear — & then they accuse me of being absent &c. But so it is I like to see these happy faces & hear the merry laugh of these rattling Jovial fellows, & here is John just come in from feeding his dogs & examining his fowl eggs.

All are well here as far as health is concerned. The old Lady [Lucy Bakewell Audubon] is as straight as an arrow — in fine health but much worried — what with her particularity in house keeping — her looking over the advertisements for maids — (plenty of trouble in these changes) & taking care of the poor old Gentleman. His is indeed a most melancholy case. I have often sat down sad & gloomy in witnessing a ruin, that I had seen in other years in order & neatness, but the ruins of a mind once bright & full of imagination is till more melancholy to the observer. The outlines of his countenance & his general robust form are there, but the mind is all in ruins. But why dwell upon it. Imagine to yourself a crabbed restless uncontroulable child — worrying & bothering every one & you have not a tythe of a description of this poor old man. He thinks of nothing but eating — scarcely sits down two minutes at a time — hides hens eggs — rings the bell every five minutes calling the people to dinner & putting the old Lady into all manner of troubles. But I turn away from the subject with a feeling of sadness.

Lucy & Harriet [Audubon grandchildren] have much improved & are really fine girls. About the little Children, Lynch who I believe has been writing to you will no doubt have told you all about them.

The weather has been quite rainy these few days & this afternoon it is quite cold. The Spring is farther advanced than I had expected to find it. The fruit trees are in full bloom & the grass is of dark green. The woods & even the yard here is full of melody with the singing of birds.

There are I think not less than twenty wood robins whose notes can be heard in this vicinity. A red start has built a nest on the cherry tree near the piazza. The Pe-wee has built in an outhouse & the robins have all found a home here. I could spend a month here with great satisfaction if I had nothing to do, but so it is—time passes & I must soon turn my face southward.

Lynch I think is not homesick as yet, indeed I rather think as long as she is with Jenny she will be content. Yesterday Mrs John & the girls went to Town & I scarcely know what are their future plans. On Saty. Morning—day after tomorrow we all go to stay in N.Y. during the meeting of the Synod. I went down on the day after our arrival, but I had not much time to pay any visits. I perceive Hope & King Philip have arrived but I have not seen them. We are I believe to stay at Erbens who I have no doubt will make some fuss & show off some New York humbug. Fri. mor[nin]g Mrs. Aud is going to Town Maid hunting & I will send this by her. I am working on the Quadrupeds.

Tell master Johny [Bachman Haskell] these little folks of all sizes sit & play all day in my room & do not touch the specimens. If the little restless, roaring, tearing dog was here he would make the fur fly as well as the heads & tails. All send love & kisses to Harriet [Eva Bachman], Catty [Catherine Bachman], Boys [William E.] Haskell & c.

Aff yrs / Jno Bachman

[Addressed to: "Miss Maria Martin / care of Dr Bachman / Charleston S.C."]

(14) 1848 Oct. 20: to Victor Audubon on a pet rat (typescript, Charleston Museum; partially published, Bachman 1888: 280–81).

20 October 1848

Dear Victor,

In regard to plate No 30 ["Sigmodon Hispidum.—Say and Ord"; Cotton Rat] you will have once more to be put in a fever—but I cant well help it. Now you will recollect I wrote you about this said Georgia pouched rat two years ago. I cannot give it a name until I am certain that no one has named it before me. I have followed your letter.

1. Pouched rat from Georgia

This animal was described by Mitchel in the New-York Medical repository Jany 1821. You must get & copy the description for me. I will send you by tomorrows steamer a living one. I have had it all summer. It is a gentle & most pleasant companion of mine eating from my hand & looking at & seeming to talk with me. If John cant figure the one you have already (both came from Florida) he must try his hand on this, but dont kill my pet if you can avoid it. I take it out by the tail & hold it in my hand. It has never attempted to bite. You perceive it has a naked tail. If Mitchell & Georgia animal has a short hairy tail then we must give this fellow a new name—Pseudostoma Floridanus Aud & Bach Southern pouched Rat [added to plate 150 as fig. 1, in the folio edition of the *Quadrupeds*]. You have I think specimens both of the Florida & Georgia species. They greatly resemble each other. So do all the species. Go to the Medical Libraries in the city & you will find Barton's description & send me a copy of it.

2. Sorex DeKayi Bach. [DeKay's Shrew] is not the same as S[orex] brevicaudus [Short-tailed Shrew]. It is quite large more than treble the size of S[orex] Brevicaudus.

3. S[orex] longirostris Bach. ["long-nosed Shrew"] is a good species. John once figured it for me—the figure in Trans. Acad. Nat Science Vol. 7 p 2nd.

4. Scalops argentatus Aud & Bach [Silvery Shrew-Mile]—you may figure it.

I suppose the Girls are thinking of returning. We have heard from them a few days ago—when they were still undecided. I find the steamer carries our letters both ways safe—but to day I will once more try the old mail. Please write me particularly what you want.

Last evening I tried my eyes by candle light for the first time. Dr Frost has me in hand. He thinks the eyes have some connexion with the state of the Stomach. He has given me two pills a day for several weeks. I at least fancy that I am better a shade, but am obliged to be Bat-like & avoid the light.

Soon I will once more go to work again on the quadrupeds. Maria [Martin] promises to write while I dictate. You must be patient. You see how I am situated. Your Father began the work before he or I was ready. I have imperative duties. My life is worth something to my children at least. I will aid you all I can but I cannot consent to endanger sight & life to oblige even you, for whom I will do more than for any other human being. When I begin again I know I cannot stop—but you must have pity on me. Whilst you are thinking of yourself, think also of me.

Yr frd. / Jno Bachman

P S Love to all.

(15) 1852 Mar. 13: to Edward Harris on completing the *Quadrupeds* (typescript, Charleston Museum; partially published, Bachman 1888: 276–78).

Charleston March 13th 1852

My Dear Sir

Rejoice with me the book is finished. I did not expect to have lived to have completed it. But Victor Aud[ubon] came on here & I made him hold the pen whilst I dictated with specimens & books before me. We went on rapidly. We worked hard & now we are at the end of our labours. I have at last prevailed on them to give the bats [planned when all mammals were to be included, but omitted from the *Quadrupeds*]. At the end of the work I intend to give a synopsis & scientific arrangement of all our American species including the Seales, whales & porpoises [also omitted from the *Quadrupeds*]. This will be included in the letter press of the 3rd Vol.

Here I will venture to consult you in regard to the publication of additional plates of species not figured in the large work [three folio volumes of fifty plates each]. A very few small Arvicolae & shrews we may not obtain & cannot be figured but nearly all are within our reach. Some of the subscribers have bound up their plates & there cannot be a sufficient number to make even the half of another volume. I propose as all these figures will be contained in the small work, that they should be inserted in the letter press of the large work. So that the subscribers by merely paying for the cost of the [six extra] small plates would have the work complete. What think you of this? What think you of Victors obtaining 129 subscribers in three days nearly, & I trust he will double the number next week. So if the large work will not pay—the small one, & this is large enough, is sure to do it.

But I had almost forgotten the main object of my writing to you. Have you a recollection of a small animal, a spermophile that resembles Says S. lateralis [Marmot-Squirrel] that you brought to me. I took it for that animal although Says description did not exactly suit it. Since then I recd. Says species & on comparing them I find that yours is a new species which I have named Spermophilus Harrisii [included in the *Quadrupeds*]. Now as you have been flying to immortality on the wings of woodpeckers & other birds, you may be unwilling to submit to the slow process of riding thither on the back of a marmot Squirrel. But you must endure it as I was once compelled to do when a shabby fellow in the back country who had never seen me—walked some miles to show me a dirty little Urchin without shoes or stockings hat or a clean face whom he had christened John Bachman. Now what do you know of the history of this little namesake of yours. Where was it procured. Did it live in communities like the rest of its species. I see it has cheek pouches.

Victor & Mrs. B. join me in kind remembrances to you & Mrs Harris. Believe me very truly

Your friend / Jno Bachman

[Addressed to: "Edward Harris / Moorestown, N.J."]

(16) 1871 Sept. 9: to John Bachman Haskell as biographer (handwritten copy and typescript, Charleston Museum).

Charleston Sep 9 / [18]71

My Dear [Grand]Son,

I am inclined to think that my letters have not received the attention that I expected. I suspect that some accident has caused my letters to miscarry inasmuch as you give us so many promises of punctuality. I have an idea that some pretty girl has turned the head of my boy—who may have some of the frailties of his grand-father. However we will try to be patient for another week, & will hope that in time a letter may make its appearance. Let me advise you, if you wish to keep in the good graces of the home folks—old & young to keep them well posted in all that interests you.

My health is still very feeble. I am thankful to a kind Providence that he has spared me so long & when the summons comes I hope to be enabled to say with one of old, "Now, Lord lettest thou thy servant depart in peace." I look to the atoning blood of my Savior for pardon & salvation. And if God will save the soul, surely we can trust in his mercy for all the rest. Remember me affectionately to all the descendants of the Audubons. I was overlooked by the biographer [of John James Audubon]; my labors connected with the quadrupeds, were not even alluded to. I think they deserve some mention, inasmuch as they cost me much labor by night & by day. You may be required to take up the cudgel in my defence when I am gone. I wrote every line which composed the latter volumes of the Quadrupeds. I have selected you as my biographer, & I know that you will discharge the duty faithfully. I mean to have said something with regard to your studies with Dr. Rude, but I will only say now—rest a little, recreate a little & you will be better able afterwards to carry on your studies successfully. Yet you must not forget that life is very short—& we have much to do in it. Recollect Rev. Thad. Boinest, buried a few days since, after a brief illness—'& what thy hand findeth to do, do it with all thy might, for there is no work or device in the grave whither thou art hastening." But John I did not intend to moralize. Still these thoughts under every circumstance come into my mind for death is always near us, & it should be our endeavor at all times to be prepared to meet it.

Yr affectionate G-father
Jno Bachman

Bibliography

Audubon, John James

1831–1839. *Ornithological Biography, or an Account of the Habits of the Birds of the United States of America; Accompanied by Descriptions of the Objects Represented in the Work Entitled the Birds of America, and Interspersed with Delineations of American Scenery and Manners.* 5 vols. Edinburgh: Adam Black et al.

1979. *Complete Audubon: A Precise Replica of the Complete Works of John James Audubon Comprising the Birds of America (1840–44) and the Quadrupeds of North America (1851–54) in Their Entirety.* 10 vols. in 5. Kent: Published for the National Audubon Society by Volair Books.

1989. *Audubon's Quadrupeds of North America: Complete and Unabridged.* Secaucus, N.J.: Wellfleet Press.

Audubon, John James, and John Bachman

1845, 1846, 1848. *Viviparous Quadrupeds of North America.* 3 vols. New York: J. J. Audubon. The first edition of the separate, folio volumes of plates. The dates are from Cahalane in Audubon 1967: xv; a copy in the College of Charleston is bound in two volumes, and both volumes have title pages dated 1845.

1846, 1851, 1854. *Viviparous Quadrupeds of North America.* 3 vols. New York: vol. 1: J. J. Audubon; vols. 2 and 3: V. G. Audubon. First octavo edition of the text without illustrations. Not seen; citation from Herrick 1938: vol. 2, 406.

1849, 1851, 1854. *Quadrupeds of North America.* 3 vols. New York: V. G. Audubon. First illustrated edition; dated from the facsimiles of title pages in Audubon 1979.

1974. *Quadrupeds of North America.* 3 vols. New York: Arno Press. Not seen; cited on WorldCat as a reprint of the 1854 edition.

Bachman, John (N.B.: numbers in brackets refer to the more complete, annotated bibliography in Waddell 2005.)

1831A. *The Funeral Discourse of the Rev. J. G. Schwartz. Delivered September 11, 1831.* Charleston: James S. Burges. 23 pp. [1]

1831B. "Mr. Audubon." *City Gazette and Commercial Daily Advertiser* 18 Oct. 1831. Unsigned; attributed to Bachman by Shuler 1995: 7. [2]

1832A. "This Distinguished Naturalist. . . ." *The Courier* 6 Jun. 1832. Signed "C" (probably for Curtius, a Bachman pseudonym); attributed to Bachman by Shuler. Photocopy in the Shuler Papers, College of Charleston. [3]

1833A1. "A Successful Method of Raising Ducks; by Experimenter." Charleston, February 1832. *Southern Agriculturalist, and Register of Rural Affairs; Adapted to the Southern Section of the United States* 6 (Mar. 1833): 130–37. The Library of Congress copy is annotated "By Revd. Mr. Bachman"; microfilm edition of the journal. [4]

1833A2. "A Successful Method of Raising Ducks by 'Experimenter.'" Charleston, February 1833." *Farmer's Register* 1 (1833): 356–59. Subtitled "From the Southern Agriculturist."

1833B1. *An Address Delivered before the Horticultural Society of Charleston at the Anniversary Meeting, July 10th, 1833.* Published by request of the Society. Charleston: A. E. Miller. 30 pp. [5]

1833B2. "An Address Delivered before the Horticultural Society of Charleston at the Anniversary Meeting, July 10th, 1833." *Southern Agriculturalist, and Register of Rural Affairs; Adapted to the Southern Section of the United States* 6 (Aug. 1833): 393–410 and (Sept. 1833): 449–60. [5a]

1834A1. *An Account of Some Experiments Made on the Habits of the Vultures Inhabiting Carolina, the Turkey Buzzard, and the Carion Crow, Particularly as It Regards the Extraordinary Powers of Smelling, Usually Attributed to Them.* Charleston: John Bachman. 16 pp. Title from offprint; publisher and place cited by Shuler, 1995: 220. [6]

1834A2. "Retrospective Criticism. Remarks in Defence of ["Mr. Audubon"] the Author of 'the ["Biography of the"] Birds of America'", by the Rev. John Bachman, Charleston, South Carolina. *Magazine of Natural History and Journal of Zoology, Botany, Mineralogy, Geology and Meteorology* 7 (1834): 164–75. [8]

1834A3. "Remarks in Defence of the Author of the 'Birds of America.'" By the Rev. John Bachman, Charleston, South Carolina. Read before the Boston Society of Natural History, Feb. 5, 1834. *Boston Society of Natural History*, 1, no. 1 (1834): 15–31. [7]

1834A4. "The Vulture." *Scientific Tracts and Family Lyceum*, n.s., 1, no. 6 (15 Mar. 1834): 165–76. [8]

1834B1. *Catalogue of Phænogamous Plants and Ferns, Native or Naturalized, Found Growing in the Vicinity of Charleston, South-Carolina*. Charleston: A. E. Miller. 15 pp. [9]

1834B2. "Catalogue of Phænogamous Plants, and Ferns Native or Naturalized Found Growing in the Vicinity of Charleston, (S.C.)." *Southern Agriculturist and Register of Rural Affairs; Adapted to the Southern Section of the United States* 8 (Apr. 1835): 189–96 and (Jun. 1835): 286–91. [9a]

1835. "Defense of Audubon." *Bucks County Intelligencer*. Cited with two paragraphs quoted in Bachman 1888: 98, 99.

1836A1. "On the Migration of the Birds of North America. Read before the Literary and Philosophical Society of Charleston, (S.C.), March 15th, 1833; by Rev. J. Bachman." [Silliman's] *American Journal of Science and Arts* 30 (Jul. 1836): 81–100. [10]

1836A2. Sur les Migrations des Oiseau de l'Amérique du Nord. *Bibliothèque Universelle de Genève*, n.s., 7 (1837), pp. 204–8. Cited by Bost 1963: 525. Translation of the 1836 English version listed as 1836A1. [17]

1836B1. "On the Habits of Insects." *Southern Literary Journal* 2 (Aug. 1836): 409–27. Partly reprinted in 1845 as 1836B2 and 1858 as 1836B3. [11]

1836B2. "The Morals of Entomology." In [William Gilmore Simms, ed.], *The Charleston Book: a Miscellany in Prose and Verse* (1845), pp. 30–40. Charleston: Samuel Hart, Sen. [35]

1836B3. "The Insect World. Morals of Entomology, etc." [De Bow's] *Commercial Review of the South and West*, 2nd n.s., 1, no. 4 (Oct. 1858): 430–35. [63]

1837A2. "Description of a New Species of Hare Found in South Carolina." *Journal of the Academy of Natural Sciences of Philadelphia*, vol. 7, part 2 (1837): 194–99 and pls. 15 and 16. [12]

1837B. "Observations, on the Different Species of Hares (Genus *Lepus*) Inhabiting the United States and Canada." *Journal of the Academy of Natural Sciences of Philadelphia*, vol. 7, part 2 (1837): 282–361 and pls. 21 and 22.

1837C1. "Some Remarks on the Genus *Sorex* [Shrews], with Monograph of the North American Species." *Journal of the Academy of Natural Sciences of Philadelphia*, vol. 7, part 2 (1837): 362–402 and pls. 23 and 24. [14]

1837C2. [Some Remarks of the Genus *Sorex*, with a Monograph of the North American Species]. *Proceedings of the Boston Society of Natural History*, vol. 1 (1841–44), pp. 40–41. Cited by Bost 1963: 526.

1837D. "Additional Note on the Genus *Lepus*." *Journal of the Academy of Natural Sciences of Philadelphia*, vol. 7, part 2 (1837): 403. [15]

1837E1. *A Sermon on the Doctrine and Discipline of the Evangelical Lutheran Church, Preached at Charleston, S.C. November 12th, 1837, by Appointment of the Synod of South-Carolina, and Adjacent States*. Charleston: J. S. Burges. 37 pp. [16]

1837E2. "A Sermon on the Doctrines and Discipline of the Evangelical Lutheran Church, Preached at Charleston, S.C., November 12th, 1837, by Appointment of the Synod of South Carolina and Adjacent States." Reprinted within an article entitled "Letters on the Catholic Doctrine of Transubstantiation, and on Protestant Errors Concerning the Holy Eucharist addressed to Rev. John Bachman, D.D." *Works of the Right Rev. John England, First Bishop of Charleston*. Edited by the Right Rev. Ignatius Aloysius Reynolds (Baltimore: John Murphy & Co., 1849; 5 vols.), vol. 1, pp. 347–474 (with the sermon reprinted on pp. 348–58). [16a]

1838A1. "Monograph of the Species of Squirrel Inhabiting North America." *Proceedings of the Zoological Society of London*, part 4 (1836): 85–130. Read 14 Aug. 1838. London: R. and J. E. Taylor. Bachman described fourteen species of North American squirrels, five of which were new species, and he referred to three other species. [18]

1838A2. "Monograph of the Genus *Sciurus*, with Descriptions of New Species and Their Varieties, as Existing in North America." *Magazine of Natural History, and Journal of Zoology, Botany, Mineralogy, Geology, and Meteorology*, 2nd ser., 3 (1839): 113–23, 154–62, 220–27, 330–37, 378–90. [18a]

1838A3. "Abstract of a Monograph of the Genus *Sciurus*, with Descriptions of Several New Species and Varieties." [Silliman's] *American Journal of Science and Arts* 37, no. 2 [Oct. 1839]: 290–310. [23]

1838A4. [Abstract of a monograph of the genus *Sciurus*, with descriptions of several new species and varieties.] *Isis, oder Encyclopädische Zeitung; von Oken* (1845), col. 376–79. Cited by Bost 1963: 525. [80]

1838B. "Observations on the Changes of Colour in Birds and Quadrupeds." *Transactions of the American Philosophical Society*, n.s., 6 (1838), article 4: 197–239. [19]

1838C. "Foreign Correspondence. Extracts of a Letter from a Lutheran Minister of the United States, Now in Europe, to the Editor. London, 26th October, 1838 (to the Rev. B. Kurtz) [and] Extract of a Letter from a Lutheran Minister of the United States, Now in Europe, to the Rev. J. G. Morris, of this City [Baltimore]." London, 26 Oct. 1838. *Lutheran Observer*, n.s., 6, no. 15 (30 Nov.

1838). Two letters by Bachman published together as one article in a newspaper. [20]

1839A. "Description of Several New Species of American Quadrupeds." *Journal of the Academy of Natural Sciences of Philadelphia* 8, part 1 (1839): 57–74.

1839B. "Additional Remarks on the Genus *Lepus*, with Corrections of a Former Paper, and Descriptions of Other Species of Quadrupeds Found in North America." *Journal of the Academy of Natural Sciences of Philadelphia* 8, part 1 (1839): 75–105.

1840A. "Notes on European Agriculture by a Charlestonian." *The Southern Cabinet* 1 (1840), no. 1 (Jan.), pp. 1–7; no. 2 (Feb.), pp. 65–68; no. 3 (Mar.), pp. 129–32; no. 4 (Apr.), pp. 193–96; no. 5 (May), pp. 257–64; no. 6 (Jun.), pp. 321–23; no. 7 (Jul.), pp. 385–89. Unsigned, but the attribution is certain from internal evidence. [24]

1840B. "The Birds of America.—From Drawings Made in the United States. By J. J. Audubon, F. R. S. J. P. Beile, Agent." *The Southern Cabinet* 1, no. 1 (Jan. 1840): 57. Unsigned review, but the attribution is all but certain from internal evidence. [25]

1840C. "North-American Herpetology; or a Description of the Reptiles Inhabiting the United States. By John Edwards Holbrook, M.D." *The Southern Cabinet* 1, no. 1 (Jan. 1840): 58–59. Unsigned review, but the attribution is all but certain from internal evidence. [26]

1840D. "On the Cultivation of the Fig Tree in Carolina." *The Southern Cabinet* 1, no. 10 (Oct. 1840): 602–3. [27]

1840E. "A Hurried Visit to Newberry District by 'a Charlestonian.'" *The Southern Cabinet* 1 (1840), no. 11 (Nov.), pp. 640–47; no. 12 (Dec.), pp. 716–24. Unsigned, but the attribution is certain from internal evidence. [28]

1842A1. "Descriptions of New Species of Quadrupeds Inhabiting North America." Coauthor: John James Audubon. *Journal of the Academy of Natural Sciences of Philadelphia* 8, part 2 (1842): 280–323. [29]

1842A2. "Descriptions of new species of quadrupeds inhabiting North America (coauthor: John James Audubon)." *Proceedings of the Academy of Natural Sciences of Philadelphia*, ser. 1, no. 8 (1843): 92–103.

1842B2. "An Address, Delivered before the Washington Total Abstinence Society of Charleston, S.C., on Wednesday Evening, July 27th, 1842." 33 pp. Charleston: Burges & James. Cited in the *National Union Catalog Pre-1956 Imprints* as only in the Library of Congress; missing from its assigned location when requested in 2004 and 2009. [78]

1842C? "Sermon against Dueling, about 1842." Cited in Bachman 1888: 435, but otherwise unknown. [79]

1842–1843A. "Observations on the Genus *Scalops*, (Shrew Moles), with Descriptions of the Species Found in North America." *Journal of the Boston Society of Natural History* 4, no. 1 (1842–43): 26–35.

1842–1843B. [Observations on the Genus *Scalops*, (Shrew Moles), with Descriptions of the Species Found in North America]. *Proceedings of the Boston Society of Natural History*, vol. 1 (1841–44), 40–41 (cited by Bost 1963: 526). [77]

1843A1. *An Inquiry into the Nature and Benefits of an Agricultural Survey of the State of South-Carolina.* Charleston: Miller & Browne. 42 pp. [31]

1843A2. "An Inquiry into the Nature and Benefits of an Agricultural Survey of the State of South-Carolina." *Southern Agriculturist*, n.s., 3 (Feb. 1843): 49–65 and (Mar. 1843): 81–96. [31a]

1843A3. "An Inquiry into the Nature and Benefits of an Agricultural Survey of the State of South-Carolina." *Southern Quarterly Review* 3, no. 6 (Apr 1843): 449–67. [31b]

1843B. "Agricultural Labor, as One of the First Conditions of a National Existence." *Magnolia, or Southern Appalachian*, n.s., 2 (1843): 16–20. "From an Agricultural Oration, Delivered in South Carolina in 1840." [32]

1843C. [Mermaid Hoax] by "No Humbug." *Charleston Mercury*, 20 Jan., 26 Jan., 1 Feb., and 7 Feb. Stephens (1983: 45, 46, 48, and 54) provides definite evidence indicating that Bachman was the author. [33]

1845 "Unity of the Races" by "C." (attributed to Bachman). *Southern Quarterly Review* 14: 372–448.

1845–48. *The Viviparous Quadrupeds of North America* [folio plates without text]. Coauthor: John James Audubon. 3 vols. New York: J. J. Audubon. The 150 plates in these three folio volumes were issued serially in thirty parts from 1842 to 1848 and were dated individually. The text was printed in separate octavo volumes listed below as 1846–54. [36]

1846–54. *The Viviparous Quadrupeds of North America* [text without plates]. Coauthor: John James Audubon. 3 vols. Vol. 1: New York: J. J. Audubon, 1846 (vol. 1 alone was also published in London by Wiley & Putnam in 1847); vols. 2 and 3: New York: V. G. Audubon, 1851 and 1854 respectively. Largely unillustrated text issued as octavo volumes to accompany the separately published folio plates listed above as 1845–48. The text and reduced plates were reproduced together from 1849 to 1854. [37]

1847. *Correspondence between the Rev. Messrs. Dana and Smyth, through the*

Mediation of the Hon. R. B. Gilchrist, and the Rev. Dr. Bachman. Charleston: T. W. Haynes. 10 pp. [38]

1848A1. "Notes on the Generation of the Virginia Opossum (*Didelphis Virginiana*)." *Proceedings of the Academy of Natural Sciences of Philadelphia* 4 (Apr. 1848): 40–47. It incorporates "Further Observations on the Generation of Opossums" (pp. 42–46) and a letter from Michel Middleton, M.D., to Bachman (pp. 46 and 47). [39]

1848A2. [Notes on the Generation of the Virginia Opossum (*Didelphis Virginiana*).] *Archiv für Naturgeschichte*, 17 (1851), pp. 161–74. Cited by Bost 1963: 525. [81]

1848B. *The Design and Duties of the Christian Ministry, Preached at a Meeting of the General Synod of the Evangelical Lutheran Church in the United States, May 14th, 1848, at New York*. Baltimore: Printed at the Publication Rooms. 23 pp. [40]

1849A. "Mammalia, or Animals Which Suckle Their Young." In George White, *Statistics of the State of Georgia: including an account of its natural, civil, and ecclesiastical history; together with a particular description of each county, notices of the manners and customs of its aboriginal tribes, and a correct map of the State*, appendix, pp. 3–5. Savannah: W. Thorne Williams. Bachman listed sixty species of mammals, fourteen of which he first described. [41]

1849B1. Minnesota Rice. *Charleston Mercury*, 17 Jan. 1849. Letter to the Editor. [42]

1849B2. "Minnesota Rice." In R. F. W. Allston, *Essay on Sea Coast Crops; Read Before the Agricultural Association of the Planting States, on Occasion of the Annual Meeting, held at Columbia, the Capitol of South-Carolina, December 3d, 1853*, pp. 45–46. Charleston: A. E. Miller. [42a]

1849–54. *Quadrupeds of North America* [text with reduced plates]. Coauthor: John James Audubon. 3 vols. New York: V. G. Audubon. This is the first illustrated octavo edition, and its three volumes were issued serially and then bound with title pages dated 1849, 1851, and 1854. [43]

1850A. *The Doctrine of the Unity of the Human Race Examined on the Principles of Science*. Charleston: C. Canning. 313 pp. Bost (1963: 434, n. 135) cites a reference to a second edition of 1856, but no other edition seems to have been published until 1876 (posthumous). [45]

1850B. "An Investigation of the Cases of Hybridity in Animals on Record, Considered in Reference to the Unity of the Human Species." *Charleston Medical Journal and Review* 5, no. 2 (Mar. 1850): 168–97. [44]

1850C. "For the *Courier* by J. B." *Charleston Daily Courier* 48 (28 May 1850). [46]

1850D. "A Reply to the Letter of Samuel George Morton, M. D., on the Question of Hybridity in Animals Considered in Reference to the Unity of the Human Species." *Charleston Medical Journal and Review* 5, no. 4 (May 1850): 466–508. [47]

1850E. "Second Letter to Samuel G. Morton on the Question of Hybridity in Animals, Considered in Reference to the Unity of Human Species." *Charleston Medical Journal and Review* 5, no. 5 (Sept. 1850): 621–60. [48]

1851A. "Additional Observations on Hybridity in Animals, and on Some Collateral Subjects." *Charleston Medical Journal and Review* 6, no. 3 (May 1851): 383–96. [49]

1851B. "Letter from Rev. John Bachman, D.D." *Charleston Medical Journal and Review* 6 (1851): 598. [50]

c. 1851–54. [Stories on natural history.] *Sunday School Visitor*. Cited in Bachman 1888: 325; no copy located. [82]

1852. "Description of a New North American Fox, Genus *Vulpes*, Cuv." Coauthor: John James Audubon. *Proceedings of the Academy of Natural Sciences of Philadelphia* 6 (1852–53): 114–16. The Large Red Fox was named *Vulpes Utah*. [51]

1853A. *A Defense of Luther and the Reformation: Against the Charges of John Bellinger, M.D., and Others; To Which are Appended Various Communications of Other Protestant and Roman Catholic Writers Who Engaged in the Controversy*. Charleston: William Y. Paxton. [52]

1853B. "Original. For the Lutheran Observer." *Lutheran Observer* 21, no. 17 (22 Apr. 1853). A letter addressed to "Br. Kurtz." [53]

1854A "Essay on the Connection of the Natural Sciences with Agriculture." Read before the Agricultural Association of the Planting States, at Columbia, December 1853. *New York Daily Times*, 1 Jun. 1854.

1854B. "Types of Mankind." *Charleston Medical Journal and Review* 9, no. 5 (Sept. 1854): 627–59. [54]

1855A. *Continuation of the Review of "Nott and Gliddon's Types of Mankind."* Charleston: James, Williams and Gitsinger. 19 pp. [55a]

1855B1. "An Examination of the Characteristics of Genera and Species as Applicable to the Doctrine of the Unity of the Human Race." *Charleston Medical Journal and Review* 10, no. 2 (Mar. 1855): 201–22. [56]

1855B2. *An Examination of the Characteristics of Genera and Species as Applicable to the Doctrine of the Unity of the Human Race*. Charleston: James, Williams & Gitsinger. 24 pp. [56a]

1855C1. "An Examination of Prof. Agassiz's Sketch of the Natural Provinces of the Animal World, and Their Relation to the Different Types of Man, with a Tableau Accompanying the Sketch." *Charleston Medical Journal and Review* 10, no. 4 (Jul. 1855): 482–534. [57]

1855C2. *An Examination of Professor Agassiz's Sketch of the Natural Provinces of the Animal World and Their Relation to the Different Types of Man, with a Tableau Accompanying the Sketch.* 54 pp. Charleston: James, Williams & Gitsinger. [57a]

1855–56. "A Chapter on Fish—Fish Ponds and Artificial Fish Breeding." Read before the State Agricultural Society of South Carolina, at Columbia, 1855. *Southern Cultivator* 13 (Dec. 1855): 362–66 and 14 (Jan. 1856): 10–16. [58]

1857A. *An Address on Education, Delivered on the Day of the Laying of the Corner-stone of Newberry College, July 15, 1857.* Charleston: James & Williams. 22 pp. [59]

1857B. "Strictures on Resolutions of the Middle Conference." *Missionary* n.s., 2, no. 46 (10 Dec. 1857). Letter to the editor dated 24 Nov. 1857. [60]

1858A1. *Report on Asiatic Goats.* Accepted and published by the Society [Southern Central Agricultural Association of Georgia], at their annual meeting, October 1857. pp. 1–16. Cited from an offprint; publisher and place uncertain. [61]

1858A2. [Report on Asiatic Goats.] "U.S. Patent office. Report. Agriculture. 1857, pp. 56–66." Washington, 1858. Cited in *National Union Catalogue Pre-1956 Imprints.* [61a]

1858B. *A Discourse, Delivered on the Forty-third Anniversary of His Ministry in Charleston.* Published by the request of the Vestry. Charleston: A. J. Burke. 18 pp. [62]

1858C? [Biographical sketch.] "In 1858, he wrote a sketch of his life for a scientific journal in Europe." Part of this sketch is quoted in Bachman 1888: 9, and it was written in English. The journal was not identified. [83]

c. 1860–62. [Coeditor of the *Southern Lutheran.*] According to Morris, Bachman was "Editor. 1861" (1876: 22). According to Stephens (1999: 832), he was coeditor from 1860 to 1862. Not seen. [84]

1861A. "A Sermon by the Rev. John Bachman, D.D., on the Day of Fasting, Humiliation and Prayer, in the War with the Northern United States, June 13, 1861." *Charleston Daily Courier*, 15 Jun. 1861, p. 4. [64]

1861B. "A Reply to the Attack of the Rev. Benjamin Kurtz, D.D. Editor of the Lutheran Observer ["concluded"] by 'J. B.'" *Lutheran Observer* 16 Nov. 1861. [65]

1862A. "The Duty of the Planter to His Family, to Society and His Country by Curtius." *Charleston Daily Courier* 60 (25 and 27 Mar. 1862). Bachman acknowledged having written as Curtius in the article listed below as 1862B. [66]

1862B. "A Few Words of Advice to Planters, Farmers, and Manufacturers, with the Opinions of Patriotic and Good Men, and the Denunciations of Scripture on Monopolizers, Forestallers, and Other Extortioners." *Charleston Daily Courier* 60 (23, 24, 25, and 30 Apr. 1862). [67]

1868. *Vindication of Rev. Dr. John Bachman, of Charleston, S.C., in Answer to Rev. E. W. Hutter: in Regard to an Article Published in the "Lutheran and Missionary," of the 27th of July, 1865.* Published by a Personal Friend. [68]

c. 1868. "The Japan Clover—*Lespedera striata*." *Charleston Courier*. Undated clipping with Jan. 1868 added to the reverse; Bachman Papers, Charleston Museum. [69]

1869A. "The Humboldt Festival." *Charleston Daily Courier*, 15 Sept. 1869. [70]

1869B. "Charge to the Associate Pastor, by Rev. John Bachman, D.D., at the Installation of Rev. W. W. Hicks, on Sunday, December 5, 1869, at St. John's Evangelical Lutheran Church, Archdale-Street." *Charleston Daily Courier* 67 (9 Dec. 1869). [71]

1869C. "Circular." Undated pamphlet, 8 pp. Bachman refers to having been with St. John's for fifty-four years. [72]

1870. "Fifty-fifth Anniversary Sermon by Rev. Dr. Bachman, St. John's Lutheran Church." *Charleston Courier*, 10 Jan. 1870. [73]

1874. "Letter to J[ared] P. Kirtland...." *Cleveland Academy of Natural Science, Proceedings . . . 1845–1859* (1874): 194–96.

1888. *John Bachman, D.D., LL.D., Ph.D.; the Pastor of St. John's Lutheran Church, Charleston*. Edited by John Bachman Haskell and C. L. Bachman. Charleston: Walker, Evans & Coggswell.

1925. "[Letter to Dr. Edmund Ravenel, 1832]. "Three Letters from the Ford and Ravenel Papers." *South Carolina Historical and Genealogical Magazine* 26 (1925): 145–49.

1929. "Some Letters of Bachman to Audubon" edited by Ruthven Deane. *Auk: A Quarterly Journal of Ornithology* 46, no. 2 (Apr. 1929): 177–85.

1962. "A Letter from Bachman, London, 1838." *South Carolina Historical Magazine* 63 (1962): 211–13.

n.d. "Address of Rev. Dr. Bachman Before the Ladies' Association to Commemorate the Confederate Dead." Cited by Bost 1963: 519. Undated newspaper clipping in the Bachman Room, St. John's Lutheran Church; not found. [85]

n.d. [Letter by Bachman to Audubon]. *Caledonian Mercury* no. 191. Cited in

Shuler Papers 4/1. This tri-weekly newspaper was published in Edinburgh from 1720 to 1867. Presumably the letter dates from 1838, when both Bachman and Audubon were in Scotland. [76]

n.d. "Rules of the Lutheran Society for Promoting Religion in South Carolina and Georgia." Cited by Morris 1876: 11. [86]

n.d. "Superstitions Concerning Insects." Undated clipping in the Charleston Museum; Natural History Manuscript Collection, box 2. A short note primarily about the Death's Head Moth. [74]

Baird, Spencer F.
1859. *Mammals of North America; the Descriptions of Species Based Chiefly on the Collections in the Museum of the Smithsonian Institution*. Philadelphia: J. B. Lippincott & Co.

Barrett, Paul H., and R. B. Freeman (eds.)
1987–89. Works of Charles Darwin. 29 vols. New York: New York University Press.

Barrett, Paul H, Peter J. Gautrey, Sandra Herbert, David Kohn, and Sidney Smith (eds.).
1987. *Charles Darwin's Notebooks, 1836–1844: Geology, Transmutation of Species, Metaphysical Enquiries*. Ithaca, N.Y.: British Museum (Natural History) and Cornell University Press.

Bernasconi, Robert (ed.)
2002. *American Theories of Polygenesis*. 7 vols. Bristol: Theommes Press.

Boehme, S. E. (ed.)
2000. John James Audubon in the West: The Last Expedition; Mammals of North America. New York: Harry N. Abrams, Inc., in association with the Buffalo Bill Historical Center.

Bost, Raymond Morris
1963. "The Reverend John Bachman and the Development of Southern Lutheranism." Dissertation, Yale University, New Haven. 549 pp. With fifty-four publications by Bachman cited in alphabetical order by title and with several others noted in the text. Contains a carefully researched summary of Bachman's scientific research and writings, pp. 425–91. Includes the transcription of

a manuscript prayer by Bachman for the Secession Convention, 20 Dec. 1860; collection of the Charleston Museum, pp. 514–15.

Cahalane, Victor H. (ed.)
  1967. *Imperial Collections of Audubon Animals: The Quadrupeds of North America.* Edited with new text by Victor H. Cahalane; foreword by Fairfield Osborn; illustrated by John James Audubon and John Woodhouse Audubon. Maplewood, N.J.: Hammond Inc. Bachman is credited on the title page.

Darwin, Charles
  1875. *Variation of Animals and Plants Under Domestication.* 2nd ed. London: John Murray. (Vols. 19 and 20 in Barrett and Freeman 1988).
  1877. *Descent of Man, in Relation to Sex.* 2nd ed. London: John Murray. (Vols. 21 and 22 in Barrett and Freeman 1989).
  1958. *Autobiography of Charles Darwin, 1809–1882, with Original Omissions Restored.* Edited by Nora Barlow. London: Collins. (In vol. 29 of Barrett and Freeman 1989.)
  1975. *Charles Darwin's Natural Selection, Being the Second Part of His Big Species Book Written from 1856 to 1858.* Edited by Robert C. Stauffer. Cambridge: Cambridge University Press.

Davidson, James Wood
  1869. "John Bachman, Ph.D., D.D., L.L.D." *Living Writers of the South*, pp. 23–25. New York: Carleton, Publisher. A typescript of this biographical sketch in the Bachman Collection at the Charleston Museum has corrections that appear to be in Bachman's handwriting.

Davis, Jefferson
  1881. *The Rise and Fall of the Confederate Government.* 2 vols. New York: D. Appleton and Company.

Elliott, Stephen
  1821. *Sketch of the Botany of South-Carolina and Georgia.* 2 vols. Charleston: J. R. Schenck.

Ficken, Col. John F.
  1924. *A Sketch of the Life and Labors of John Bachman, D.D., LL.D.* Charleston: Brotherhood of St. Johns Lutheran Church. 12 pp.

Godman, John D.

1826–28. *American Natural History*. 3 vols. Philadelphia: H. C. Carey and I. Lea.

1831. *American Natural History: Part 1.—Mastology*. 3 vols. Philadelphia: Key and Mielkie.

Happoldt, Christopher

1960. *The Christopher Happoldt Journal: His European Tour with the Rev. John Bachman (June–December, 1838)*. Edited with introductory material by Claude Henry Neuffer. Contributions from the Charleston Museum, vol. 13. Charleston: Charleston Museum. Includes lengthy biographical sketches by Neuffer of Bachman as well as Happoldt. Although in private possession when published, the original two-volume journal has since been acquired by the Southern Historical Collection, University of North Carolina, Chapel Hill.

Harlan, Richard

1825. *Fauna Americana: Being a Description of the Mammiferous Animals Inhabiting North America*. Philadelphia: A. Finley.

[Haskell, John B.]

1874A "John Bachman: The Death of the Distinguished Naturalist and Divine; The Story of His Life and Labors." *Charleston Courier*, 25 Feb. 1874. Detailed obituary attributed to Haskell by Shuler 1995: 221. Lists fourteen publications; refers to another specifically, and mentions others; largely the same list published in 1888.

Herrick, Francis Hobart

1938. *Audubon the Naturalist: A History of His Life and Time*. 2nd ed. 2 vols. New York: D. Appleton-Century Co., Inc. Reprinted, New York: Dover Publications, 1968.

Holbrook, John Edwards

1842. *North American Herpetology; or, a Description of the Reptiles Inhabiting the United States*. 5 vols. Philadelphia: J. Dobson.

1976. *North American Herpetology*. Lawrence, Kans.: Society for the Study of Amphibians and Reptiles.

Holifield, E. Brooks
1978. *The Gentlemen Theologians: American Theology in Southern Culture, 1795-1860.* Durham, N.C.: Duke University Press.

Horn, E. T.
1884. "St. John's Evangelical Lutheran Church." *Yearbook—1884. City of Charleston, So. Ca.*, pp. 262-79. Charleston: News and Courier Book Presses. Horn wrote that the vestry of St. John's asked Rev. Dr. Philip F. Mayer of Philadelphia to recommend a preacher, and Mayer recommended Bachman.

Lyell, Charles
1845. *Travels in North America in the Years 1841-2; with Geological Observations on the United States, Canada, and Nova Scotia.* New York: Wiley and Putnam.
1850. *Second Visit to the United States of North America.* New York: Harper & Brothers.

Martin, B[enjamin] N[icolas].
1874. A Review of His [Bachman's] Life and Labors. *Proceedings of the Lyceum of Natural History in the City of New York*, 2nd ser., pp. 117-18.

Mazyck, William G.
1908. "Charleston Museum, Its Genesis and Development." *Yearbook 1907.* Charleston: City of Charleston.

Nott, J. C., and George R. Gliddon (eds.)
1854. *Types of Mankind: or, Ethnological Researches, Based Upon the Ancient Monuments, Paintings, Sculptures, and Crania of Races, and Upon Their Natural Geographical, Philological, and Biblical History; Illustrated by Selections from the Inedited Papers of Samuel George Morton, M.D.* Edited by J. C. Nott and George R. Gliddon. Philadelphia: Lippincott, Grambo & Co.

Peattie, Donald C.
1928. "Bachman, John." *Dictionary of American Biography.* Edited by Allen Johnson, vol. 1, pp. 466-67. New York: Charles Scribner's Sons.

Peck, R. M.
2000. "Audubon and Bachman: A Collaboration in Science." In Boehme 2000: 71–115.

Porcher, Frederick Adolphus
1946. [Notes on Bachman]. "Memoirs of Frederick Adolphus Porcher," edited by Samuel Gaillard Stoney. *South Carolina Historical and Genealogical Magazine* 47 (1946): 151, 218, and 219. Porcher was a fellow professor at the College of Charleston. He briefly but incisively discusses Bachman's teaching as professor of natural history, his personality, and some of his most strongly held opinions.

Richardson, John
1829. *Fauna Boreali-Americana; or the Zoology of the Northern Parts of British America: Containing Descriptions of the Objects of Natural History Collected on the Late Northern Land Expeditions, under Command of Captain Sir John Franklin, R.N.* Assisted by William Swainson and William Kirby. London: John Murray. In the Quadrupeds, a citation to Bachman was intended to be to Richardson: "Bach. Fauna Bor. Am., vol. i, p. 45." This reference is part of Bachman's description of *Putorius Pusillus* (1989: 200), and "vol. i, p. 45" refers to the entry in the *Fauna Boreali-Americana* for "Mustela (Putorius) Vulgaris. (Lin.). "Bach." is probably a mistake made by Victor Audubon, who edited proofs that Bachman did not see.

Sanders, Albert E., and William D. Anderson, Jr.
1999. *Natural History Investigations in South Carolina from Colonial Times to the Present.* Columbia: University of South Carolina.

Sanders, Albert E., and Warren Ripley
1985. *Audubon: The Charleston Connection.* Contributions from the Charleston Museum, vol. 16. Charleston: Charleston Museum.

Shuler, Jay
1995. *Had I the Wings: The Friendship of Bachman and Audubon.* Athens: University of Georgia Press. 233 pp. A dual biography with some separate biographical sections; the bibliography lists ten items by Bachman, but many others are noted in Shuler's research in the College of Charleston Library.

Stanton, William Ragan

1960. *The Leopard's Spots: Scientific Attitudes Toward Race in America, 1815–59.* Chicago: University of Chicago Press.

Stephens, Lester D.

1989. "Scientific Societies in the Old South," in *Science and Medicine in the Old South,* edited by Ronald L. Numbers and Todd L. Savitt (Baton Rouge: Louisiana State University Press, 1989), pp. 55–78.

1999. "Bachman, John." *American National Biography.* Edited by John A. Garraty and Mark C. Carnes, vol. 1, pp. 831–33. New York: Oxford University Press.

2000. *Science, Race, and Religion in the American South: John Bachman and the Charleston Circle of Naturalists, 1815–1895.* Chapel Hill: University of North Carolina Press.

2003. "Literary and Philosophical Society of South Carolina: A Forum for Intellectual Progress in Antebellum Charleston." *South Carolina Historical Magazine* 104 (2003): 154–75.

Tyler, Ron

2000. "The Publication of *The Viviparous Quadrupeds of North America.*" In Boehme 2000: 119–82.

Waddell, Gene

2005. "A Bibliography for John Bachman." *Archives of Natural History* 32–1: 53–69.

Wilson, Don E., and Sue Ruff (eds.)

1999. *Smithsonian Book of North American Mammals.* Washington: Smithsonian Institution Press in association with the American Society of Mammalogists.

Index

Agassiz, Louis, 1, 3, 12, 351; polygenesis controversy, 18, 252
agricultural societies, need for, 186, 189–92
agricultural survey of S.C.: article on, 177–216; need to publish survey results cheaply, 193–94; qualifications needed to conduct survey, 188–89
agriculture school: most needed to improve agriculture, 196; subjects to be taught (in order of importance): chemistry, 49, 50, 186–87, 196–98; geology, 198–99; vegetable physiology, 199–200; math, 201; physics (mechanical philosophy), 201; domesticated animals, 202–3; botany, 203–6; ornithology, 209–14
Allegheny Mountains, 5
American Association for the Advancement of Science, 12
American Beaver: article on, 217–37. *See also* Beaver, American
Aristotle, unity of the human species, 252
Audubon, John James: beginning of friendship with JB, 7, 172; defended by JB, 18; drinking problem, 327–29; experiments with vultures, 70–75; friendship with JB, 7, 10, 172; "indefatigable," 95; letters by JB to, 320–22, 324–27, 329–40, 345–48; senility, 352–53; specialized in larger mammals, 218; western expedition criticized by JB, 346–48. See also

Viviparous Quadrupeds of North America
Audubon, John Woodhouse (artist; son of John James Audubon), 10, 328
Audubon, Lucy Bakewell (Mrs. John James Audubon), 10, 352; letter by JB to, 318–19
Audubon, Maria Rebecca Bachman (Mrs. John Woodhouse Audubon; dau. of JB), 10, 328
Audubon, Mary Eliza Bachman (Mrs. Victor Gifford Audubon; dau. of JB), 10
Audubon, Victor Gifford (publisher; son of John James Audubon), 10, 324, 355; letters by JB to, 327–29, 340–45, 348–51, 353–55

Bachman, Catherine L. (dau. and biographer of JB), 19
Bachman, Eva (mother of JB), 5
Bachman, Harriet Martin (first Mrs. John Bachman), 18–19, 329, 338, 345, 346
Bachman, Jacob (father of JB), 5
Bachman, John (1790–1874): "accustomed to hard work," 340; bilingual, 14; b. Rhinebeck, N.Y., 5; controversies engaged in, 18; degrees conferred on, 17; education, 12–13, 16–17; European travels in 1838, 9, 105, 177; friendship and collaboration with John James Aubudon (*q.v.*); health problems, 14, 19, 350–51; initially named Johannes, 5; injured during Civil War, 312–16;

Bachman, John (*continued*)
library destroyed in 1865, 312; minister of Saint John's Evangelical Lutheran Church, 1815-1870, 14-15; moved to Charleston, S.C., in 1815, 14; offered presidency of South Carolina College, 336-37; selected letters by, 317-57; specialized in smaller mammals, 218; studied botany with William Bartram, 7; studied ornithology with Alexander Wilson, 5-6; vindication by, 299-316; writing schedule, 220. *See also* Luther, methodologies, polygenesis controversy, *Vivaparous Quadrupeds of North America*
Bachman, Julia (dau. of JB), last days described, 348-50
Bachman, Maria (second Mrs. JB; sister of first Mrs. JB). *See* Martin, Maria
Bachman, Maria Rebecca (dau. of JB; Mrs. John Woodhouse Audubon), 10
Bachman, Mary Eliza (dau. of JB; Mrs. Victor Gifford Audubon), 10
Bacon, Francis: "book of nature," 3; inductive approach, 3
Bakewell, Lucy. *See* Audubon, Lucy Bakewell
Banks, Joseph, 43, 51, 157, 197
Barton, Benjamin Smith, 335
Bartram, William, taught Bachman botany, 6, 7, 234
bats (*Vespertilio*), intended for inclusion in *Viviparous Quadrupeds of North America*, 338
Beaver, American (*Castor fiber*): article on, 217-37; article rewritten, 340; color, 225-26; compared with European Beaver, 236; dams built by, 227-29; described, 225; distribution, 234-35; habits, 226-34; lodges built by, 228-29
Bell, John G., 337-38, 341-42, 345
Bellinger, John, 264, 266, 272
birds: article on color changes, 134-76 *passim*; article on migration of, 82-104; bird songs indicating different species, 8; value of birds for agriculture and horticulture, 48-49, 207-9
blacks, 14, 20, 35, 242, 303; same species as whites, 253. *See also* polygenesis controversy; slavery
"book of nature," 3
Braun, Anthony Theodore, 13
Brewer, Thomas Mayo, 331
Buffon, George-Louis Leclerc, 251-52, 294

Cahalane, Victor H., 2, 20
Carion Crow, 70. *See also* vultures
Carolina Parakeet: changing colors, 139-40; rarely seen by 1837, 139
Castor fiber Linn., article on, 217-37. *See also* Beaver, American
Catesby, Mark, 234
Catholicism. *See* Roman Catholic Church; Reformation, Protestant
Charleston Museum, developed from College of Charleston Museum, 12
Charleston, S.C.: group of naturalists in, 11; plundered after the Civil War, 299, 304-6
Civil War: JB's experiences in, 299-316; JB ministered to soldiers of both sides, 307; raids by Sherman's Army, 307-17
College of Charleston: JB on board of trustees, 18; JB on faculty of, 28; museum contributed to by JB, 12
color changes of animals, article on 134-76
Cooper, Thomas, 279
criminals, five-sixths uneducated, 292
Cuvier, Frédéric, use of dentition for classification, 221, 235, 294

Darwin, Charles, 9, 20; influence by JB on, 105-6; praised JB's article on bird migration, 82-83

Davis, Jefferson, published JB's Civil War experiences, 299
DeKay, James E., 331, 343
dentition, use for classification, 221
Didelphis Virginiana (opossums), article on reproduction of, 238–50

education: address on, 279–98; criminals, five-sixths uneducated, 292. *See also* College of Charleston, Hartman College, Newberry College, South Carolina College
Elliot, Stephen, 11–12, 65
Elliott Society of Natural History, 12, 118
England, John, 16, 271
ermine (*Mustela erminea* Bonaparte), 5, 165–70
experiments, with vultures, 70–81

fevers, stagnant water drained to prevent, 193
Franklin, Benjamin, 196, 291
Frémont, John C., 344
fruit grown in S.C., 60–62; oranges, 210–11

Geomys. *See* gopher
German Friendly Society, Charleston, 27, 35
Germany, education in, 285, 296–97
Gibbes, Lewis Reeve, 11
Gildersleeve, Basil, 264
God, as Creator, 16, 85, 149, 284, 287
Godman, John, 2, 165n
gopher (*Geomys*), Pouched Rat, 332, 336, 339, 348, 353
Gray, John Edward, 324, 330
gypsum, use as fertilizer, 199, 208n

Happoldt, Christopher, 337
hares (*Lepus*): *Americanus* (American Hare), 165; *articus* Ross a.k.a. *glacialis* Leach (Polar Hare), 163–64, 171;

sylvaticus, 322; *townsendii*, 342; *variabilis*, 171; *Virginianus* (Virginia or Northern Hare), 164. *See also* rabbits
Harlan, Richard, 2, 323, 326, 330, 339
Harris, Edward, 106, 323, 325, 337; letter by JB to, 355–56; marmot squirrel named for, 356
Hartman College, 13
Haskell, Harriet Bachman (Mrs. William E. Haskell; dau. of JB), 334, 346
Haskell, John Bachman (grandson and biographer of John Bachman), 19, 318, 353; letter by JB to, 356–57
Herbemont, Nicholas, 63
Hessian fly, attempts to identify, 51–52
Holbrook, John Edwards, 11, 331, 334–35
Holmes, Francis Simmons, 11–12
Horticultural Society of Charleston, address to, 41–69
horticulture: address on, 41–69; main purposes of, 43–44
human race: article on the unity of, 251–62; human beings as domesticated animals, 4. *See also* polygenesis controversy
Humboldt, Alexander von, 6, 9, 17, 253
Humboldt, Wilhelm von, 17
Hutter, E. W., 299–300, 312, 315
hybrids, not separate species, 4

Indians, American, 221, 231, 234, 260, 347; mounds, 365; observe birds to determine when to plant, 102; Oneida visited by JB, 6
Irving, Washington, JB accompanies to Oneida, 6

Johnson, Samuel, 283–84

Kalm, Peter, 49, 207
Knickerbocker, Diedrich, pen name for Washington Irving, 6
Kunze, John Christopher, 13

Lamarck, Jean-Baptist, 16
Lavosier, Antoine-Laurent de, 50, 198, 196
LeConte, John Eatton, 333–34, 336, 343
Legomys princeps Richardson (Little Chief Hare; renamed *Ochtona princeps*), 322
Lehman, George, 7, 318
Lepus, comprehensive study of by JB, 218. See also hares; rabbits
Lewis and Clarke Expedition, 74, 127, 346
Lichtenstein, M. H. C., 344
Liebig, Justus von, 192–93, 196
Linnaeus, Carl, 52, 294
Literary and Philosophical Society of Charleston, 11, 12, 17; presentations by JB for, on an agricultural survey for S.C., 177–216, and on bird migration, 82–104
Luther, Martin, 4, 12, 21; defended by JB, 263–66, 278, 287
Lutheranism, effects on freedom of speech and education, 3. See also Saint John's Evangelical Lutheran Church
Lyell, Charles, visited JB twice in Charleston S.C., 9
Lynch, Patrick Neeson, 272, 275–77

mammals. See common names for general and species; *Viviparous Quadrupeds of North America*
marmot squirrel (*Spermophilus*), 343
marsupial. See opossum (*Didelphis*)
Martin, Harriet (first Mrs. JB). See Bachman, Harriet Martin
Martin, Maria (second Mrs. JB; sister of first Mrs. JB), 19, 319, 334, 338, 343, 350; letter by JB to, 351–53; takes dictation for JB, 354; taught painting by Audubon, 19, 319
Mayer, Philip Frederic, 12–13
McCrady, John, 11
methodologies, used by JB: accounting for all relevant evidence (as for coloration, 134–76); comprehensive studies (as for entire genera such as *Sciurus* [squirrels], 105–33); experiments (as for vultures, 70–81); inductive reasoning (as for establishing the unity of the human race, 251–62); reliance on primary sources (as on Luther and the Reformation, 263–78); reliance on science rather than on assumptions, authoritative assertions, religion, or theory (as for establishing the unity of the human race, 251–62); systematic observations over long periods (as on opossums, 238–50)
Michel, Middleton, letter by on opossums, 248–50
migration of the birds, article on, 82–104
Milton, John, 43, 294, 350
Morton, Samuel George, 252, 261; letter by JB to, 322–24
Mustela erminea Bonaparte (ermine), 165–70

Native Americans. See Indians, American
Neotoma. See rats
Newberry College, 17, 20, 279, 280, 289; address on education prepared by JB for, 279–298; JB president of trustees, 280; JB's library intended for, 312
New York State, 5, 13–14, 45, 53, 177; agricultural survey, 182; poplar trees destroyed, 211; seeds coated to prevent birds from eating, 218n; use of gypsum for agriculture, 199
Nuttall, Thomas, 172, 173n, 322

Ochtona. See *Legomys princeps* Richardson
olives, experiments growing, 62–63
Oneida Indians, 6
opossum (*Didelphis*), article on reproduction, 238–50
orange trees, insect damage to in S.C., 210–11

Ord, George, 137, 147, 154, 155, 160
oriole, color changes, 140–41
ornithology. *See* birds
Owen, Richard, 244, 248, 334–35

Painted Bunting, color changes, 141–42
Passenger Pigeon, migration of, 93
pewee (bird), 101, 353
Pickering, Charles, 322–23, 331
pine trees, insect damage to, 212–13
polygenesis controversy, 18, 251–54. *See also* Agassiz, Louis
poplar trees, destroyed in N.Y. State, 211–12
Porcher, Francis Peyre, 11, 60
Porcher, Frederick, 18
Pouched Rat (*Geomys*), 332, 336, 339, 348, 353

quadrupeds. See *Viviparous Quadrupeds of North America*; and common names such as squirrels, shrews, and rabbits
Quitman, Frederick Henry, 13–14

rabbits (formerly *Lepus*; renamed *Sylviagus*): *bachmani* Waterhouse (Bush Rabbit), named for Bachman, 131–33, depicted on dustjacket; *palustris* Bachman (Marsh Hare), 165, 322. *See also* hares
races: unity of the human race, 251–62; as varieties of the human species, 4
Rafinesque, Constantine Raphael, 332
rats (*Neotoma*), types needed for study, 332–33. *See also* Pouched Rat (*Geomys*)
Ravenel, Edmund, 11
Reformation, Protestant, 21, 263–78 *passim*
religion. *See* Luther, Martin; Lutheranism; Reformation, Protestant; Roman Catholic Church
reproduction, marsupial (opossum), 238–50

Rice Bird, color changes, 155–56, 176
Richardson, John, 2, 87, 102, 106; on beavers, 227, 235; on pouched rats, 332, 339
Roman Catholic Church, 16; use of the confessional condemned by JB, 263–78
Ruffin, Edmund, 189, 196, 199

Saint John's Evangelical Lutheran Church, JB its minister from 1815 to 1870, 14–15
Say, Thomas, 326
Schlegel, Frederick, on Luther, 263
Schwartz, John G., eulogy by JB, 23–40; letter by Schwartz quoted, 38–39
Sciurus, article on, 105–33. *See also* squirrels
Scotland: major supplier of seeds, 191; progress of, 295
Seabrook, William, 196, 214
seeds: coated to prevent birds eating, 208n; requirements to maintain quality of, 191
Semler, Johann Salomo, 13
Sherman, William T., 301; raids by his soldiers in S.C., 307–16
shrews (*Sorex*), comprehensive study of by JB, 218
slavery: in N.Y. State, 6; in S.C., 19
Sorex (shrews), comprehensive study of by JB, 218
South Carolina: agricultural survey funded for, 179; importing food, 185, 214. *See also specific individuals, places, topics*
South Carolina College, 17, 27, 279; JB offered presidency of, 336–37
Spanish fly, 53
species: definition of, 251; in-vitro fertilization for distinguishing species and varieties, 220
Spermophilus (marmot squirrel), 342

squirrels (*Sciurus*): article on, 105–33; species noted: *Auduboni* Bachman (Larger Louisiana Black Squirrel), 124; *aureogaster aureogaster* F. Cuvier, 111–13; *capistratus* Bosc. (Boscot Fox Squirrel), 107–9; *carolinensis* Gmelin (Little Carolina Gray Squirrel), 119–21; *cinereus* Gmelin (cat squirrel), 113–16; *Douglasii* (Gray Squirrel or Oppocepoce), 125–27; *fuliginosus* Bachman (Sooty Squirrel), 124–25; *Hudsonius* Pennant (Chickaree Hudson's Bay Squirrel), 127; *lanuginus* Bachman (Downy Squirrel), 129–31; *leucotis* (Northern Gray Squirrel), 116–19; *niger* Linnaeus (Black Squirrel), 121–23; *nigrescens*, 121; *Richardsonii* Bachman (Columbia Pine Squirrel), 127–29; *Sciurus Colliœi*, 121; *subauratus* Bachman (Golden-bellied Squirrel), 110–11; *Texianus* Bachman, (Texian Squirrel), 109–10
Stephens, Lester D., 11
Strobel, B. B., 74, 77
swallows: supposed hibernation of, 98–100; usually safe to plant after their return, 103
Switzerland, 5, 295
Sylviagus. See rabbits; *see also* hares

teeth (dentition), use for classification, 221
theory: repudiated by JB, 2, 7, 134, 149
Tomato Worm, life history, 212
Townsend, John Kirk, 331–32
trees, types of ornamental, 65–66
Trudeau, James, 233, 326
tuberculosis, illness of JB, 19
Tuomey, Michael, 12
turkey, domestication and changing colors of, 145–56
Turkey Buzzard, 70

University of South Carolina. *See* South Carolina College

varieties, distinguished from species, 3
Viviparous Quadrupeds of North America, article on beavers from, 223–37; collaboration of JB and Audubon family, 9–11, 217–23, 317, 322–56 *passim*
vultures, experiments with, 70–81

Ward, Henry, 7, 75
Waterhouse, George Robert, 131–33, 324, 326, 330; hare named for JB, 106
West Point (military college), study of by JB, 17
White, Blanco, 273–74
Whooping (Hooping) Crane, 92, 142–43
Wilson, Alexander, 94; JB studied with, 5–8, 13; reduced number of bird species, 172
women, to be college educated, 291
woodpeckers, wantonly destroyed, 212–13
Wright, John, 345–46

Young, Arthur, 190, 196

Zoological Society of London, 9, 105, 325–26, 330
zoology. *See Viviparous Quadrupeds of North America; specific common names of animals*

The Publications of the Southern Texts Society
Books published by the University of Georgia Press

A DuBose Heyward Reader
Edited and with an Introduction by James M. Hutchisson

To Find My Own Peace: Grace King in Her Journals, 1886–1910
Edited by Melissa Walker Heidari

The Correspondence of Sarah Morgan and Francis Warrington Dawson, with Selected Editorials Written by Sarah Morgan for the Charleston News and Courier
Edited by Giselle Roberts

Shared Histories: Transatlantic Letters between Virginia Dickinson Reynolds and Her Daughter, Virginia Potter, 1929–1966
Edited by Angela Potter

Princes of Cotton: Four Diaries of Young Men in the South, 1848–1860
Edited by Stephen Berry

Mary Telfair to Mary Few: Selected Letters, 1802–1844
Edited by Betty Wood

In Black and White: An Interpretation of the South
Lily Hardy Hammond, edited by Elna C. Green

Pioneering American Wine: Writings of Nicholas Herbemont, Master Viticulturist
Edited by David S. Shields

John Bachman: Selected Writings on Science, Race, and Religion
Edited by Gene Waddell

www.ingramcontent.com/pod-product-compliance
Lightning Source LLC
Chambersburg PA
CBHW011754220426
43672CB00018B/2950